Perspectives in Neural Computing

Springer
London
Berlin
Heidelberg
New York
Barcelona
Hong Kong
Milan
Paris
Santa Clara
Singapore
Tokyo

Also in this series

J.G. Taylor
The Promise of Neural Networks
3-540-19773-7

Maria Marinaro and Roberto Tagliaferri (Eds)
Neural Nets - WIRN VIETRI-96
3-540-76099-7

Adrian Shepherd
Second -Order Methods for Neural Networks
3-540-76100-4

Jason Kingdon
Intelligent Systems and Financial Forecasting
3-540-76098-9

Dimitris C. Dracopoulos
Evolutionary Learning Algorithms for Neural Adaptive Control
3-540-76161-6

M. Kárný, K. Warwick and V. Kůrková (Eds)
Dealing with Complexity: A Neural Networks Approach
3-540-76160-8

John A. Bullinaria, David W. Glasspool and George Houghton (Eds)
4th Neural Computation and Psychology Workshop, London,
9-11 April 1997: Connectionist Representations
3-540-76208-6

Maria Marinaro and Roberto Tagliaferri (Eds)
Neural Nets - WIRN VIETRI-97
3-540-76157-8

L.J. Landau and J.G. Taylor (Eds)
Concepts for Neural Networks: A Survey
3-540-76163-2

Gustavo Deco and Dragan Obradovic
An Information-Theoretic Approach to Neural Computing
0-387-94666-7

Thomas Lindblad and Jason M. Kinser
Image Processing using Pulse-Coupled Neural Networks
3-540-76264-7

L. Niklasson, M. Bodén and T. Ziemke (Eds)
ICANN 98
3-540-76263-9

Maria Marinaro and Roberto Tagliaferri (Eds)

Neural Nets WIRN VIETRI-98

Proceedings of the 10th Italian Workshop on Neural Nets, Vietri sul Mare, Salerno, Italy, 21-23 May 1998

Società Italiana Reti Neuroniche (SIREN)
•
International Institute for Advanced Scientific Studies
"E.R. Caianiello" (IIASS)
•
IEEE NNC
•
INNS - SIG Italia
•
Department of Computer Science and Applications,
"R.M. Capocelli", University of Salerno
•
Department of Information Science, University of Milan
•
Department of Physics "E.R. Caianiello", University of Salerno
•
Elsag Bailey (Genova)
•
Istituto Italiano per gli Studi Filosofici (Naples)
•
Provincia di Salerno

Springer

Professor Maria Marinaro
Dipartimento di Scienze Fisiche "E.R. Caianiello"
Università di Salerno, 84081 Baronissi (SA), Italy

Dr Roberto Tagliaferri
Dipartimento di Informatica ed Applicazioni "R.M. Capocelli"
Università di Salerno, 84081 Baronissi (SA), Italy

Series Editor
J.G. Taylor, BA, BSc, MA, PhD, FlnstP
Centre for Neural Networks,
Department of Mathematics, King's College,
Strand, London WC2R 2LS, UK

ISBN-13: 978-1-4471-1208-2 Springer-Verlag London Berlin Heidelberg

British Library Cataloguing in Publication Data
A catalogue record for this book is available from the British Library

Library of Congress Cataloging-in-Publication Data
Italian Workshop on Neural Nets (10th : 1998 : Vietri sul Mare, Italy)
Neural nets, WIRN Vietri-98 : Tenth Italian Workshop on Neural Nets, Vietri sul Mare, Salerno, 21-23 May 1998 / Maria Marinaro and Roberto Tagliaferri, eds. ; [jointly organized by] Società italiana reti neuroniche (SIREN) ...[et al.].
p. cm. -- (Perspectives in neural computing)
Includes bibliographical references.
ISBN-13: 978-1-4471-1208-2 e-ISBN-13: 978-1-4471-0811-5
DOI: 10.1007/978-1-4471-0811-5
1. Neural networks (Computer science)--Congresses. I. Marinaro, M. II. Tagliaferri, Roberto. III. Società italiana reti neuroniche. IV. Series.
QA76.87.I85 1998 98-38064
006.3'2--dc21 CIP

Typesetting: Camera ready by contributors

34/3830-543210 Printed on acid-free paper

Preface

This volume contains the Proceedings of the *10th Italian Workshop on Neural Nets WIRN VIETRI-98*, jointly organized by the International Institute for Advanced Scientific Studies "Eduardo R. Caianiello" (IIASS), the Società Italiana Reti Neuroniche (SIREN), the IEEE NNC Italian RIG and the Italian SIG of the INSS.

This volume is dedicated to Marco Protasi, a leading scientist in the fields of Algorithmics, Computability, Discrete Mathematics and Neural Networks, who died on 1st February 1998. A biographical sketch of Professor Protasi by Stefano Fanelli is included.

Following the tradition of previous years, we invited two foreign scientists to the workshop, Professor P. Baldi and Professor D. Saad, who respectively presented the lectures "Software Foundation Libraries for Intelligent Systems" and "Theoretical Approaches to On-Line Learning in Smooth Multilayer Networks". In addition, a review talk was presented, dealing with a very up-to-date topic: "Combining Neural Networks and Hidden Markov Models for Speech Recognition".

A large part of the book contains original contributions approved by the referees as oral or poster presentations which have been assembled for reading convenience into 3 sections: Architectures and Algorithms, Mathematical Models, Pattern Recognition and Signal Processing.

The editors would like to thank the invited speakers, the review lecturer and all the contributors whose highly qualified papers helped with the success of the workshop.

Finally, special thanks go to the referees for their accurate work.

Maria Marinaro
Roberto Tagliaferri

Organizing - Scientific Committee:

B. Apolloni *(Univ. Milano)*, A. Bertoni *(Univ. Milano)*, N.A. Borghese *(CNR Milano)*, P. Campadelli *(Univ. Milano)*, D.D. Caviglia *(Univ. Genova)*, A. Colla *(ELSAG BAILEY - Genova)*, A. Esposito *(IIASS)*, M. Frixione *(IIASS)*, C. Furlanello *(IRST - Trento)*, M. Gori *(Univ. Siena)*, G.M. Guazzo *(IIASS)*, F. Lauria *(Univ. Napoli)*, M. Marinaro *(Univ. Salerno - IIASS)*, F. Masulli *(Univ. Genova)*, F.C. *Morabito (Univ. Reggio Calabria)*, P. Morasso *(Univ. Genova)*, G. Orlandi *(Univ. Roma)*, T. Parisini *(Univ. Trieste)*, E. Pasero *(Politecnico Torino)*, A. Petrosino *(INFM)*, V. Piuri *(Politecnico di Milano)*, S. Rampone *(IIASS)*, R. Serra *(CRA Montecatini - Ravenna)*, F. Sorbello *(Univ. Palermo)*, R. Tagliaferri *(Univ. Salerno)*.

Referees:

Apolloni B., Bertoni A., Biella G., Borghese N.A., Burattini E., Burrascano P., Campadelli P., Caviglia D., Chella A., Colla A.M., Costa M., De Gregorio M., Di Claudio E.D., Esposito A., Frattale Mascioli F.M., Frisone F., Frixione M., Furlanello C., Giuliani D., Gori M., Lauria F., Malcheoni D., Marinaro M., Masulli F., Morabito F.C., Morasso P., Orlandi G., Palmieri F., Parisi R., Parisini T., Pasero E., Petrosino A., Piuri V., Rampone S., Serra R., Sperduti A., Tagliaferri R.

The sponsorship and support of:

- Società Italiana Reti Neuroniche (SIREN)
- International Institute for Advanced Scientific Studies "E.R. Caianiello" (IIASS)
- IEEE NNC
- INNS - SIG Italia
- Department of Computer Science and Applications, "R.M. Capocelli", University of Salerno
- Department of Information Science, University of Milan
- Department of Physics "E.R. Caianiello", University of Salerno
- Elsag Bailey (Genova)
- Istituto Italiano per gli Studi Filosofici (Naples)
- Provincia di Salerno

are gratefully acknowledged

Contents

Section 1
Marco Protasi Lectures

In Memory of Marco Protasi

Stefano Fanelli (*)

(*) Dipartimento di Matematica
Università di Roma "Tor Vergata" , Roma (ITALY)
fanelli@mat.uniroma2.it

On February 1st , 1998 Marco Protasi's life was defeated by a long and terrible illness that took him away from his family , his friends, his colleagues, his students .
Marco was born in Spoleto (Umbria) in 1950 and received the Dr. degree (laurea)in Mathematics in 1973 from the University of Rome "La Sapienza".
In 1983 he became Associate Professor of Computer Science at the University of L'Aquila and in 1987 Full Professor of Computer Science at the University of Rome "Tor Vergata" .
He was the chief of many important research projects c/o the National Research Centre and , several times ,Visiting Professor at the International Computer Science Institute in Berkeley (CA) .
He was also the co-ordinator of a joint project MURST/British Council on "Approximate and Learning Algorithms" and main responsible of a European Community project regarding summer schools on "Approximate Algorithms" , "Randomised Algorithms" and "Dynamical Algorithms".
Marco was author of five text books in Algorithmics , Computability and Discrete Mathematics.
In the field of Neural Networks , he was the leader of an operational unit of the strategic project "Reti Neurali" c/o the National Research Centre.
His papers in this area concern mainly :

A) *Efficient and Generalised BackPropagation (BP) Algorithms for MLP-networks*
B) *Direct Methods for Training MLP-networks*
C) *Algorithms for Automatic Speech Recognition using Recurrent Networks*
D) *Links between Computational Complexity and Learning in Neural Networks*
E) *Applications of Neural Networks to Numerical Analysis Problems*

The works of groups A) and B) received the support of myself and of some brilliant graduates of the Department of Mathematics of the University of Rome "Tor Vergata" , while those belonging to groups C) , D) and E) were developed in the frame of a collaboration with some researchers of the Department of Computer Science of the University of Florence and myself .
A) (see [1],[2],[4],[8])
The papers in this group regard the implementation of advanced versions of

BackPropagation Algorithms for training MLP-networks.
More precisely , [1] contains a critical analysis of some heuristic algorithms known in the literature and of the corresponding learning rules (Momentum terms (MT) , Delta-Bar-Delta (DBD), Extended Delta-Bar-Delta (EDBD)) .
It is shown that the application of the classical Steepest Descent Algorithm (SDA) in conjunction with Armijo's method for the determination of a suitable step-size produces better results in comparison with MT , DBD and EDBD.
In [2] are reported some numerical experiments concerning the use of several variants of the Conjugate Gradient Method in the Fletcher and Reeves version (FRCG) .
In particular, it is described a new algorithm , called *Adaptive Parameters Conjugate Gradient BP (APCGBP)* , which matches the well known efficiency of FRCG with the excellent properties of a special form of Vogl's method .
[4] deals with the performances of BP methods by assuming various environmental hypotheses (on-line or off-line implementations , redundancy of patterns or target data) . The paper examines off-line BP-algorithms using patterns with a high level of redundancy and on-line BP-algorithms with a low level of redundancy , thereby completing the analysis performed in the literature .
Finally , [8] shows an extension of a quasi-newtonian memory-less algorithm (One Step Secant (OSS)) , described by Battiti . Although the latter method utilises some suitable information of Second Order , it has the peculiarity of requiring O(N) operations. However , since the one-dimensional optimisation performed in the line search is a crucial point to obtain an effective algorithm ,the paper studies alternative heuristic procedures . Once again , by combining a variant of Vogl's method with OSS , the extended algorithm , named *OSSV* , is able to obtain the best results both from a convergence point of view and in terms of precision .

B) (see [3],[5],[6],[10])
The works belonging to this group concern the study of some alternative (to BP) methods for the training of MLP-networks.
These methods are based on a direct determination of the matrices of weights , by solving in the classical or in the least squares sense a set of systems of linear equations , derived by the condition of perfect learning for a three-layered feedforward neural network .
The main advantages of this approach are the following ones :

i) there is no risk of being trapped in local minima of the error function during the training phase
ii) weights adapt quickly to the patterns presented to the network

The fundamental idea of these "direct methods" lies essentially in an extension and a generalisation of the Single-Value Decomposition (SVD) algorithm for the evaluation of the matrices of weights .
A preliminary algorithm (named *ICGSVD*) in the simplest case of a network having a unique output unit with binary values was described in [3] . Computational experiences were , in fact , particularly encouraging in the latter case . However , in [5] it was shown that the algorithm could be successfully extended to networks having an arbitrary

number of output units , by determining a suitable "target vector" in the numerical procedure. Since other authors , using similar methods , had suggested different computational schemes to evaluate the latter vector , investigations to determine the best criterion were performed in this paper .
Moreover , in order to compare carefully the performances of the other direct methods known in the literature , several benchmark problems were implemented in [6] and in [10] . The deep and detailed analysis carried out in [10] is extremely sharp and allows to estimate the behaviour of the direct methods both in the training phase and in the testing one .In particular , it is important to emphasise the comparative study developed in [10] , which characterises the performances of the algorithms in terms of relationships among the cardinalities of hidden nodes , patterns , input and output units .

C) (see [7],[9])
These two papers describe novel algorithms for the Recognition of Isolated Words (IWR) , by using Recurrent Neural Networks (RNN) with a "K" (a-priori Knowledge) or "K-L" (a-priori Knowledge and Learning) architecture .
The innovative algorithms , which require an explicit knowledge of the IW in terms of a Finite State Automaton (FSA) , are based on the application of the Reduced Gradient (RG) Method . Since it can be proved that the automaton rules are realised in terms of linear inequalities on the weights , the learning process is modelised as a Non-Linear Programming problem with Linear Constraints (NLPLC) . It is well known that in this situation RG is able to determine efficiently a Kuhn-Tucker (KT) point , which , however , is not in general an optimal solution (i.e. the global minimum) if the objective function is not convex (as in the case of the Error Function) . If the latter KT point is neither the global minimum nor a satisfactory suboptimal solution for the problem , it is necessary to escape from its basin of attraction in order to continue the optimisation process .
[7] describes a heuristic method based on a combined use of RG and a skilled Back Propagation Through Time (BPTT) procedure . The essential idea of the method , named *FMPRG* , is to perform a prescribed number of iterations of the BPTT algorithm , whenever RG has reached a KT point being not a constrained local minimum for the problem .
In [9] , by using a deterministic global algorithm , acronymed TRUST , a more efficient method, named *RGTRUST1*, was proposed to jump out of the neighbourhood of undesirable solutions .
The general idea is to utilise a procedure based on a single iteration of TRUST for each local optimisation phase of RG . More precisely , during the learning process performed by RG , the implementation of TRUST takes place whenever a suitable set of "local stopping criteria" is satisfied . The latter set of conditions has the aim of establishing the right point of restarting the error dynamics in the neighbourhood of the KT point.

D) (see [11],[12],[13],[16])
These papers deal with optimal learning and provide a unified viewpoint of most significant results in the field . The focus is mainly on the problem of local minima in the

cost function of a neural network , which , in general affects any learning algorithm of BP-type .
The main aim of these contributions is to introduce a novel concept , called "*non-suspiciousness*" , by showing some intriguing links between optimal learning and the computational complexity of the loading problem associated to the network .
Moreover , it is proved that all the problems giving rise to unimodal error functions belong to the same class , defined on the basis of computational requirements .
The new theory is developed from different point of views. In particular, in [11] and [13] the interest is concentrated on the theoretical results ensuring local minima free error surfaces under suitable hypotheses on the network and on the learning environment . The identification of similar conditions guarantees the convergence to the global minimum just by using classical gradient descent algorithms .
On the other hand , it is shown that there exists a "*canonical form of gradient descent learning*" , derived by Zak's theory of Terminal Attractors , which allows to compute the optimal solution under some suitable "*non-suspiciousness conditions*" , thereby generalising the concept of unimodality (or convexity) for the objective function .
The paper [12] deals with the interpretation of the concept of non-suspiciousness in the frame of the theory of the initialisation oracles . More exactly , it is shown that , by considering a neural network as a parametrical system charged of learning concepts , any loading problem can be properly described inside this representation . Consequently , non-suspiciousness and the complementary concept of "*suspiciousness*" can be directly linked with the computational complexity of the loading problem and the shape of the associated error surface .
Finally , paper [16] is included in these Proceedings .

E) (see [14],[15],[17])
This last group of papers deals with the application of the theory of non-suspiciousness to the resolution of linear systems .
[14] describes a novel terminal attractor algorithm , named *ELISA*, which is based on a special neural network continuous form of gradient descent . The proposed method approximates the minimum of a quadratic function in a constant time , depending solely on the initial value of the residual function .
Under general hypotheses it is proven that ELISA has O(N**2) as computational complexity and therefore is theoretically optimal .
Numerical experiments clearly show that ELISA is particularly recommended if one desires a fast efficient resolution of well-conditioned linear systems with undefined matrices of large size N . For instance , by dominant operation counting , the threshold of convenience w.r.t. classical Gauss direct algorithm is around the value N=1700 , while in terms of total CPU-time the algorithm is already competitive when N=800 .
[15] is essentially a survey on the possible relationships between the concept of non-suspiciousness and Smale's theory of Computational Complexity for Continuous Numerical Analysis problems . It is shown that the operational inequality , proved in [14] , connecting the norm of the gradient computed by ELISA and the condition number of the associated matrix , can be seen as a particular case of Smale's inequality , involving the complexity of a polynomial algorithm defined by a continuous machine .

Other connections with classical results by Wilkinson and Demmel are underlined. More in general , the paper tries to discover a possible way of overcoming the dichotomy between Computer Science and Numerical Analysis by applying the theory of Terminal Attractors .
The last paper [17] is in preparation and will be dedicated to the memory of Marco .

Marco Protasi's References in the field of Neural Networks

[1] S. Fanelli - R. Lanuti - M. Protasi
"True steepest descent methods in MLP-networks"
(IV Italian Workshop on Par. Arch. and Neural Networks , 53-59 , 1991)

[2] A. Di Medio - S. Fanelli - M. Protasi
"Is classical Back Propagation really inefficient?"
(IJCNN `92 , Beijing , II , 368-374 ,1992)

[3] M. Di Martino - S. Fanelli - M. Protasi
"An efficient algorithm for the binary classification of patterns using MLP-networks"
(IEEE Int. Conf. on Neural Networks , San Francisco , II , 936-943 , 1993)

[4] A. Di Medio - S. Fanelli - M. Protasi
"Off-line and on-line Back Propagation methods with various levels of redundancy"
(SPIE Conf. on OE/Aerospace and RS , Orlando , V.1966 , 296-302 , 1993)

[5] M. Di Martino - S. Fanelli - M. Protasi
"A new improved on-line algorithm for multi-decisional problems based on MLP-networks using a limited amount of information"
(IJCNN`93 , Nagoya , I , 617-620 ,1993)

[6] M. Di Martino - S. Fanelli - M. Protasi
"Computational experiences of new direct methods for the on-line training of MLP-networks with binary outputs"
(ICANN'94 , Sorrento , I , 627-630 , 1994)

[7] S. Fanelli - A. Manzo - M. Protasi
"On the application of the reduced gradient algorithm in a recurrent neural network environment for automatic speech recognition"
(INNC'94 , Dublin , 55-60 , 1994)

[8] S. Fanelli - P. Paparo - M. Protasi
"Improving performances of Battiti-Shanno's quasi newtonian algorithms for learning in feed-forward neural networks"
(ANZIIS '94 , Brisbane , 115-119 , 1994)

[9] M. Bianchini - S. Fanelli - M. Gori - M. Maggini - M. Protasi
"A new heuristic global algorithm for automatic speech recognition using recurrent neural networks"
(ICANN'95 , Paris , I , 601-605 , 1995)

[10] M. Di Martino - S. Fanelli - M. Protasi
"Exploring and comparing the best direct methods for the efficient training of MLP-networks"
(IEEE Trans. on Neural Networks , (7) , 6 , 1497-1502 , 1996)

[11] M. Bianchini - S. Fanelli - M. Gori - M. Protasi
"Unimodal loading problems"
(MANNA: Models , Algorithms and Applications , 107-112 , Kluwer Acad. ,1997)

[12] S. Fanelli - P. Frasconi - M. Gori - M. Protasi
"Suspiciousness of loading problems"
(IEEE Int. Conf. on Neural Networks , Houston , 2 , 1240-1245 , 1997)

[13] M. Bianchini - S. Fanelli - M. Gori - M. Protasi
"Terminal attractor algorithms and the class of unimodal loading problems"
(IMACS , Berlin , 4 , 277-282 , 1997)

[14] M. Bianchini - S. Fanelli - M. Gori - M. Protasi
"Solving linear systems by a neural network canonical form of efficient gradient descent"
(ICONIP' 97 , Dunedin , I , 531-534, 1997)

[15] M. Bianchini - S. Fanelli - M. Gori - M. Protasi
"Non-suspiciousness : a generalisation of convexity in the frame of foundations of numerical analysis and learning"
(IJCNN' 98 , Anchorage , II , 1619-1623, 1998)

[16] S. Fanelli - P. Frasconi - M. Gori - M. Protasi
"Computational suspiciousness of learning in artificial neural nets"
(X Italian Workshop on Par. Arch. and Neural Networks , 1998 , to appear)

[17] M. Bianchini - S. Fanelli - M. Gori - M. Protasi
"A terminal attractor algorithm for solving linear systems"
(in preparation)

Computational Suspiciousness of Learning in Artificial Neural Nets

S. Fanelli*, P. Frasconi◇, M. Gori‡, and M. Protasi*

*Dip. di Matematica
Università Tor Vergata, Roma (Italy)
(fanelli,protasi)@mat.utovrm.it
◇Dip. Sistemi e Informatica
Università di Firenze (Italy)
paolo@mcculloch.ing.unifi.it
‡Dip. Ingegneria dell'Informazione
Università di Siena (Italy)
marco@ing.unisi.it

Abstract

In this paper we study the time complexity of learning in terms of continuous-parametric representations (CPR) of concept classes, that underly most of the connectionist approaches to machine learning. We introduce the notion of non-suspect concept classes in CPR. Suspiciousness essentially qualifies the shape of the error function associated to a CP representation when the learner adopts an optimization algorithm for searching in the parameter space. We show that a concept class $\mathcal{F}$ is non-suspect in a CPR only if $\mathcal{F}$ is efficiently learnable. Under some further assumptions, that are met for example in the case of the class of linearly separable patterns, we claim that there exists an *optimal* algorithm for learning non-suspect concept classes in the representation of multilayered neural networks, whose time complexity is $\Theta(mp)$, being m the number of training examples and p the dimension of the parameter space.

I. Introduction

The computational complexity of loading a set of training examples into connectionist learning models is a topic that has been sparkling the attention of numerous researchers, see e.g. [1],[2], who have mainly investigated the intractability of learning.

In this paper we study the time complexity of learning from the perspective of the particular representation of concepts that is commonly embraced by connectionist models: the continuous parametric representation (CPR). Concept classes in this representation are described by a *parameter* vector (e.g. connection weights) *continuously* varying in a region $\Omega \subset R^p$. Searching in the concept class is accomplished by exploring the parameter region. Neural networks and hidden Markov models are typical examples of systems that rely on such CP representations. It is straightforward to extend the definition of loading problem given for neural networks to every CP representation. An eventual solution to the loading problem is seeked by defining a scalar "cost" function E (such as error with respect to supervision, or negative log-likelihood) and minimizing E with respect to the parameters. If a given concept class is hard to learn in a given CPR (i.e. loading the examples in that CPR is intractable) we can intuitively expect that the corresponding shape of the cost function will be hard to climb for any optimization algorithm that adapts the parameters. However, to the best of our knowledge, no theoretical framework has been developed so far to formalize the relationship between the computational complexity of learning and the hardness of the associated continuous optimization. This paper seeks a contribution in this direction.

We introduce the notion of non-suspect concept classes in CPR. According to the definition that will be given in the next sections, suspiciousness depends on the shape of the cost surface in terms of "almost" stationary points. Informally, if the cost function is plagued by local minima and flat regions (plateaus), then there is the "suspect" that optimization algorithms cannot output a consistent solution in polynomial time. Indeed, we prove that a concept class $\mathcal{F}$ matches our definition of non-suspiciousness in a CPR only if it is efficiently learnable. An interesting aspect of this theory is that the above result can be given in a stronger form in the case of common backpropagation neural network. Under some further assumptions, that are met, for example, in the case of linearly separable patterns, concept classes that are non-suspect in the single-layer neural network representation are not only learnable in polynomial time, but are also learnable with *optimal* time complexity $O(mp)$, being m the number of training examples and p the dimension of the parameter space.

II. CPR of concept classes

We are interested in learning models in which the learner relies on a real-valued adaptive function $a(\cdot)$ defined on the instance space $\mathcal{X}$. The output $y = a(x, w) \in R$ depends on the input $x \in \mathcal{X}$ through a continuous parameter vector $w \in \Omega \subset R^p$. We require that $a(\cdot)$ be continuous and has continuous and differentiable first derivatives with respect to w. $\mathcal{X}$ may be a continuous space (e.g., R^n) or a discrete space (e.g., $\{0,1\}^n$). A decision function $d(\cdot) : R \to \{0,1\}$ on top of $a(\cdot)$ maps the analog output y into a Boolean variable $d(y)$, so that for each $w \in \Omega$, a concept $h = \{x \in \mathcal{X} | d(a(x,w)) = 1\}$ on the domain $\mathcal{X}$ is obtained. The hypothesis class $\mathcal{H} = \{h | w \in \Omega\}$ is the class of concepts obtained by letting w to vary in Ω. The *continuous parametric representation* of concept class $\mathcal{H}$ will be denoted by $\mathcal{R}_{\mathcal{H}}$. For any concept $h \in \mathcal{H}$, $\mathcal{R}_{\mathcal{H}}(h)$ is the pair $r = (N, w)$, where N is a name in $\{0,1\}^*$ that identifies $d(a(x, \cdot))$ and $\forall x \in \mathcal{X},\ \ d(a(x,w)) = h(x)$.

Throughout this paper, in order to get rid of unuseful representations that only "memorize" information, we assume that the length of representation $\mathcal{R}_{\mathcal{H}}$ is a polynomial of the input dimension. Common architectures such as sigmoidal feedforward neural networks, radial basis functions, recurrent networks, and (hierarchical) mixtures of experts, all fit into the above assumptions when applied to classification problems. The learner is supposed to be a PAC learning algorithm A for a target class of concepts $f \in \mathcal{F}$ in terms of a continuous parametric representation $\mathcal{R}_{\mathcal{H}}$. Following standard notation, A takes as input a precision parameter $\epsilon > 0$, a confidence parameter δ, and a size parameter n. A is given access to a set $\mathcal{L}_m \doteq \{(x_1, f(x_1)), (x_2, f(x_2)), \ldots, (x_m, f(x_m))\}$ of labeled examples, and is required to output $r \in \mathcal{R}_{\mathcal{H}}(h)$ for some $h \in \mathcal{H}$ such that h is ϵ-consistent with f with probability at least $1-\delta$. We assume that $\mathcal{F}$ is efficiently sample-learnable, i.e. m grows only polynomially with $1/\epsilon$, $1/\delta$, and n. The time complexity, however, is also a function of the representation. Hence, the time complexity depends on the particular choice of functions $a(\cdot)$ and $d(\cdot)$ that define the particular continuous-parametric architecture.

A well known result in computational learning theory states that any concept class $\mathcal{H}$ can be efficiently PAC learned in a representation $\mathcal{R}$ if there exists a polynomial *fitting* algorithm Q for $\mathcal{H}$ in $\mathcal{R}$ (i.e., an algorithm Q that takes $\mathcal{L}_m$ as input and outputs a parameter vector w such that $\mathcal{L}_m \subset graph(d(a(\cdot, w)))$) [3]. Fitting algorithms for connectionist models are typically based on optimization of a continuous cost function E, as described below.

Let $t : \{0, 1\} \rightarrow R$ be a function such that $t(1) > t(0)$, associating an analog target to (Boolean) concept membership. Let $e : \{0, 1\}^2 \rightarrow R$ be a real-valued function such that $\forall a, c$:

$$\begin{array}{l} e(a,c) \geq 0; \quad e(a,c) = 0 \text{ or } e_a(a,c) = 0 \quad \rightarrow d(a) = c; \\ \exists A : e(a,c) < A \text{ iff } d(a) = c; \quad sgn[e_a(a,c)] = sgn[a - t(c)], \end{array} \tag{1}$$

where $'$ denotes differentiation with respect to the argument a. The cost function is then defined as

$$E(w, \mathcal{L}_m) = \frac{1}{m} \sum_{i=1}^{m} e(a(x_i, w), c(x_i)). \tag{2}$$

This class of cost functions includes most of the error criteria commonly employed with connectionist approaches to concept learning.

III. Forced Gradient Descent: A canonical reduction

In this section we calculate the computational complexity of learning in CPRs. First, we describe a continuous time dynamical system that can solve the cost optimization problem in any assigned constant time. Let $E(w)$ be a positive C^2 function of w such that $\exists w^* : E(w^*) = 0$. Let us consider the following differential equation:

$$\frac{dw(t)}{dt} = -\gamma(t) \nabla E(w(t)), \tag{3}$$

and choose $\gamma(t)$ as

$$\gamma(t) \doteq \frac{\Psi(E)}{\| \nabla E(w(t)) \|^2}, \tag{4}$$

being Ψ a non-negative continuous function of E. Based on this choice of the "learning rate", the dynamics of the error function becomes

$$\frac{dE}{dt} = \nabla E(w)^T \frac{dw}{dt} = \nabla E(w)^T \left(-\frac{\Psi(E)}{\| \nabla E(w) \|^2} \nabla E(w) \right) = -\Psi(E), \tag{5}$$

which makes the error function continuously decreasing to zero. In this paper, we are interested in finding terminal attractors and, particularly, in minimizing the time σ required to approach the optimal solution. For this reason, a choice as simple as $\Psi(E) = E_0/\sigma$ fulfills our needs. In this way, the error function decreases linearly to zero and the terminal attractor is approached for $t = \sigma$ *independently of the problem at hand.* The corresponding weight updating equation becomes

$$\frac{dw}{dt} = -\frac{E_0}{\sigma}\frac{\nabla E(w)}{||\nabla E(w)||^2}. \tag{6}$$

for $t \in [0, \sigma)$.

In order to establish links between continuous optimization and and complexity, let us consider a discrete version of the dynamical system described above. Let us consider the following difference equation obtained by applying Euler's method to approximate equation (6),

$$w_{k+1} \leftarrow w_k - \tau\frac{E_0}{\sigma}\frac{\nabla E(w_k)}{||\nabla E(w_k)||^2}, \tag{7}$$

where the discrete time index k is related to the continuous time variable t by $t = k\tau$, being τ the sampling step in the Euler approximation.

The following definition formalizes the approximation of the discretization process.

Definition 1 *Difference equation (7) is said to be a* consistent approximation *of differential equation (6) if* $\forall\epsilon_a > 0\ \exists\tau > 0$ *such that* $\forall k$, $|E(w(k\tau)) - E(w_k)| < \epsilon_a$.

Definition 2 *A parameter vector* $w \in \Omega$ *is called* ϵ_s*-stationary for* $E(w)$ *if*

$$\left|\frac{\partial E(w)}{\partial w_j}\right| \leq \epsilon_s \quad \forall j = 1 \ldots p. \tag{8}$$

Definition 3 *An algorithm* $\mathcal{A}$ *is an* initialization oracle *for equation (7) if:*

1. $\mathcal{A}$ *takes as input a representation* $r(E)$ *for the error function* $E(w)$ *and an approximation threshold* ϵ_a*;*

2. $\mathcal{A}$ *predicts an initial parameter vector* w_0 *and a positive threshold* ϵ_s *such that the conditions "*w_k *is* ϵ_s*-stationary", and and "*$E(w_k) \geq \epsilon_a$*" do not hold simultaneously for the discrete trajectory* $\{w_k, k = 0, 1, \ldots\}$.

If moreover:

1. *the running time of* $\mathcal{A}$ *is a polynomial* $p(|r|, 1/\epsilon_a)$*, being* $|r|$ *the length of the representation* $r(E)$*, and*

2. $1/\epsilon_s$ *is a polynomial* $q(|r|, 1/\epsilon_a)$

we say that A *is a* polynomial initialization oracle.

Once an initialization oracle is available, algorithm (1) allows us to reach a consistent approximation of differential equation (6).

Algorithm 1 `FORCED-GD`

Input:
a representation $r(E)$ for the error function;
an approximation threshold ϵ_a.
Output:
A parameter vector w^* such that $E(w^*) < \epsilon_a$.
begin
invoke an initialization oracle on inputs $r(E)$ and ϵ_a;
$E_0 \leftarrow E(w_0)$;
$h \leftarrow \max_{i,j=1\ldots,p} \frac{\partial^2 E(w)}{\partial w_i \partial w_j}$
$K \leftarrow \left\lceil \frac{hE_0^2}{2\epsilon_a \epsilon_s^2} \right\rceil$
for $k \leftarrow 1, \ldots, K$ **do**
$w_{k+1} \leftarrow w_k - \frac{E_0}{K} \frac{\nabla E(w_k)}{\|\nabla E(w_k)\|^2}$;
return w_K;
end

Theorem 1 (Correctness of FORCED-GD**)** *If an initialization oracle exists for $E(w)$, then for any input,* FORCED-GD *outputs a parameter vector w^* such that $E(w^*) < \epsilon_a$.*

Proof The proof can easily be derived from [4].

□

The time complexity of FORCED-GD cannot be evaluated without further assumptions about E. The following fitting algorithm can be derived by using algorithm (1)

Algorithm 2 CPR-FITTING
Input:
A CP representation (N, w);
a set of m training examples $\mathcal{L}_m$.
Output:
A parameter vector w^* such that $d(a(\cdot, w^*))$ is consistent with $\mathcal{L}_m$.
begin
$\epsilon_a \leftarrow A/(2m)$;
build a formula F for $E(w)$;
return FORCED-GD(F, ϵ_a)
begin

The proof of the following result trivially follows from equations (1, 2):

Theorem 2 (Correctness of CPR-FITTING**)** *The parameter vector w^* computed by* CPR-FITTING *induces a hypothesis $h = d(a(\cdot, w^*))$ which is consistent with $\mathcal{L}_m$.*

IV. Suspiciousness of learning in CPR

In this section, we propose a way for assessing the complexity of learning concept classes that is inspired by their CPR representation. Informally, when the associated error function is severely plagued by local minima and/or plateaux, it is very unlikely that an efficient initialization oracle be discovered.

Definition 4 *A concept class $\mathcal{F} \subset \mathcal{H}$ is said to be* suspect *in terms of the polynomial CP representation $\mathcal{R}_{\mathcal{H}}$ of $\mathcal{H}$ if for some $f \in \mathcal{F}$, there exists $\mathcal{L}_m \subset graph(f)$ such that we cannot find $\mathcal{H}(h)$ with a polynomial initialization oracle for $E(\mathcal{R}_{\mathcal{H}}(h), \mathcal{L}_m)$ and h consistent with $\mathcal{L}_m$.*

We denote by $\mathcal{S}(\mathcal{R}_{\mathcal{H}})$ the set of all concept classes that are suspect in terms of $\mathcal{R}_{\mathcal{H}}$. We also define (understanding CPR) $\mathcal{S} = \{\mathcal{F} | \, \forall \mathcal{H} : \; \mathcal{F} \in \mathcal{S}(\mathcal{R}_{\mathcal{H}})\}$. The complements of $\mathcal{S}(\mathcal{R}_{\mathcal{H}})$ and $\mathcal{S}$ are denoted by $\mathcal{NS}(\mathcal{R}_{\mathcal{H}})$ and $\mathcal{NS}$, respectively.

Theorem 3 *A concept class $\mathcal{F}$ belongs to $\mathcal{NS}$ only if $\mathcal{H}$ is polynomial-time learnable in terms of the CP representation $\mathcal{R}_{\mathcal{H}}$.*

Proof (*sketch*) Whenever $\mathcal{F} \in \mathcal{NS}$, there exists a CPR such that algorithm `CPR-FITTING` is a polynomial fitting for $\mathcal{F}$ in $\mathcal{R}_{\mathcal{H}}$, and, thus $\mathcal{F}$ is polynomial-time learnable.

□

As an application of the theory, we consider a standard multilayered perceptron for concept learning. It is well-known that the concept class of linearly-separable patterns $\mathcal{F}_{LS}$ is efficiently learnable in the representation of perceptrons, as the fitting problem can be solved by linear programming, which has polynomial complexity. Using our theory, this result turns out to be a particular case, since this concept class is proven to be polynomially-time learnable also in the multilayer perceptron CPR.

Theorem 4 *The class $\mathcal{F}_{LS}$ of linearly separable functions is polynomial-time learnable in terms of the continuous parametric representation of single-layer networks.*

This theorem is proven by showing that there exists a CPR for $\mathcal{F}_{LS}$ in which the concept class is non-suspect(see an extended version of the paper [5]). A single-layer perceptron can be chosen, since in this case it possible to use a trivial initialization oracle which returns a random initial point. This is essentially due to the absence of local minima of the associated error functions [6].

Most importantly, it can be proven that for class $\mathcal{F}_{LS}$, the number of iterations K of algorithm (1) is independent of the problem dimension [5]. Consequently the computational complexity of algorithm (1) can be evaluated directly by inspecting $\nabla E(w, \mathcal{L}_m)$. Since Backpropagation is an optimal algorithm for computing the gradient we conclude that this is one of those cases in which algorithm (1) turns out to be optimal.

Acknowledgments

We thank M. Bianchini, M. Maggini, F. Scarselli, H.T. Siegelmann, and F. Schoen for fruitful discussions on the topic addressed in this paper.

References

[1] J. Judd, *Neural Network Design and the Complexity of Learning.* Cambridge, London: The MIT Press, 1990.

[2] A. Blum and R. Rivest, "Training a 3-node neural net is NP-complete," in *Advances in Neural Information Processing Systems* (D. Touretzky, ed.), vol. 1, pp. 494–501, Morgan Kaufmann, San Mateo, 1989.

[3] B. Natarajan, *Machine Learning: A Theoretical Approach.* San Mateo, CA: Morgan Kaufmann Publishers, Inc., 1991.

[4] P. Frasconi, S. Fanelli, M. Gori, and M. Protasi, "Suspiciousness of loading problems," in *IEEE International Conference on Neural Networks*, pp. II 1240–1245, IEEE Press, 9–12 June 1997.

[5] S. Fanelli, P. Frasconi, M. Gori, and M. Protasi, "Computational suspiciousness of learning in artificial neural networks," Tech. Rep. TR–DII–13–98, Dipartimento di Ingegneria dell'Informazione, Univ. di Siena, 1998.

[6] E. Sontag and H. Sussman, "Backpropagation separates when perceptrons do," in *International Joint Conference on Neural Networks*, vol. 1, (Washington DC), pp. 639–642, IEEE Press, June 1989.

Section 2
Invited Papers

Software Foundation Libraries for Intelligent Systems

Pierre Baldi
Net-ID and Netcore
Los Angeles, CA 90042
pfbaldi@netid.com

Yves Chauvin
Net-ID and Netcore
San Francisco, CA 94107
yves@netid.com

Van Mittal-Henkle
Net-ID and Netcore
Los Angeles, CA 90025
van@netid.com

Abstract

We describe a set of object-oriented software foundation libraries for the design and modeling of complex and intelligent systems. The basic thesis underlying the libraries is that the language required to represent and design all such systems in a scalable way is the language of sparse recursive graphs with the corresponding local message passing propagation algorithms. We provide some formal arguments to support this thesis going from decision theory, to Bayesian statistics, to probabilistic graphical models. The current version of the libraries consist of 6 different modules: ObjectCost, ObjectStat, ObjectNet, ObjectComp, ObjectData, and ObjectGraph. The libraries support most of the standard models used in artificial intelligence and machine learning, such as neural networks, hidden Markov models, and Bayesian networks. The libraries are being used in real data mining applications and products, and are actively being developed and extended.

1 Introduction

At the dawn of the information age, scientists and engineers from numerous disciplines are already grappling with the challenges of designing complex and intelligent systems, and of navigating and mining large data sets. Indeed, rapid progress in sensors, data storage and data base technology is leading to an explosion of the volume of data in many different fields, from biology to finance. The goal of this article is to outline a general purpose software foundation for the modeling and design of intelligent systems. The basic thesis we put forward is that *the language needed to represent and design complex and intelligent systems is the language of sparse recursive graphs and the related local message passing algorithms.* We provide some supporting evidence for the thesis, and describe the main consequence: its embodiment into a set of object-oriented software foundation libraries for the rapid design and deployment of sparse recursive graphs and algorithms based on local message passing iterations. Loosely speaking, recursive means that graphs can be combined in modular and hierarchical ways to reflect, for instance, different scales in a given domain. A node at one level of the hierachy can be "opened" up into a graph at the level below. Similarly, a graph is sparse if the total number of connections is only a small fraction of the total number of possible connections. In this sense, biological neural circuits are both recursive and sparse, in spite of their exceedingly complex connectivity. Recursivity and sparseness reflect properties of the world and are needed to ensure tractability.

At some level, the basic thesis is intuitively clear. In fact, sparse graphs with a hierarchical recursive structure seem ubiquitous in complex systems such as telephone switches, computer circuits, social organizations, and many biological networks from gene regulation, to cell metabolism, to neural networks. When two scientists convene to discuss a scientific problem, almost invariably they will start out by scribbling some small graphs on the legendary "back of the envelope". This intuitive view, however, is only a weak form of the thesis. As such, alternative descriptions based on a different formalism could and have been put forward. What we are really after is a strong version of the thesis, namely that the sparse recursive graph approach is almost *inevitable*. While of course such a statement cannot really be proven, we will try in the next sections to provide a stronger theoretical foundation that goes beyond simple intuitive ideas.

The theoretical foundation of our approach has three components: decision theory, Bayesian probability theory, and the theory of probabilistic graphical models. Graph algorithms and the implementation of many conventional algorithms by graph propagation are also important. In some form or another, the foundations for the first two theories have been known for at least half a century. The theory of probabilistic graphical models is perhaps a little younger. But it is only in the past few years that the overarching framework has reached a mature state and begun to have significant practical applications. Only now are computers beginning to reach the level of computing power required to deploy and exploit this general framework. Indeed, it is the confluence of three factors: computing power, data volume, and theoretical framework that is supporting our long-term approach.

In the next sections, we first give a very brief introduction to decision theory, Bayesian probability theory, and graphical models. None of these topics can be treated in reasonable depth in a few pages. Numerous pointers to the literature are provided for the interested reader. For completeness, we have also added a more technical appendix on probabilistic graphical models. We then provide a general overview of our object-oriented software libraries for intelligent systems which support and implement the theoretical framework in a *scalable* way, from the data to the user interface. The libraries are particularly suited for the rapid and integrated development of machine learning applications.

2 Decision theory and Bayesian probability theory

In business, engineering, and many other situations individuals and organizations are very often faced with the problem of reasoning and taking rational decisions in the presence of uncertainty and incomplete information. Over half a century of research in decision theory, game theory and statistics have firmly established that there is essentially a unique methodology for addressing such problems. All other approaches are at best approximations to the unique solution provided by what can be called Bayesian decision theory [43, 21, 9, 35]. More precisely, consider a decision maker that is faced with a number of possible actions, each action having a given set of consequences depending on the state of the world. Typically the state of the world is not entirely known to the decision maker, who can gather information, for instance, by sampling or experimentation. It can then be shown that if the decision maker is willing to scale both his preferences for the possible consequences and his judgments concerning the uncertain states, and accept two simple principles of consistent behavior, then his strategy is mathematically determined in a unique way. In particular, it can be shown that: (1) his judgments concerning the uncertain states of the world must obey the rules of probability theory; and (2) his final goal of rational behavior must be equated to maximizing the expected utility of his actions.

While this is not the place to present the axioms for consistent behavior and derive the unique methodology for rational decisions, or discuss any of the related philosophical issues, it is still useful to introduce the basic formalism. So consider that

- The states of the world are represented by the symbol s (s could of course be a large array of numbers).
- The actions are represented by the symbol a.
- The data about the world are represented by the symbol d.
- $\pi(s|d)$ represents our judgment, or degree of confidence, that the state of the world is s given data d.
- A cost, risk, return, reward or utility $c(a, s)$ is associated with each action and each state of the world (more complex actions/functions that include the cost of gathering data can also be considered).
- A decision rule r is simply a function that associates each data d with an action a.

Under the small set of axioms describing consistent behavior, it results that:

1. Once rescaled, the degrees of confidence $\pi(s|d)$ must obey all the rules of probability calculus. Thus we shall write $\mathbf{P}(s|d)$ for the rescaled version. In fact, this approach can be used to define probabilities, the so- called "subjectivist" approach, and construct Bayesian statistics as a special chapter of Bayesian decision theory. An alternative set of axioms [4], known as the Cox-Jaynes axioms can be used to derive Bayesian statistical theory alone from scratch, without its overarching decision component. This is of interest to scientists in quest of pure knowledge regardless of decisions. In all cases, the degrees of belief satisfy Bayes theorem:

$$\mathbf{P}[s|d] = \frac{\mathbf{P}[d|s]\mathbf{P}[s]}{\mathbf{P}[d]} \tag{1}$$

This is the fundamental relation that allows inversion between cause and effects, data and hypothesis. It applies to any hypothesis s, and in particular to any complex model $M(w)$ parameterized by a set of parameters w.

2. A decision rule is rational or optimal if it minimizes the expected cost or risk, or maximizes the expected utility or return. That is, the goal of the decision maker is to find decision rule r that extremizes the expectation

$$\mathbf{E}[r] = \int c(r(d), s)\mathbf{P}(s|d)ds \tag{2}$$

that is the average cost of the actions, across states of the world given the data. Thus conceptually, we can see that there is an overarching decision framework (Bayesian statistical decision theory), that contains a submodule associated with probability theory and Bayesian statistics. As we shall see, this hierarchical organization is closely reflected in the structure of the libraries, which include a general decision module (ObjectCost) and a general statistical module (ObjectStat)to be described below.

Finally, the reader should be aware that an even more general theory is provided by game theory [43], where the uncertain environment includes other agents or players interested in maximizing their own utility functions. Decision theory is a game against nature. Thus axiomaticaly game theory contains decision theory, which in turn contains Bayesian probability theory.

Once actions and costs are defined most of the remaining work has to do with statistical calculations: estimating probabilities and expectations associated with the state of the world and the corresponding models. This typically requires dealing with very high dimensional distributions of the form $\mathbf{P}(x_1, \ldots, x_n)$ over the space of parameters, data, hypothesis, and so on. Such a treatment is impossible in its generality and often requires simplifying assumptions or approximations. By far the most important class of simplifying assumptions are conditional independence assumptions leading to the theory of graphical models. Probabilistic graphical models are at the core of Bayesian methods and are the key ingredient both in terms of descriptive and compositional abilities, and computational efficiency.

3 Graphical Models

3.1 Probabilistic Graphical Models and Independence

In most situations, a given variable depends only on a small cluster of variables and, conditioned on the values in the cluster, is independent of all the remaining variables. These independence relationships can often be represented in terms of a graph, where the variables are associated with the nodes, and a missing edge represents a particular independence relationship (precise definitions can be found in the Appendix). See, for instance, [34, 29, 44, 12, 40, 11] for general reviews, treatments, or pointers to the large literature on this topic. The independence relationships result in the fundamental fact that the global high-dimensional probability distribution $\mathbf{P}(x_1, \ldots, x_n)$ over all variables, can be factored into a product of simpler local probability distributions over lower-dimensional spaces associated with smaller clusters of variables. The clusters are again reflected in the structure of the graph. Inference and learning are based on message passing propagation algorithms in the graphs. Thus the third and fourth essential modules of the libraries (ObjectNet and ObjectComp) have to do with general graphical (or network) models, and more specifically with probabilistic graphical models. Most of the models routinely used in artificial intelligence and machine learning fall in this category. This is the case, for instance, of neural networks, Bayesian networks, hidden Markov models, decision trees, cellular automata, and influence diagrams.

Probabilistic graphical models can be subdivided into two broad categories depending on whether the edges of the associated graph are directed or undirected. Undirected edges are typical in problems where interactions are considered to be symmetric, such as in statistical mechanics or image processing [24, 17, 31]. In the undirected case, in one form or another, these models are called random Markov fields, undirected probabilistic independence networks, Boltzmann machines, Markov networks, and log-linear models.

Directed models are used in cases where interactions are not symmetric and reflect causal relationships or time irreversibility [34, 22]. This is typically the case in expert systems and in all problems based on temporal data. The Kalman filter, for instance, a tool widely used in signal processing and control, can be viewed in this framework. In the case of temporal series, the independence assumptions are often those used in Markov models. Hidden Markov models and artificial neural networks are examples of graphical models with directed edges. Typical names for such models in the literature are Bayesian networks, belief networks, directed probabilistic independence networks, causal networks, and influence diagrams (see also [7] for a generalization). It is also possible to develop a theory for the mixed case [44], where both directed and undirected edges are present. Such mixed graphs are also called chain independence graphs. The basic theory of graphical models is reviewed in the Appendix.

One fundamental observation is that in most applications the resulting graphs are *sparse*, with at least some degree of structure. Thus the global probability distribution factors into a number of relatively small local distributions. And this is key to the implementation of efficient computational structures for learning and inference, based on the local propagation of information between clusters of variables in the graph. Sparseness is a property of many systems which reflects a certain degree of organization and is essential to their function, and not surprisingly to the tractability of the corresponding models. While it is not possible to prove such a general statement, many "proofs" exist in special cases showing that a particular graphical problem is not tractable in general, but has a polynomial-time solution if the graph is sparse. A good example is the NP completeness of general inference in Bayesian networks [13] which becomes tractable if the graph is singly connected. The recursive construction and manipulation of sparse graphs and sparse-graph-algorithms is the essence of ObjectNet and ObjectComp and is the central engine of the libraries. Related modeling techniques that cannot be discussed here include the use of hidden variables, hierarchical modeling, hybrid modeling and neural network reparameterization [5, 4], and the use of the exponential family of distributions.

3.2 Other Graphs

The libraries support the representation of arbitrary graphs, whether they are sparse or not, whether they come with probabilistic or physical semantics attached to them or not, and whether they make use of machine learning techniques or not. Dense graphs however are rarely seen in applications and very often turn out to be intractable. In a few situations, analytical approximate solutions can be found using some kind of mean field theory. There are also situations where the graphical formalism applies but where there is no underlying probabilistic model. This could be for many reasons ranging from the case of deterministic variables to the case of situations where the probabilistic components of the problem have not yet been introduced and are deferred to later more mature stages. Graphs can also be constructed with or without a physical semantics. In a graph with physical semantics, networks are used directly as models of particular physical entities. For instance, in life- cycle analysis, networks are used to represent interconnected industrial processes associated with the production of a good. Nodes are associated with processes, and edges are associated with flows of energy, materials, and goods. At a finer level of analysis, nodes can be recursively expanded into sub-networks. While ultimately there are statistical fluctuations associated with the flows, a first approximation often used in current life-cycle analysis simulators is obtained by considering only deterministic quantities. Another contrasting example is the case of hidden units in artificial neural networks. These "computational" nodes are typically associated with deterministic hidden variables devoid of any specific physical interpretation. All these different types of graphs can be represented and manipulated in the libraries and differ by the way and degree to which they make use of the statistical modules and propagation algorithms. Finally, for a significant number of graphical models, machine learning techniques are particularly important.

4 Machine Learning

Machine learning is just a modern term for the older discipline of statistical model fitting. Its expansion is closely related to the double expansion in data volume and computing power. Machine learning techniques are best suited for domains characterized by the availability of large amounts of data and the lack of good explanatory theories– not an uncommon situation these days, for instance in biology or finance. The basic idea is to extract the "theory" automatically from the data via a process of learning from examples. Thus the first typical machine learning step in the libraries is the construction of a parameterized probabilistic model for the data with an underlying graphical structure. Again typical examples of such models would include neural networks, hidden Markov models, or Bayesian networks. Models with 10-100 thousand parameters are routinely handled on today s computers. Depending on the situation, a number of different algorithms can be used in learning or model fitting, inference, or optimization, from gradient descent, to EM (expectation-maximization), to dynamic programming, to Monte Carlo methods such as Gibbs sampling and simulated annealing [4]. In most situations, such algorithms correspond to a propagation mechanism in an underlying graph, which can be implemented using simple message passing methods between neighbours in the graph. As learning progresses, models (i.e. graphs) can be combined in many ways from parallel, to hierarchical, to hybrid in order to improve pattern recognition and data mining performance.

5 Overview of the Libraries

The libraries contain a hierarchy of fundamental classes that can be used, over and over, to recursively construct nodes and connections associated with complex networks. In the libraries, the user can use the class hierarchy that is adapted to the current problem and network, which for instance may or may not include training. The classes support diverse computations associated with nodes, and propagation of information associated with connections, at all hierarchical levels. They support the creation of network models for specific application domains, as well as statistical graphical models for pattern recognition and data mining, such as neural networks, hidden Markov models, and Bayesian networks. The classes also support several learning and optimization algorithms, as well as decision modules.

Like most software products, the libraries can be subdivided into three major components: data access and management, the computational engine, and the graphical user interface (GUI). These three components are implemented by the following six modules: ObjectCost, ObjectStat, ObjectNet, ObjectComp, ObjectData, and ObjectGraph. ObjectGraph is associated with the GUI and ObjectData with data access and management. The remaining modules are associated with the engine and cover concepts in decision theory and statistics, as well as all graph algorithms and manipulations. It is our belief that such a uniform and integrated approach is needed in at least two different essential ways. First, a good engine without a good connection to data and a good user interface is useless. For example, a substantial portion of any machine learning project is often spent on the data alone. Second, both academic and commercial engines exist for specific sub-classes of networks and algorithms. But such engines cannot scale up to the construction of large software projects and the design of complex intelligent systems where networks of different sorts must be combined and composed in a recursive fashion over several different scales, and where algorithmic complexity and data volume expand rapidly. Our experience suggests that a flexible, unified, and theoretically well grounded object-oriented foundation is a key element in enabling the efficient construction and maintenance of large scale intelligent systems. Naturally, the straightforward use of such an approach involves some tradeoffs between implementation and maintenance efficiency versus runtime efficiency. The class hierarchy of the core libraries, however, is quite shallow with only a few layers. This is what allows computational efficiency and lends support to the general thesis. Thus in many cases execution speed is not an issue. But even when it is, the libraries offer the ability to dynamically compile a high-level model into a representation optimized for computational speed.

Currently, all components except ObjectGraph are written in C++ to optimize speed and flexibility. ObjectGraph is written in Java for portability across platforms. Each component consists of a set of files containing the corresponding classes and methods. The libraries are being used in applications that run under Windows 95/NT, Solaris, and IRIX. They also make use of the following third-party components:

- the STL (Standard Template Library)
- Systems¡Toolkit¿ (from ObjectSpace)
- ObjectStore, the leading object-oriented data base (from Object Design)

6 Modules of the Libraries

6.1 ObjectCost

ObjectCost is the highest level decision module in the hierarchy. It contains classes and methods related to decision analysis, actions and, depending on the problem, the associated costs, risks, losses, or gains and returns. The main goal is to evaluate the expected cost of a strategy and to find strategies of lowest expected cost. Alternative sub-optimal strategies such as MiniMax (minimize the maximum possible loss) are also possible. ObjectCost also contains classes and methods for a number of financial objects. ObjectCost allows the integration of costs and benefits with computational networks. Decision trees often include costs/benefits at branch points (nodes). Influence diagrams also include costs at decision points or during external events. Decision networks may also include costs/benefits in the connections. ObjectCost allows such costs and benefits to occur as a function of time. Through ObjectCost, costs, which may depend on long distance or long term interactions, can be represented in local dependencies over hierarchical networks and then globally optimized.

6.2 ObjectStat

ObjectStat is the module that contains the classes and methods related to probability distributions and statistical objects supporting Bayesian analysis. This is where random variables can be defined and associated with the nodes of a probabilistic graphical model. Probability distributions and densities can also be defined and manipulated together with several basic statistical concepts such as sample, expectation, variance, moments, and tests. ObjectStat includes classes and methods such as:

- random variables, continuous and discrete, one-dimensional and vectorial
- samples
- distributions of various types, especially in the exponential family such as normal, Poisson, and Dirichlet distributions.
- statistics
- statistical tests and tables
- moments and characteristic functions
- simple and multiple correlation and regression
- analysis of variance and covariance

- maximum likelihood, MAP estimation
- regularization and prior estimations

Decoupling probabilistic and graphical concepts is important for flexibility and to handle graphs devoid of probabilistic semantics.

6.3 ObjectNet and ObjectComp

ObjectNet and ObjectComp are the core modules of the libraires which allow the recursive manipulation of graphs and graph algorithms, regardless of their probabilistic interpretation. ObjectNet deals with graph theoretical objects such as nodes, connections, orientation, and flows. ObjectNet classes are used to represent the structural information of networks (graphs) independently of algorithms or values. ObjectComp classes are layered on top of ObjectNet and used to encapsulate algorithmic and computational information.

ObjectNet classes and methods, such as "node" or "connection" support the recursive creation of arbitrary graphs, including graphs associated with probabilistic graphical models. ObjectComp methods and classes support message passing propagation algorithms, such as inference, learning, dynamic programming, optimization, and Monte Carlo algorithms.

ObjectNet provides additional specific support for three special classes of graphical models and associated algorithms: Bayesian networks, neural networks, and hidden Markov models which are useful in a large number of applications. General Bayesian networks are emerging as one of the main tools in artificial intelligence and expert systems. Neural networks, as they are typically used in classification and regression, can be viewed as a special case of Bayesian network where: (1) each node in a given layer has a deterministic relationship with its parents; (2) the outputs of the output units have a true probabilistic interpretation, for instance by adding gaussian noise in regression. Hidden Markov models are also a special class of Bayesian networks with a particularly simple architecture. The hierarchical recursive nature of ObjectNet and ObjectComp naturally facilitates the creation of auxiliary or dual graphs (such as the moral graph and the junction tree for Bayesian networks) or complex networks consisting of multiple modules, for instance in mixture of experts or hybrid architectures [5].

6.4 ObjectGraph

ObjectGraph is the module that contains all the classes and methods required to visually represent and manipulate graphs, and the corresponding user interface. ObjectGraph consists of graphical classes and methods to display various properties associated with graphs and graph computations. ObjectGraph is written in Java and linked to the other C++ modules using the Java Native Interface (JNI) mechanism. JNI provides a clean, cross-platform mechanism of calling back and forth between C++ and Java. can display node Examples of view features include:

- a scrollable view area for the display of large networks
- single and multiple selection of graphical components

- zooming operations for both detailed and overview displays of network components
- multiple network views in splitter windows for simultaneous display of various parts of a selected network or of various levels of a hierarchical network.
- various drag and drop facilities
- various node layouts (e.g., stars, matrices)

Network components can be created, selected, and deleted in the ObjectGraph graphical view. ObjectGraph allows the selection of single or multiple nodes and supports a full set of object alignment and resizing commands, allowing the user to choose the display properties of individual nodes.

6.5 ObjectData

ObjectData contains classes and methods related to the reading and manipulation of data. The data, in object-oriented format, is used by the other modules, for instance during model fitting (machine learning) and other inference processes.

ObjectData implements the link between sources of data (for training or analysis) and the models which operate on the data. It provides facilities for online data validation (data scrubbing) and the management of data sets for training adaptive models. The architecture of the ObjectData module is flexible enough to accommodate a variety of data sources from simple flat file formats to full blown OODBMS (Object Oriented Database Management Systems). As an example, ObjectData has been used as the foundation for linking biological data stored in the ObjectStore OODBMS (from ObjectDesign) to a hidden Markov model simulator [6] for biological sequence analysis built on top of ObjectNet and ObjectComp.

ObjectStat statistical objects can be layered on top of ObjectData, allowing ObjectNet components to get access to data either directly through ObjectData or through ObjectStat. ObjectData hides the underlying data base representation from the ObjectStat and ObjectNet users.

7 Conclusion

Sometime early in the 21st century, humankind should achieve two very significant milestones: (1) the completion of the human genome sequencing project; (2) the development of computers that will reach and surpass the level of operations per second of the human brain. Together with other developments, such as the Internet, these milestones raise profound questions about the nature and boundaries of life and intelligence.

Here, we have put forward the thesis that the language needed to deploy intelligent system in a scalable way is the language of sparse recursive graphs with the associated message passing algorithms. We have provided some theoretical arguments in favor of such a thesis using the theory of graphical models within Bayesian decision theory and statistics. Other arguments using complementary ideas related to formal grammars, and in particular graph grammars, are under investigation [7]. The basic idea would be to develop a kind of graph grammar for intelligent system that is the equivalent of standard formal grammars for language.

More importantly perhaps, we have developed a set of object-oriented software foundation libraries implementing the thesis. By necessity, we have provided only a very broad outline of the libraries. Additional information can be obtained by contacting the authors directly. The libraries are being used and tested in real applications and products, and continue to be actively developed and extended, so far without encountering any major signs that would contradict the central reasoning behind them.

Finally, on a much more speculative note, it should not pass unnoticed that if true, the strong version of our thesis would have non-trivial repercussions in many areas of neurobiology and cognitive sciences. If the language of recursive sparse graphs is essential for intelligence and is somehow its archetypical pre-verbal form, then mature brain circuits must be capable of implementing such language in some form or another. If this thesis were confirmed, it would significantly impact our ideas about brain wiring, developmental stages, representations, and learning.

8 Appendix: Probabilistic Graphical Models

8.1 Notation and Preliminaries

In this appendix, which follows [4], we review the basic theory of probabilistic graphical model and the corresponding factorization of high-dimensional probability distributions. First, a point of notation. If X and Y are two independent random variables, we write $X \perp Y$. Conditional independence on Z is denoted by $X \perp Y | Z$. This means that $\mathbf{P}(X, Y|Z) = \mathbf{P}(X|Z)\mathbf{P}(Y|Z)$. It is important to note that conditional independence implies neither marginal independence nor the converse. Let $G = (V, E)$ denote a graph with a set V of vertices, and a set E of edges. The vertices are numbered $V = \{1, 2, ..., n\}$. If the edges are directed, we write $G = (V, \vec{E})$. In all the graphs to be considered, there is at most one edge between any two vertices, and there are no edges from a vertex to itself. In an undirected graph, $N(i)$ represents the sets of all the neighbors of vertex i, and $C(i)$ represents the set of all the vertices that are connected to i by a path. So, for instance,

$$N(i) = \{j \in V : (i, j) \in E\}. \tag{3}$$

If there is an edge between any pair of vertices, a graph is said to be *complete.* The *cliques* of G are the subgraphs of G that are both complete and maximal. The *clique graph* G^C of a graph G is the graph consisting of a vertex for each clique in G, and an edge between two vertices, if and only if the corresponding cliques have a nonempty intersection.

In a directed graph, the direction of the edges will often represent causality or time irreversibility. We use the obvious notation $N^-(i)$ and $N^+(i)$ to denote all the parents of i and all the children of i, respectively. Likewise, $C^-(i)$ and $C^+(i)$ denote the ancestors, or the "past," and the descendants of i, or the "future," of i. All these notations are extended in the obvious way to any set of vertices I. So, for instance, for any $I \in V$,

$$N(I) = \{j \in V : i \in I \quad \text{and} \quad (i,j) \in E\} - I. \tag{4}$$

This is also called the boundary of I. In an undirected graph, a set of vertices I is *separated* from a set J by a set K if and only if I and J are disjoint and any path from any vertex in I to any vertex in J contains a vertex in K.

We are interested in high-dimensional probability distributions of the form $\mathbf{P}(X_1, ..., X_n)$, where, for instance, the X variables represent both hidden and observed variables. In particular, we are interested in the factorization of such distributions into products of simpler distributions, such as conditionals and marginals. Obviously, it is possible to describe the joint distribution using the marginals

$$\mathbf{P}(X_1, ..., X_n) = \prod_{i=0}^{n-1} \mathbf{P}(X_{i+1}|X_1, ..., X_i). \tag{5}$$

The set of complete conditional distributions $\mathbf{P}(X_i|X_j : j \neq i)$ also defines the joint distribution in a unique way, *provided they are consistent* (or else no joint distribution can be defined) [10, 3]. The complete set of marginals $\mathbf{P}(X_i)$ is in general highly insufficent to define the joint distribution, except in special cases (see factorial distributions below). The problem of determining a multivariate joint distribution uniquely from an arbitrary set of marginal and conditional distributions is examined in [16]. As we shall see, graphical models correspond to joint distributions that can be expressed economically in terms of local conditionals, or joint distributions over small clusters of variables. Probabilistic inference in such models allows one to approximate useful probabilities, such as posteriors. A number of techniques are typically used to carry inference approximations including: probability propagation, Monte Carlo methods, statistical mechanics, variational methods, and inverse models.

For technical reasons [44], we assume that $\mathbf{P}(X_1, ..., X_n)$ is positive everywhere, which is not restrictive for practical applications because rare events can be assigned very small but nonzero probabilities. We are going to consider graphs of the form $G = (V, E)$, or $G = (V, \vec{E})$, where each variable X_i is associated with the corresponding vertex i. We let X_I denote the set of variables $X_i : i \in I$, associated with a set I of indices. For a fixed graph G, we will denote by $\mathcal{P}(G)$ a family of probability distributions satisfying a set of independence assumptions embodied in the connectivity of G. Roughly speaking, the absence of an edge signifies the existence of an independence relationship. These independence relationships are defined precisely below, in the two main cases of undirected and directed graphs. In modeling situations, the real probability distribution may not belong to the set $\mathcal{P}(G)$, for any G. The goal then is to find a G and a member of $\mathcal{P}(G)$ as close as possible to the real distribution—for instance, in terms of relative entropy.

8.2 The Undirected Case: Markov Random Fields

In the undirected case, the family $\mathcal{P}(G)$ corresponds to the notion of Markov random field, or Markov network, or probabilistic independence network, or, in a slightly different context, Boltzmann machine [24, 1]. Symmetric interaction models are typically used in statistical mechanics—for instance, Ising models and image processing [17, 31], where associations are considered to be more correlational than causal.

8.2.1 Markov Properties

A Markov random field on a graph G is characterized by any one of the following three equivalent Markov independence properties. The equivalence of these properties is remarkable, and its proof is left as an exercise.

1. **Pairwise Markov Property.** Nonneighboring pairs X_i and X_j are independent, conditional on all the other variables. That is, for any $(i,j) \notin E$,

$$X_i \perp X_j | X_{V-\{i,j\}}. \tag{6}$$

2. **Local Markov Property.** Conditional on its neighbors, any variable X_i is independent of all the other variables. That is, for any i in V,

$$X_i \perp X_{V-N(i)\cup\{i\}} | X_{N(i)}. \tag{7}$$

3. **Global Markov Property.** If I and J are two disjoint sets of vertices, separated by K, the corresponding set of variables is independent conditional on the variables in the third set:

$$X_I \perp X_J | X_K \tag{8}$$

These independence properties are equivalent to the statement

$$\mathbf{P}(X_i|X_{V-\{i\}}) = \mathbf{P}(X_i|X_{N(i)}). \tag{9}$$

8.2.2 Factorization Properties

The functions $\mathbf{P}(X_i|X_j : j \in N(i))$ are called the *local characteristics* of the Markov random field. It can be shown that they uniquely determine the global distribution $\mathbf{P}(X_1, ..., X_n)$, although in a complex way. In particular, and unlike what happens in the directed case, the global distribution is not the product of all the local characteristics. There is, however, an important theorem that relates Markov random fields to Boltzmann–Gibbs distributions. It can be shown that, as a result of the local independence property, the global distribution of a Markov random field has the functional form

$$\mathbf{P}(X_1, ..., X_n) = \frac{e^{-f(X_1,...,X_n)}}{Z} = \frac{e^{-\sum_C f_C(X_C)}}{Z}. \tag{10}$$

where Z is the usual normalizing factor. C runs over all the cliques of G, and f_C is called the *potential* or clique function of clique C. It depends only on the variables X_C occurring in the corresponding clique. f is also called the energy. In fact, $\mathbf{P}$ and G determine a Markov random field if and only if (10) holds [41].

It is easy to derive the local characteristics and marginals from the potential clique functions, by applying the definition in combination with the Boltzmann–Gibbs representation. The potential functions, on the other hand, are not unique. The determination of a set of potential functions in the general case is more elaborate. But there are formulas to derive the potential functions from the local characteristics. There is an important special case that is particularly simple. This is when the graph G is *triangulated.* A graph G is triangulated if any cycle of length greater than or equal to 4 contains at least one chord. A singly connected graph (i.e. a tree) is an important special case of a triangulated graph. A graph is triangulated if and only if its clique graph has a special property called the *running intersection* property, which states that if a vertex of G belongs to two cliques C_1 and C_2 of G, it must also belong to all the other cliques on a path from C_1 to C_2 in the clique graph G^C. The intersection of two neighboring cliques C_1 and C_2 of G—that is, two adjacent nodes of G^C—is called a *separator.* In a triangulated graph, a separator of C_1 and C_2 separates them in the probabilistic independence sense defined above.

Another important characterization of triangulated graphs is in terms of *perfect numbering.* A numbering of the nodes in V is perfect if for all i, $N(i) \cap \{1, 2, ..., i-1\}$ is complete. A graph is triangulated if and only if it admits a perfect numbering (see [42], [29], and references therein).] The key point here is that for Markov random fields associated with a triangulated graph, the global distribution has the form

$$\mathbf{P}(X_1, ..., X_n) = \frac{\prod_C \mathbf{P}(X_C)}{\prod_S \mathbf{P}(X_S)}, \tag{11}$$

where C runs over the cliques, and S runs over the separators, occurring in a *junction tree*, that is, a maximal spanning tree [] of G^C. $\prod_C \mathbf{P}(X_C)$ is the marginal joint distribution of X_C. The clique potential functions are then obvious.

A very special case of the Markov random field is when the graph G has no edges. This is the case when all the variables X_i are independent, and $\mathbf{P}(X_1, ..., X_n) = \prod_{i=1}^n \mathbf{P}(X_i)$. Such joint distributions or Markov random fields are called *factorial.* Given a multivariate joint distribution P, it is easy to see that among all factorial distributions, the one that is closest to P in relative entropy is the product of the marginals of P.

8.3 The Directed Case: Bayesian Networks

In the directed case, the family $\mathcal{P}(G)$ corresponds to the notions of Bayesian networks, belief networks, directed independence probabilistic networks, directed Markov fields, causal networks, influence diagrams, and even Markov meshes [34, 44, 12, 11, 22]. As already mentioned, the direction on the edges usually represents causality or time irreversibility. Such models are more common, for instance, in the design of expert systems.

In the directed case, we have a directed graph $G = (V, \vec{E})$. The graph is also assumed to be *acyclic*, that is, with no directed cycles. This is because it is not possible to consistently define the joint probability of the variables in a cycle from the product of the local conditioning probabilities. That is, in general the product $\mathbf{P}(X_2|X_1)\mathbf{P}(X_3|X_2)\mathbf{P}(X_1|X_3)$ does not consistently define a distribution on X_1, X_2, X_3. An acyclic directed graph represents a partial ordering. In particular, it is possible to number its vertices so that if there is an edge from i to j, then $i < j$. In other words, the partial ordering associated with the edges is consistent with the numbering. This ordering is also called a topological sort. We will assume that such an ordering has been chosen whenever necessary, so that, for instance, the past of i $C^-(i)$ is included in $\{1, 2, ..., i-1\}$, and the future $C^+(i)$ is included in $\{i+1, ..., n\}$. The *moral* of $G = (V, \vec{E})$ is the undirected graph $G^M = (V, E+M)$ obtained by removing the direction on the edges of G and by adding an edge between any two nodes that are parents of the same child in G (if they are not already connected, of course). The term "moral" was introduced in [29] and refers to the fact that all parents are "married." We can now describe the Markov independence properties of graphical models with an underlying acyclic directed graph.

8.3.1 *Markov Properties*

A Bayesian network on a directed acyclic graph G is characterized by any one of a number of equivalent independence properties. In all cases, the basic Markov idea in the directed case is that, conditioned on the present, the future is independent of the past or, equivalently, that in order to predict the future, all the relevant information is contained in the present.

8.3.2 *Pairwise Markov Property*

Nonneighboring pairs X_i and X_j with $i < j$ are independent, conditional on all the other variables in the past of j. That is, for any $(i, j) \notin \vec{E}$ and $i < j$,

$$X_i \perp X_j | X_{C^-(j)-\{i\}}. \tag{12}$$

In fact, one can replace $C^-(j)$ with the larger set $\{1, ..., j-1\}$. Another equivalent statement is that, conditional on a set of nodes I, X_i is independent of X_j if and only if i and j are *d- separated*, that is, if there is no d-connecting path from i to j [12]. A d-connecting path from i to j is defined as follows. Consider a node k on a path from i to j. The node k is called linear, divergent, or convergent, depending on whether the two edges adjacent to it on the path are incoming and outgoing, both outgoing, or both incoming. The path from i to j is d-connecting with respect to I if and only if every interior node k on the path is either (1) linear or diverging and not a member of I, or (2) converging, and $[k \cup C^+(k)] \cap I \neq \emptyset$. Intuitively, i and j are d-connected if and only if either (1) there is a causal path between them or (2) there is evidence in I that renders the two nodes correlated with each other.

8.3.3 *Local Markov Property*

Conditional on its parents, a variable X_i is independent of all other nodes, except for its descendants. Thus

$$X_i \perp X_j | X_{N^-(i)}, \tag{13}$$

as long as $j \notin C^+(i)$ and $j \neq i$.

8.3.4 Global Markov Property

If I and J are two disjoint sets of vertices, we say that K separates I and J in the directed graph G if and only if K separates I and J in the moral undirected graph of the smallest ancestral set containing I, J, and K [28]. With this notion of separation, the global Markov property is the same—that is, if K separates I and J,

$$X_I \perp X_J | X_K. \tag{14}$$

It can be also be shown [44] that the directed graph G satisfies all the Markov independence relationships of the associated moral graph G^M. The converse is not true in general, unless G^M is obtained from G by removing edge orientation only, that is, without any marriages. Finally, any one of the three Markov independence properties is equivalent to the statement

$$\mathbf{P}(X_i | X_{C^-(i)}) = \mathbf{P}(X_i | X_{N^-(i)}). \tag{15}$$

In fact, $C^-(i)$ can be replaced by the larger set $\{1, ..., i-1\}$.

8.3.5 Factorization Properties

It is not difficult to see, as a result, that the unilateral local characteristics $\mathbf{P}(X_i | X_{N^-(i)})$ are consistent with each other, and in fact uniquely determine a Bayesian network on a given graph. Indeed, we have

$$\mathbf{P}(X_1, ..., X_n) = \prod_i \mathbf{P}(X_i | X_{N^-(i)}). \tag{16}$$

This property is fundamental. The local conditional probabilities can be specified in terms of lookup tables. This is often impractical, due to the size of the tables. A number of more compact but also less general representations are often used, such as noisy OR- [34] or NN-style representations, such as sigmoidal belief networks [33] for binary variables, where the characteristics are defined by local connection weights and sigmoidal functions, or the obvious generalization to multivalued variables using normalized exponentials. Having a local NN at each vertex to compute the local characteristics is another example of hybrid model parameterization.

8.3.6 Learning and Propagation

There are several levels of learning in graphical models in general, and Bayesian networks in particular, from learning the graph structure itself to learning the local conditional distributions from the data. These will not be discussed here, but reviews and pointers to the literature can be found in [11, 22]. Another fundamental operation with Bayesian networks is the propagation of evidence, that is, the update of the probabilities of each X_i conditioned on the observed node variables. Evidence propagation is NP-complete in the general case [13]. But for singly connected graphs (no more than one path between any two nodes in the underlying undirected graph), propagation can be executed in time linear with n, the number of nodes, using a simple message-passing approach [34, 2]. In the general case, all known exact algorithms for multiply connected network rely on the construction of an equivalent singly connected networks, the junction tree, by clustering the original variables, according to the cliques of the corresponding triangulated moral graph [34, 29, 37], with refinements in [26]. A similar algorithm for the estimation of the most probable configuration of the variables X_i is given in [14]. Schachter et al. [38] show that all the known exact inference algorithms are equivalent in some sense to the algorithms in [26] and [14].

A number of well known algorithms in very different fields, originally derived using specialized techniques have been shown to be special cases of belief propagation and typically of Pearl's algorithm. Such reductions usually have simplified and illuminated the original derivation. Examples of this phenomena include Fourier and Hadamard transforms, a number of decoding algorithms (Gallager-Tanner-Wiberg, Turbo, etc.) [2, 30, 18], Kalman filters (the Rauch-Tung- Streibel smoother), the EM algorithm for hiddden Markov models [39], and the inside-outside algorithm for stochastic context-free grammars [7]. Both emprirical and theoretical evidence suggest that at least in some situations the simple message-passing algorithm of [34] yields reasonably good approximations in the multiply connected case. This is the case in coding theory where the decoding algorithm used with Turbo codes is an instance of belief propagation in belief networks with loops (see [30] for details).

While belief propagation in general remains NP complete, approximate algorithms can often be derived using Monte Carlo methods, such as Gibbs sampling [20, 46], and variational methods, such as mean field theory [36, 25, 18], sometimes leveraging the particular structure of a network (see also [23, 15, 45] for an interesting learning algorithm for a particular class of Bayesian networks). Gibbs sampling is particularly attractive for Bayesian networks, both because of its simplicity and generality.

8.3.7 Gibbs Sampling

Assuming that we observe the values of the variables associated with some of the visible nodes, we want to sample the value of any other node i according to its conditional probability, given all the other variables. From the factorization (16),

$$\mathbf{P}(X_i|X_{V-\{i\}}) = \frac{\mathbf{P}(X_V)}{\mathbf{P}(X_{V-\{i\}})} = \frac{\prod_j \mathbf{P}(X_j|X_{N^-(j)})}{\sum_{x_i} \mathbf{P}(X_1, \ldots, X_i = x_i, \ldots, X_n)}, \tag{17}$$

which yields, after simplifications of common numerator and denominator terms,

$$\mathbf{P}(X_i|X_{V-\{i\}}) = \frac{\mathbf{P}(X_i|X_{N^-(i)}) \prod_{j \in N^+(i)} \mathbf{P}(X_j|X_{N^-(j)})}{\sum_{x_i} \mathbf{P}(X_i = x_i|N^-(i)) \prod_{j \in N^+(i)} \mathbf{P}(X_j|X_{N^-(j)})}. \tag{18}$$

As expected, the conditional distributions needed for Gibbs sampling are local, and depend only on i, its parents, and its children. Posterior estimates can then be obtained by averaging simple counts at each node, which requires very little memory. Additional precision may be obtained by averaging the *probabilities* at each node (see [32] for a partial discussion). As in any Gibbs sampling situation, important issues are the duration of the procedure (or repeated procedure, if the sampler is used for multiple runs) and the discarding of the initial samples ("burn-in"), which can be nonrepresentative of the equilibrium distribution.

8.3.8 Generality

It is worth noting that many models commonly used are instances of Bayesian networks. Artificial feed-forward NNs are Bayesian networks where the local conditional probability functions are delta functions. Likewise HMMs and Markov systems in general have a very simple Bayesian network representation. In fact, HMMs are a special case of both Markov random fields and Bayesian networks. We leave as a useful exercise for the reader to derive these representations, as well as the Bayesian network representation of many other concepts such as mixtures, hierarchical priors, Kalman filters and other state space models, and so on. The generality of the Bayesian network representation is at the root of many new classes of models currently under investigation. This is the case, for instance, of several generalizations of HMMs, such as input-output HMMs [8], tree-structured HMMs [27], and factorial HMMs [19].

References

[1] D. H. Ackley, G. E. Hinton, and T. J. Sejnowski. A learning algorithm for Boltzmann machines. *Cognitive Science*, 9:147–169, 1985.

[2] S. M. Aji and R. J. McEliece. The generalized distributive law. Technical Report, Department of Electrical Engineering, California Institute of Technology, 1997.

[3] B. C. Arnold and S. J. Press. Compatible conditional distributions. *Journal of the American Statistical Association*, 84:152–156, 1989.

[4] P. Baldi and S. Brunak. *Bioinformatics: the machine learning approach*. MIT Press, Cambridge, MA, 1998. In press.

[5] P. Baldi and Y. Chauvin. Hybrid modeling, HMM/NN architectures, and protein applications. *Neural Computation*, 8(7):1541–1565, 1996.

[6] P. Baldi, Y. Chauvin, T. Hunkapillar, and M. McClure. Hidden Markov models of biological primary sequence information. *Proc. Natl. Acad. Sci. USA*, 91:1059–1063, 1994.

[7] P. Baldi and P. Frasconi. Parsing graphical models. Net-ID/DSI Technical Report, April 1998.

[8] Y. Bengio and P. Frasconi. An input-output HMM architecture. In J. D. Cowan, G. Tesauro, and J. Alspector, editors, *Advances in Neural Information Processing Systems*, volume 7. Morgan Kaufmann, San Francisco, CA, 1995. (IEEE).

[9] J. O. Berger. *Statistical decision theory and Bayesian analysis*. Springer-Verlag, New York, 1985.

[10] J. Besag. Spatial interaction and the statistical analysis of lattice systems. *J. Royal Statis. Soc. B*, 36:192–225, 1974.

[11] W. Buntine. A guide to the literature on learning probabilistic networks from data. *IEEE Trans. on Knowledge and Data Eng.*, 8:195–210, 1996.

[12] E. Charniak. Bayesian networks without tears. *AI Mag.*, 12:50–63, 1991.

[13] G. F. Cooper. The computational complexity of probabilistic inference using Bayesian belief networks. *Art. Intell.*, 42:393–405, 1990.

[14] A. P. Dawid. Applications of a general propagation algorithm for probabilistic expert systems. *Stat. and Comp.*, 2:25–36, 1992.

[15] P. Dayan, G. E. Hinton, R. M. Neal, and R. S. Zemel. The Helmholtz machine. *Neural Computation*, 7(5):889-904, 1995.

[16] A. Gelman and T. P. Speed. Characterizing a joint probability distribution by conditionals. *Journal of the Royal Statistical Society B*, 55(1):185–188, 1993.

[17] S. Geman and D. Geman. Stochastic relaxation, Gibbs distributions and the Bayesian restoration of images. *IEEE Trans. Pattern Anal. and Machine Intell.*, 6:721–741, 1984.

[18] Z. Ghahramani. Learning dynamic Bayesian networks. In M. Gori and C. L. Giles, editors, *Adaptive Processing of Temporal Information. Lecture Notes in Artifical Intelligence.* Springer Verlag, Heidelberg, 1998. In press.

[19] Z. Ghahramani and M. I. Jordan. Factorial hidden Markov models. *Machine Learning*, 1997.

[20] W. R. Gilks, A. Thomas, and D. J. Spiegelhalter. A language and program for complex Bayesian modelling. *The Statistician*, 43:69–78, 1994.

[21] John C. Harsanyi. *Rational behavior and bargaining equilibrium in games and social situations.* Cambridge University Press, Cambridge, UK, 1977.

[22] D. Heckerman. Bayesian networks for data mining. *Data Mining and Knowl. Discov.*, 1:79–119, 1997.

[23] G. E. Hinton, P. Dayan, B. J. Frey, and R. M. Neal. The wake-sleep algorithm for unsupervised neural networks. *Science*, 268(5214):1158–1161, 1995.

[24] V. Isham. An introduction to spatial point processes and Markov random fields. *Internat. Statist. Rev.*, 49:21–43, 1981.

[25] T. S. Jaakkola and I. Jordan. Recursive algorithms for approximating probabilities in graphical models. In M. C. Mozer, M. I. Jordan, and T. Petsche, editors, *Advances in Neural Information Processing Systems*, volume 9, pages 487–493. MIT Press, Cambridge, MA, 1997.

[26] F. V. Jensen, S. L. Lauritzen, and K. G. Olesen. Bayesian updating in causal probabilistic networks by local computations. *Comput. Statist. Quart.*, 4:269–282, 1990.

[27] M. I. Jordan, Z. Ghahramani, and L. K. Saul. Hidden Markov decision trees. In M. C. Mozer, M. I. Jordan, and T. Petsche, editors, *Advances in Neural Information Processing Systems*, volume 9, pages 501–507. MIT Press, Cambridge, MA, 1997.

[28] S. L. Lauritzen, A. P. Dawid, B. N. Larsen, and H. G. Leimer. Independence properties of directed Markov fields. *Networks*, 20:491–505, 1990.

[29] S. L. Lauritzen and D. J. Spiegelhalter. Local computations with probabilities on graphical structures and their application to expert systems. *J. Royal Statis. Soc. B*, 50:157–224, 1988.

[30] D. J. C. MacKay, R. J. McEliece, and J. F. Cheng. Turbo decoding as an instance of Pearl's belief propagation algorithm. *IEEE J. Sel. Areas Commun.*, 1997. In press.

[31] J. M. Modestino and J. Zhang. A Markov random field model-based approach to image interpretation. *IEEE Trans. Pattern Anal. Machine Intell.*, 14:606–615, 1992.

[32] R. M. Neal. Probabilistic inference using Markov chain Monte Carlo methods. Technical report. Department of Computer Science, University of Toronto, 1993.

[33] Radford M. Neal. Connectionist learning of belief networks. *Artificial Intelligence*, 56:71–113, 1992.

[34] J. Pearl. *Probabilistic reasoning in intelligent systems.* Morgan Kaufmann, San Mateo, CA, 1988.

[35] J. W. Pratt, H. Raiffa, and R. Schlaifer. *Introduction to statistical decision theory.* MIT Press, Cambridge, MA, 1995.

[36] L. K. Saul and M. I. Jordan. Exploiting tractable substructures in intractable networks. In D. S. Touretzky, M. C. Mozer, and M. E. Hasselmo, editors, *Advances in Neural Information Processing Systems*, volume 8, pages 486–492. MIT Press, Cambridge, MA, 1996.

[37] R. D. Schachter. Probabilistic inference and influence diagrams. *Operation Research*, 36:589–604, 1988.

[38] R. D. Schachter, S. K. Anderson, and P. Szolovits. Global conditioning for probabilistic inference in belief networks. In *Proceedings of the Uncertainty in AI Conference*, pages 514–522, San Francisco, CA, 1994. Morgan Kaufmann.

[39] P. Smyth, D. Heckerman, and M. I. Jordan. Probabilistic independence networks for hidden Markov probability models. *Neural Computation*, 9:227–267, 1997.

[40] D. J. Spiegelhalter, A. P. Dawid, S. L. Lauritzen, and R. G. Cowell. Bayesian analysis in expert systems. *Stat. Sci.*, 8:219–283, 1993.

[41] F. Spitzer. Markov random fields and Gibbs ensembles. *Am. Math. Monthly*, 78:142–154, 1971.

[42] R. E. Tarjan and M. Yannakakis. Simple linear-time algorithms to test the chordality of graphs, test acyclicity of hypergraphs. and selectively reduce acyclic hypergraphs. *SIAM Journal on Computing*, 13(3):566–579, 1984.

[43] J. von Neumann and O. Morgenstern. *Theory of games and economic behavior.* Princeton University Press, Princeton, NJ, 1953. third edition (first edition published in 1944).

[44] J. Whittaker. *Graphical models in applied multivariate statistics.* John Wiley & Sons, New York, 1990.

[45] L. Xu. A unified learning scheme: Bayesian-Kullback Ying-Yang machine, 1996.

[46] J. York. Use of the Gibbs sampler in expert systems. *Artif. Intell.*, 56:115–130, 1992.

Theoretical Approaches to On-line Learning in Smooth Multilayer Nets

David Saad and Magnus Rattray
Neural Computing Research Group, Aston University
Birmingham, UK

Abstract
On-line learning is one of the leading training techniques for neural networks. In this paper we will review recent theoretical approaches, based on methods from non-equilibrium statistical mechanics, for analysing on-line learning in multilayer neural network, concentrating on new approaches for defining optimal learning parameters or rules and on incorporating second order information into the learning process using natural gradient descent.

1 Introduction

Feed-forward neural networks have been extensively applied during the last decade for a variety of classification, regression, prediction and control tasks and are the most commonly used neural network architecture due to their ability to learn random mappings to any desired accuracy, give a sufficient number of hidden units. The most common learning methods for training feed-forward networks may be divided into two main paradigms: batch and on-line learning (and variations between these two extremes). On-line learning is arguably the most efficient way of training large feed-forward networks, especially when the task to be learnt is non-stationary, and is based on instantaneous modifications of the network parameters calculated according to only the latest in a sequence of training examples. This process is inherently stochastic because a new training example is selected at random each time the training error is determined. In batch learning, on the other hand, all training examples are used to determine the training error, leading to a deterministic algorithm.

We will first focus on deriving a compact description of the dynamics using methods from statistical mechanics (section 2) which will then be employed to derive, in a principled way, optimal training parameters and training rules (sections 3, 4 and 5). Once this framework is in place we will examine the efficacy of a novel second order training method, natural gradient descent (section 6) throughout the learning process. In section 7 we will point out the main remaining open questions and future research directions.

2 Learning in soft committee machines

For setting up the basic framework [1, 2] we will consider a learning scenario whereby a feed-forward neural network model, the 'student', emulates an unknown mapping, the 'teacher', given examples of the teacher mapping (in this case another feed-forward neural network) which may be corrupted by noise. This provides a rather general learning scenario since both student and teacher can represent a very broad class of functions [3]. Student performance is typically measured by the generalization error, which is the student's expected error on an unseen example. The object of training is to minimize the generalization error by adapting the student network's parameters appropriately.

We consider a student mapping from an N-dimensional input space $\boldsymbol{\xi} \in \Re^N$ onto a scalar function $\sigma(\mathbf{J}, \boldsymbol{\xi}) = \sum_{i=1}^{K} g(\mathbf{J}_i \cdot \boldsymbol{\xi})$, which represents a soft Committee machine (SCM) [1], where $g(x) \equiv \mathrm{erf}(x/\sqrt{2})$ is the activation function of the hidden units, $\mathbf{J} \equiv \{\mathbf{J}_i\}_{1 \le i \le K}$ is the set of input-to-hidden adaptive weights for the K hidden nodes and the hidden-to-output weights are set to one. The activation of hidden node i in the student under presentation of the input pattern $\boldsymbol{\xi}^\mu$ is denoted $x_i^\mu = \mathbf{J}_i \cdot \boldsymbol{\xi}^\mu$. This general configuration represents most properties of a general multi-layer network and can easily be extended to accommodate adaptive hidden-to-output weights [4].

Training examples are of the form $(\boldsymbol{\xi}^\mu, \zeta^\mu)$ where $\mu = 1, 2, \ldots$ labels each independently drawn example in a sequence. Components of the independently drawn input vectors $\boldsymbol{\xi}^\mu$ are uncorrelated random variables with zero mean and unit variance. The corresponding output ζ^μ is given by a teacher which may be corrupted by output noise and is of a similar configuration to the student except for a possible difference in the number M of hidden units: $\zeta^\mu = \sum_{n=1}^{M} g(\mathbf{B}_n \cdot \boldsymbol{\xi}^\mu) + \rho^\mu$, where $\mathbf{B} \equiv \{\mathbf{B}_n\}_{1 \le n \le M}$ is the set of input-to-hidden adaptive weights for teacher hidden nodes and ρ^μ is zero mean Gaussian noise of variance σ^2. The activation of hidden node n in the teacher under presentation of the input pattern $\boldsymbol{\xi}^\mu$ is denoted $y_n^\mu = \mathbf{B}_n \cdot \boldsymbol{\xi}^\mu$. We will use indices i, j, k, l to refer to units in the student network and n, m for units in the teacher network.

The error made by the student is given by the quadratic deviation,

$$\epsilon(\mathbf{J}^\mu, \boldsymbol{\xi}^\mu) \equiv \frac{1}{2}\left[\, \sigma(\mathbf{J}^\mu, \boldsymbol{\xi}^\mu) - \zeta^\mu \,\right]^2 = \frac{1}{2}\left[\sum_{i=1}^{K} g(x_i^\mu) - \sum_{n=1}^{M} g(y_n^\mu) + \rho^\mu \right]^2 . \qquad (1)$$

This training error is then used to define the learning dynamics via a gradient descent rule for the update of student weights $\mathbf{J}_i^{\mu+1} = \mathbf{J}_i^\mu + \frac{\eta}{N} \delta_i^\mu \boldsymbol{\xi}^\mu$, where $\delta_i^\mu \equiv g'(x_i^\mu) \left[\sum_{n=1}^{M} g(y_n^\mu) - \sum_{j=1}^{K} g(x_j^\mu) + \rho^\mu\right]$ and the learning rate η has been scaled with the input size N. Performance on a typical input in the absence of output noise defines the generalization error $\epsilon_g(\mathbf{J}) \equiv \langle \epsilon(\mathbf{J}, \boldsymbol{\xi}) \rangle_{\{\xi\}}|_{\sigma=0}$ through an average over all possible input vectors $\boldsymbol{\xi}$.

Expressions for the generalization error and learning dynamics have been obtained [2] in the thermodynamic limit ($N \to \infty$), and can be represented by a set of macroscopic variables (order parameters) of the form: $\mathbf{J}_i \cdot \mathbf{J}_k \equiv Q_{ik}$, $\mathbf{J}_i \cdot \mathbf{B}_n \equiv R_{in}$, and $\mathbf{B}_n \cdot \mathbf{B}_m \equiv T_{nm}$, measuring overlaps between student and teacher vectors. The overlaps R and Q become the dynamical variables of the system while T is defined by the task. The learning dynamics is then defined in terms of differential equations for the macroscopic variables with respect to the normalized number of examples $\alpha = \mu/N$ playing the role of a continuous time variable:

$$\frac{\mathrm{dR_{in}}}{\mathrm{d}\alpha} = \eta\ \phi_{in}\ , \qquad \frac{\mathrm{dQ_{ik}}}{\mathrm{d}\alpha} = \eta\ \psi_{ik} + \eta^2\ \upsilon_{ik}\ , \tag{2}$$

where $\phi_{in} \equiv \langle \delta_i y_n \rangle_{\{\xi\}}$, $\psi_{ik} \equiv \langle \delta_i x_k + \delta_k x_i \rangle_{\{\xi\}}$ and $\upsilon_{ik} \equiv \langle \delta_i \delta_k \rangle_{\{\xi\}}$. The explicit expressions for ϕ_{in}, ψ_{ik}, υ_{ik} and ϵ_g depend exclusively on the overlaps Q, R and T and are given in Ref. [2]. The equations of motion, depending on a closed set of parameters, can be integrated and iteratively solved, providing a full description of the order parameters evolution from which the evolution of the generalization error can be derived. Although the dynamical equations considered here are only strictly valid in the large N limit, they have been shown to describe mean behaviour accurately for systems of realistic size [5].

3 Optimal learning rate

On-line methods are often sensitive to the choice of learning parameters and for gradient descent in particular the choice of learning rate can be critical. If the learning rate is chosen too large then the learning process may diverge, but if the learning rate is too low then convergence can take an extremely long time; moreover, in either case the algorithm may get trapped at a sub-optimal fixed point. The appropriate learning rate will also vary substantially over time and may require annealing towards the end of the learning process. Existing analytical results for defining optimal learning rates, based on either the stochastic approximation [6] or the order parameters approach [7] concentrate on the asymptotic regime where the system may be linearized.

The naive approach to learning rate optimization is to consider the fastest rate of decrease in generalization error as a measure of optimality. It is then straightforward to find the locally optimal learning rate by determining the value of η that minimizes $\mathrm{d}\epsilon_g/\mathrm{d}\alpha$, using Eqs.(2) and the fact that the change in generalization error over time depends exclusively on the overlaps. The expression obtained for the locally optimal learning rate [8] may be useful for some phases of the learning process but it is likely to be useless for others. For example, the lowest generalization error for the symmetric phase[2], a pronounced transient phase characterized by lack of differentiation between the student nodes, is achieved by gradually reducing the learning rate towards zero; however, decaying the learning rate in the symmetric phase will prevent the system from escaping the symmetric fixed point, thus resulting in a sub-optimal solution.

A more appropriate measure of optimality is the total reduction in generalization error over the entire learning process [10]. With this measure one can then define the *globally optimal* learning rate in a given time-window $[\alpha_0, \alpha_1]$ to be that which provides the largest decrease in generalization error between these two times. We write the change in generalization error as an integral,

$$\Delta\epsilon_g(\eta) \;=\; \int_{\alpha_0}^{\alpha_1} \frac{\mathrm{d}\epsilon_g}{\mathrm{d}\alpha}\,\mathrm{d}\alpha \;=\; \int_{\alpha_0}^{\alpha_1} \mathcal{L}(\eta,\alpha)\,\mathrm{d}\alpha\ . \tag{3}$$

This functional of η will be minimized by a variational calculation.

Since the generalization error depends solely on the overlaps Q, R and T, which are the dynamical variables (T remains fixed here), we can expand the integrand in terms of these variables,

$$\begin{aligned}\mathcal{L}(\eta,\alpha) &= \sum_{in} \frac{\partial \epsilon_g}{\partial R_{in}} \frac{\mathrm{dR_{in}}}{\mathrm{d}\alpha} + \sum_{ik} \frac{\partial \epsilon_g}{\partial Q_{ik}} \frac{\mathrm{dQ_{ik}}}{\mathrm{d}\alpha} \\ &- \sum_{in} \mu_{in} \left(\frac{\mathrm{dR_{in}}}{\mathrm{d}\alpha} - \eta\, \phi_{in} \right) - \sum_{ik} \nu_{ik} \left(\frac{\mathrm{dQ_{ik}}}{\mathrm{d}\alpha} - \eta\, \psi_{ik} - \eta^2\, \upsilon_{ik} \right) .\end{aligned} \tag{4}$$

The last two terms in equation (4) force the correct dynamics using sets of Lagrange multipliers μ_{in} and ν_{ik} corresponding to the equations of motion for R_{in} and Q_{ik} respectively.

Variational minimization of the integral in equation (3) with respect to the dynamical variables leads to a set of coupled differential equations for the Lagrange multipliers,

$$\begin{aligned}\frac{\mathrm{d}\mu_{\mathrm{jm}}}{\mathrm{d}\alpha} &= -\eta \sum_{in} \mu_{in} \frac{\partial \phi_{in}}{\partial R_{jm}} - \eta \sum_{ik} \nu_{ik} \frac{\partial \left(\psi_{ik} + \eta\, \upsilon_{ik} \right)}{\partial R_{jm}} , \\ \frac{\mathrm{d}\nu_{\mathrm{jl}}}{\mathrm{d}\alpha} &= -\eta \sum_{in} \mu_{in} \frac{\partial \phi_{in}}{\partial Q_{jl}} - \eta \sum_{ik} \nu_{ik} \frac{\partial \left(\psi_{ik} + \eta\, \upsilon_{ik} \right)}{\partial Q_{jl}} ,\end{aligned} \tag{5}$$

along with a set of boundary conditions,

$$\mu_{in}(\alpha_1) = \left. \frac{\partial \epsilon_g}{\partial R_{in}} \right|_{\alpha_1} \quad \text{and} \quad \nu_{ik}(\alpha_1) = \left. \frac{\partial \epsilon_g}{\partial Q_{ik}} \right|_{\alpha_1} . \tag{6}$$

Then taking variations with respect to η we find a simple expression for the globally optimal learning rate,

$$\eta = -\frac{\sum_{in} \mu_{in} \phi_{in} + \sum_{ik} \nu_{ik} \psi_{ik}}{2 \sum_{ik} \nu_{ik} \upsilon_{ik}} . \tag{7}$$

Equations (5), (6) and (7) determine necessary conditions for η to maximize the reduction in generalization error over the interval $[\alpha_0, \alpha_1]$. The boundary conditions correspond to the locally optimal solution, reflecting the fact that at α_1 the choice of η does not affect the dynamics at other times. To find the learning rate which satisfies this set of conditions we use gradient descent on the functional derivative of $\Delta\epsilon_g$ with respect to η,

$$\begin{aligned}\eta(t+1) &= \eta(t) - \Theta \frac{\delta \Delta \epsilon_g}{\delta \eta} , \\ \frac{\delta \Delta \epsilon_g}{\delta \eta} &= \sum_{in} \mu_{in} \phi_{in} + \sum_{ik} \nu_{ik} \left(\psi_{ik} + 2\eta\, \upsilon_{ik} \right) ,\end{aligned} \tag{8}$$

where t is the iteration index and Θ is the step-size for the iteration process. In order to choose an appropriate value for Θ one may also employ second order variations. Standard heuristics can be used to ensure that the iteration process does not diverge if the second order variations become negative, or close to zero.

To update η using Eq.(8), one should first solve Eqs.(5) along with the end conditions (Eq.6), determined by the order parameter values at α_1. For updating η one should therefore: 1) assume some randomly chosen initial conditions for the overlaps and integrate the equations for the overlaps forward using Eqs.(2), starting at α_0 and ending at α_1. 2) Then, calculate the end conditions (Eq.6) using the overlap values at α_1. 3) Solve Eqs.(5) numerically by integrating them backwards using the boundary conditions of Eq.(6). 4) Update the time dependent learning rate using Eq.(8). 5) Use the current time-dependent learning rate to integrate Eqs.(2) forward and iterate the whole process until convergence.

This algorithm converges within a few iterations and results in an exact function for the optimal learning rate over the given time-window. We will now demonstrate the usefulness of the method for several learning scenarios.

4 Numerical results

The theory presented in the previous section and the resulting recipe are completely general and may be employed for any number of teacher and student hidden units and for any teacher correlation matrix T. However, in this section we will focus on students and teachers of equal complexity ($K = M$) and on isotropic teachers ($T_{nm} = \delta_{nm}$). Throughout this paper, unless stated otherwise, initial conditions for the overlaps R_{in} and $Q_{i \neq k}$ are taken randomly from a uniform distribution $U[0, 10^{-6}]$ while the vector lengths Q_{ii} are taken from $U[0, 0.5]$, a choice which corresponds to an input dimension of about $N \simeq 10^{12}$. The optimal η is independent of the initial conditions which mainly affect the length of the symmetric phase.

4.1 Realizable rules

In our first example we consider a realizable ($K = M = 3$) noiseless training task. The time-window is $0 \leq \alpha \leq 600$ and the learning rate is initially fixed at some arbitrary value. The update in equation (8) is then iterated until convergence and figure 1 shows results for the dynamics using the globally optimal learning rate. Figures 1(a)–(d) show the optimal learning rate, generalization error and overlaps respectively. After a short initial transient both the learning rate and generalization error stabilize at almost constant values, corresponding to a symmetric phase in which the student nodes have not yet specialized to particular teacher nodes, as required to learn perfectly. The overlaps also exhibit a plateau within this phase and figure 1(c) shows that the student-teacher overlaps are almost indistinguishable (the indices have been ordered *a posteriori* so that student node i eventually specializes to teacher node i). The learning rate takes a value of about $\eta \simeq 0.97$ within the symmetric phase, which is in close agreement to the optimal value obtained numerically [11]. Eventually, the

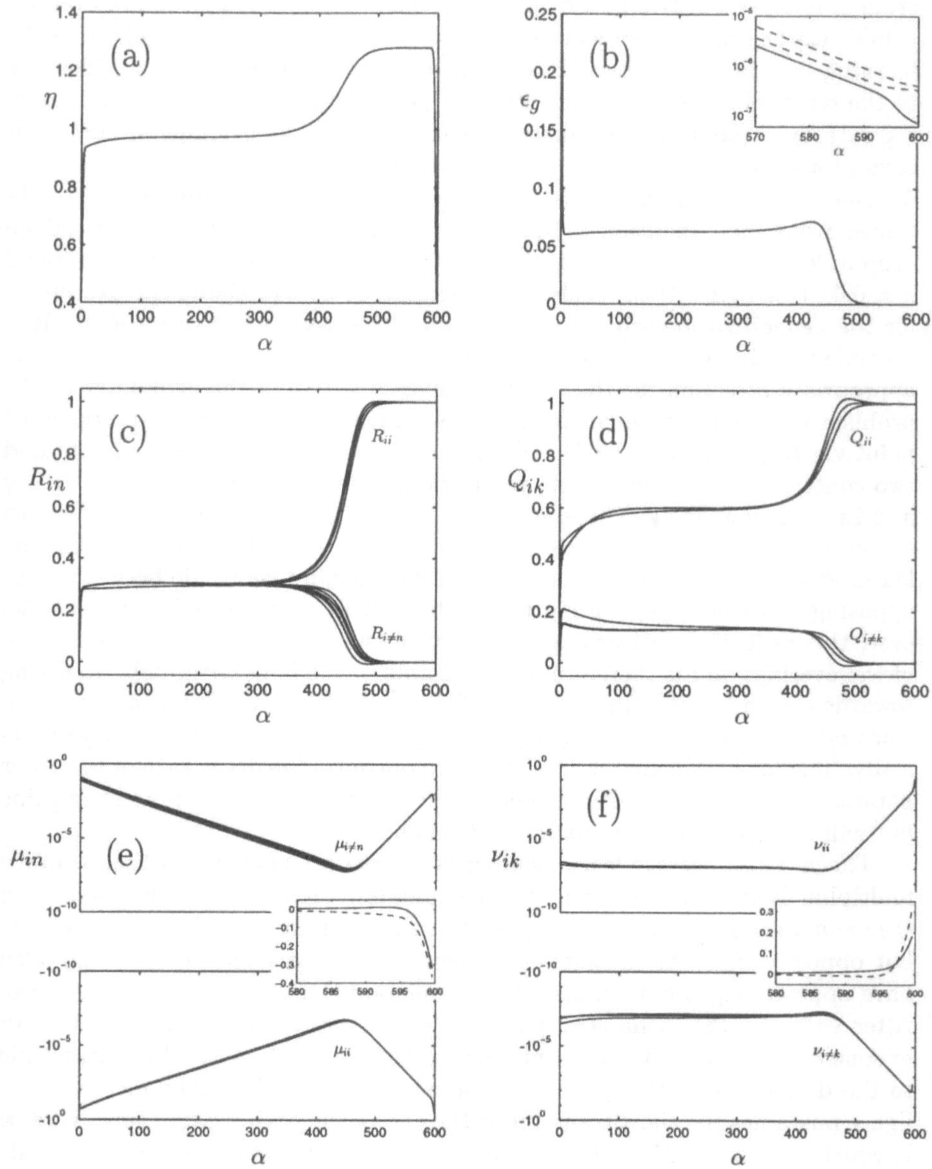

Figure 1: Results are presented for a three hidden node student trained to emulate an isotropic teacher of the same configuration. The globally optimal learning rate is shown in (a) along with the corresponding evolution of the generalization error and order parameters in (b),(c) and (d). The inset of (b) shows the generalization error (solid line) and the magnitude of the opposing contributions to the leading term (dashed lines – upper proportional to $2r - q$, lower proportional to $2s - c$). The Lagrange multipliers are shown in (e) and (f) using a log scale, with the later stages magnified in each inset (dashed line for a curve associated with the lower figure).

student escapes the symmetric phase and the generalization error and overlaps exhibit exponential convergence towards their respective optimal values, as the learning rate increases towards another constant value of $\eta = 1.28$, identical to the result obtained independently [11] for the asymptotically optimal learning rate by expanding the dynamical equations for the overlaps around their asymptotic fixed point.

One significant difference between the results obtained here and those obtained from the expansion around the asymptotic fixed point is the sudden drop in learning rate towards the end of the time-window to a value of around $\eta = 0.41$ (see figure 1(a)). This can be explained by examining the expression for the generalization error in the vicinity of its asymptotic fixed point. It is possible to gain an immediate reduction in generalization error by choosing an appropriate direction for the decay eigenvectors. Using the symmetry of the problem we expand the expression for the generalization error around the fixed point via $R_{in} = \delta_{in}(1-r) + (1-\delta_{in})s$ and $Q_{ik} = \delta_{ik}(1-q) + (1-\delta_{ik})c$ to find two contributions to the leading term of opposite sign, proportional to $2r - q$ and $2s - c$ respectively. These are shown in the inset to figure 1(b) along with the corresponding generalization error for $570 \leq \alpha \leq 600$. By reducing the learning rate it is possible to reduce the difference in magnitude between these opposing contributions, leading to a reduction in generalization error. However, this reduction in learning rate slows down the exponential convergence of the overlaps and is therefore unsustainable in the long term, only occurring towards the end of the time-window. This example shows how locally optimal learning does not necessarily give good long-term performance, even asymptotically. The long-term goal in this case is to optimize the decay rate of the order parameters, while changes in the decay direction can provide short-term gains but will eventually lead to poorer performance.

The various phases of learning described above are mirrored by the Lagrange multiplier dynamics shown in figures 1(e) and (f). Figure 1(e) shows how during the symmetric phase μ_{ii} and $\mu_{i \neq n}$ decay exponentially with similar magnitude but opposite sign. At the same time figure 1(f) shows that ν_{ii} and $\nu_{i \neq k}$ also have opposite signs but remain almost constant during the symmetric phase. After escaping the symmetric phase all the Lagrange multipliers exhibit an exponential growth with the same constant rate, which is equal in magnitude to the decay rate of the generalization error at this point. The inset to each figure magnifies the short transient at the end of the optimization time-window in which the exponential growth is interrupted as each Lagrange multiplier finds the appropriate boundary value.

Notice that the dynamics of the overlaps and Lagrange multipliers forms a small number of bundled similar trajectories, reflecting symmetries in the task. By exploiting these symmetries the dimensionality of the system can be reduced significantly, allowing a compact description for arbitrary K and T. This dimensionality reduction has already been used to study the different phases of learning in [2, 11, 8], obtaining generic results and scaling properties of learning parameters and training times with respect to K and T.

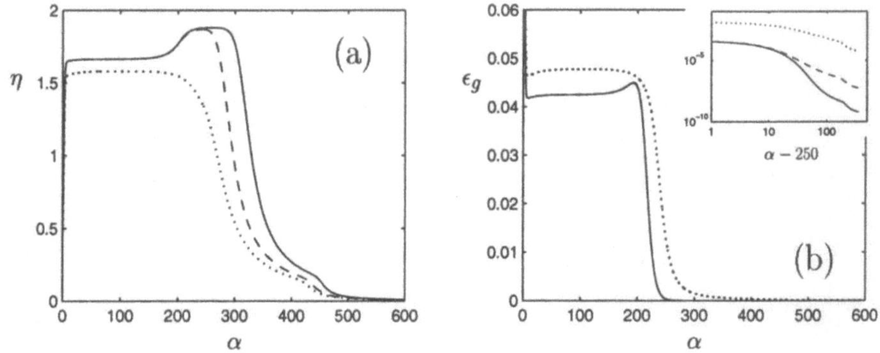

Figure 2: A two hidden node student is trained on noise corrupted examples generated by an isotropic teacher of the same configuration. The optimal learning rate is shown in (a) for three noise levels $\sigma^2 = 10^{-2}$, 10^{-5}, 10^{-7} (from left to right) over a fixed time-window $0 \leq \alpha \leq 600$. The corresponding generalization error is shown in (b) with the inset showing a log-log plot for the decay at late times, which indicates that the dashed and solid lines are only split after $\alpha \simeq 300$.

4.2 Noise corrupted examples

In our second example we consider an unrealizable learning scenario by introducing additive uncorrelated Gaussian noise of zero mean and variance σ^2 to the teacher's output. Qualitatively similar results are obtained for structural unrealizability ($K < M$). The picture that emerges, shown in figure 2 for various noise levels ($\sigma^2 = 10^{-2}$, 10^{-5} and 10^{-7}), is initially similar to the realizable case but changes dramatically as the system escapes the symmetric phase. As the system begins convergence towards zero generalization error[1], as shown in figure 2(b), the optimal learning rate shown in figure 2(a) begins to fall and slowly approaches a decay inversely proportional to α, proved to be optimal for linear systems (e.g., [12] and references therein), until reaching a greedy phase (after the kink around $\alpha = 440$ in figure 2). The rapid reduction in generalization error after this point is achieved by changing the decay direction, as for the realizable learning scenario described above.

Figures 2(a) and (b) show a log-log plot of the learning rate and generalization error respectively as a function of α for optimization over time-windows of varying length. One observes that both the learning rate and generalization error approach a decay proportional to $1/\alpha$ and that the curves lie on top of one another until the greedy phase which occurs towards the end of each time

[1]Recall that this is the generalization error in the absence of noise. The prediction error for a noisy teacher has an additive constant contribution equal to half the noise variance.

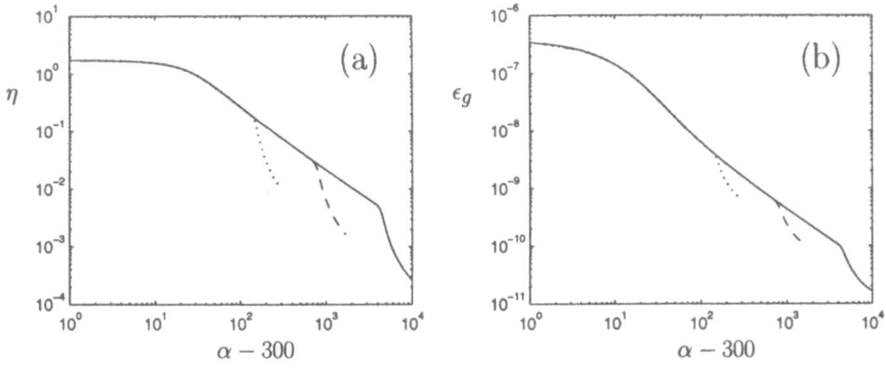

Figure 3: As in figure 2, a two node student is trained on noisy examples from a teacher of the same configuration. Here the noise level is fixed at $\sigma^2 = 10^{-7}$ while the optimization process is carried out over different time windows $0 \leq \alpha \leq \alpha_1$ with $\alpha_1 = 600$, 2000 and 10^4 (from left to right). The asymptotic decay of the learning rate (a) and generalization error (b) are shown for each case. The initial behaviour is similar to that presented in figure 2.

window. However, unlike the realizable case where the drop-off in learning rate occurs over a relatively short time, here the final greedy phase increases in length as the total learning time-window increases and this phase always takes a significant proportion of the learning time. This is simply a reflection of the slower decay time-scale for this problem.

On a practical note, our results suggest that as symmetry breaks one should gradually modify the decay rate from a constant until it is proportional to $1/\alpha$. However, it may take a prohibitively long time until the $1/\alpha$ decay becomes optimal, making it irrelevant in many instances. Moreover, if one decays the learning rate at a fixed rate it may take an extremely long time before losses, incurred due to the use of sub-optimal learning rates in earlier stages of the dynamics, can be recovered. Annealing the learning rate during the symmetric phase could even lead to trapping, since the length of the symmetric phase scales inversely to η for small η [2].

4.3 Further analysis

The overlap dynamics is often dominated by fixed points (the symmetric and convergence phases) which makes it amenable to analysis by linear expansion. Analysing the variational algorithm in the neighbourhood of these fixed points enables us to gain valuable insight into how the algorithm works as well as scaling laws for the optimal parameters and the length of the symmetric phase [8].

We simplify the problem by focusing on realizable scenarios ($K = M$) and isotropic tasks ($T_{nm} = T\delta_{nm}$), whereby the problem dimensionality can be reduced, for determining generic behaviour for variable T and K.

Using the reduced dimensionality equations of motion makes it possible to find the optimal learning rate for arbitrary K, since we avoid the increase in computation time necessary to deal with large systems. We can therefore run our algorithm for various values of K and T in order to deduce scaling laws for the optimal learning rate within the two phases of learning. However, our algorithm simply performs gradient descent on the relevant eigenvalue within each phase [8] and the results should therefore compare closely to results obtained by optimizing with respect to this eigenvalue directly [11].

Comparing the optimal learning rate determined by our method to the value found by a direct numerical eigenvalue analysis [11] for various values of K we find that the results agree well. This outcome is rather fortuitous, as the scaling laws previously determined for each phase in [11] also describe the scaling behaviour for the globally optimal learning rate. This was not obvious *a priori* since it was not known that the optimal time-dependent learning rate would be dominated by two plateaus, each with a constant learning rate.

It is interesting to summarize some of the scaling laws deduced in [11] (for fixed T). During the symmetric phase the optimal learning rate scales as $K^{-5/3}$ for large K and the trapping time as $K^{8/3}$. During the convergence phase the optimal learning rate scales as $1/K$. The maximal learning rate in each phase, above which perfect learning is impossible, scales in the same way as the optimal learning rate. One therefore finds that using a learning rate which is optimal asymptotically will be very bad during the transient, either leading to trapping within the symmetric phase or divergence of the student weight vector norms.

4.4 Limitations to the variational approach

Before moving on one should point out the possible limitation of our variational algorithms if eigenvalues are multi-modal, as might be the case if the equations of motion involved higher powers, or non-linear functions, of the parameters being optimized. In this case the algorithm may get stuck in functional local minima, as it effectively carries out gradient descent/ascent on the dominant eigenvalue.

Certain initial settings for η may also result in convergence to functional local minima which are sub-optimal. For example, if η is very small initially and the time-window is not sufficiently long for the system to leave the symmetric phase then the algorithm may anneal η in an attempt to optimize performance without leaving the symmetric subspace. As with any differential method we are at the mercy of local minima; our conditions for the globally optimal learning rate are necessary but not sufficient. From careful study of the dynamics we are satisfied that all the solutions presented in this work are globally optimal.

5 Optimal learning rule

So far we concentrated on optimizing the learning parameters keeping the learning rule similar to standard gradient descent. This section will focus on optimizing the update rule itself, restricted to the general form $\mathbf{J}_i^{\mu+1} = \mathbf{J}_i^{\mu} + \frac{1}{N} F_i^{\mu}(\boldsymbol{x}^{\mu}, \zeta^{\mu})\boldsymbol{\xi}^{\mu}$, where $\mathbf{F} = \{F_i\}$ depends only on the student activations and the teacher's output. Note that gradient descent takes this general form, as does Hebbian learning and other training algorithms commonly used in discrete machines. The optimal $\mathbf{F}$ can also depend on the self-averaging statistics which describe the dynamics, since we know how they evolve in time. Some of these would not be available in a practical application, although for some simple cases the unobservable statistics can be deduced from observable quantities [13, 14]. This is therefore an idealization rather than a practical algorithm and provides a bound on the performance of a real algorithm.

The general equations of motion can easily be derived for this case,

$$\frac{\mathrm{dR_{in}}}{\mathrm{d}\alpha} = \langle F_i y_n \rangle \qquad \frac{\mathrm{dQ_{ik}}}{\mathrm{d}\alpha} = \langle F_i x_k + F_k x_i + F_i F_k \rangle \ . \tag{9}$$

Angled brackets denote averages over activations, replacing the averages over inputs, which cannot be calculated explicitly without knowing the explicit form of the rule. We will make use of this general formulation for deriving an expression for the globally optimal learning rule, found by maximizing the total reduction in generalization error shown by Eq.(3) where the expression for the integrand is of the form

$$\begin{aligned} \mathcal{L}(\mathbf{F},\alpha) &= \sum_{in} \frac{\partial \epsilon_g}{\partial R_{in}} \frac{\mathrm{dR_{in}}}{\mathrm{d}\alpha} + \sum_{ik} \frac{\partial \epsilon_g}{\partial Q_{ik}} \frac{\mathrm{dQ_{ik}}}{\mathrm{d}\alpha} - \sum_{in} \lambda_{in} \left(\frac{\mathrm{dR_{in}}}{\mathrm{d}\alpha} - \langle F_i y_n \rangle \right) \\ &- \sum_{ik} \nu_{ik} \left(\frac{\mathrm{dQ_{ik}}}{\mathrm{d}\alpha} - \langle F_i x_k + F_k x_i + F_i F_k \rangle \right) \ . \end{aligned} \tag{10}$$

The expression for $\mathcal{L}$ still involves averages over $\mathbf{x}$ and $\mathbf{y}$ (angled brackets), so taking variations in $\mathbf{F}$, which may depend on $\mathbf{x}$ and ζ but not on $\mathbf{y}$, we find an expression for the optimal rule in terms of the Lagrange multipliers $\mathbf{F} = -\mathbf{x} - \boldsymbol{\nu}^{-1}\boldsymbol{\lambda}\overline{\mathbf{y}}/2$, where $\boldsymbol{\nu} = [\nu_{ij}]$ and $\boldsymbol{\lambda} = [\lambda_{in}]$. We define $\overline{\mathbf{y}}$ to be the teacher's expected field given the teacher's output and the student activations, which are observable quantities:

$$\overline{\mathbf{y}} = \int \mathbf{dy}\, \mathbf{y}\, p(\mathbf{y}|\mathbf{x}, \zeta) \ . \tag{11}$$

Taking variations w.r.t. to the dynamical variables we find a set of differential equations for the Lagrange multipliers,

$$\begin{aligned} \frac{\mathrm{d}\lambda_{\mathrm{km}}}{\mathrm{d}\alpha} &= -\sum_{in} \lambda_{in} \frac{\partial \langle F_i y_n \rangle}{\partial R_{km}} - \sum_{ij} \nu_{ij} \frac{\partial \langle F_i x_j + F_j x_i + F_i F_j \rangle}{\partial R_{km}} \\ \frac{\mathrm{d}\nu_{\mathrm{kl}}}{\mathrm{d}\alpha} &= -\sum_{in} \lambda_{in} \frac{\partial \langle F_i y_n \rangle}{\partial Q_{kl}} - \sum_{ij} \nu_{ij} \frac{\partial \langle F_i x_j + F_j x_i + F_i F_j \rangle}{\partial Q_{kl}} \ , \end{aligned} \tag{12}$$

where **F** takes its optimal value . The boundary conditions for the Lagrange multipliers are as before (Eq.6), and are found by minimizing the rate of change in generalization error at α_1. If the above expressions do not yield an explicit formula for the optimal rule then the rule can be determined iteratively by gradient descent on the functional $\Delta\epsilon_g(\mathbf{F})$, employing the same iterative procedure as in section 3.

In order to apply the above result we must be able to carry out the average in Eq. (11) and then in Eqs. (12). These averages are also required to determine the locally optimal learning rule [14], so that the present method can be extended to any of the problems which have already been considered under the criteria of local optimality. Here we present an example where the averages can be computed in closed form.

In this example of an over-realizable task the teacher is an analogue perceptron ($M = 1$) while the student is a SCM with an arbitrary number (K) of hidden nodes. We choose the activation function $g(x) = \mathrm{erf}(x/\sqrt{2})$ for both the student and teacher. The locally optimal rule for this scenario has recently been determined [15].

Since the teacher is invertible, the expected teacher activation $\overline{y}$ is trivially equal to the true activation y. This leads to a particularly simple form for the dynamics (the n suffix is dropped since there is only one teacher node),

$$\frac{\mathrm{dR_i}}{\mathrm{d}\alpha} = b_i T - R_i \qquad \frac{\mathrm{dQ_{ik}}}{\mathrm{d}\alpha} = b_i b_k T - Q_{ik} \ , \tag{13}$$

where we have defined $b_i = -\sum_j \nu_{ij}^{-1}\lambda_j/2$ and the optimal rule is given by $F_i = b_i y - x_i$. The Lagrange multiplier dynamics in eqs. (5) then show that the relative ratio of each Lagrange multiplier remains fixed over time, so that b_i is determined by its boundary value (see equation (6)). It is then straightforward to find solutions for long times, since the b_i approach limiting values for very small generalization error (there are a number of possible solutions with similar performance due to symmetries). For example, one possible solution is to have $b_1 = 1$ and $b_i = 0$ for all $i \neq 1$, which leads to an exponential decay of weights associated with all but a single node. This shows how optimal performance is achieved when the complexity of the student matches that of the teacher.

Figure 4 shows results for a three node student learning an analogue perceptron. Clearly, the locally optimal rule performs poorly in comparison to the globally optimal rule. In this example the optimal rule arrived at was one in which two nodes became correlated with the teacher while a third became anti-correlated, showing another possible variation on the optimal rule. The locally optimal rule gets caught in a symmetric plateau (as in section 3) and also displays a slower asymptotic decay.

It should be pointed out that the optimal rules derived here will often require knowledge of macroscopic properties related to the teacher's structure and therefore do not provide practical algorithms as they stand (although some of the required macroscopic properties may be evaluated or estimated on the basis of data gathered as the learning progresses). However, they do provide

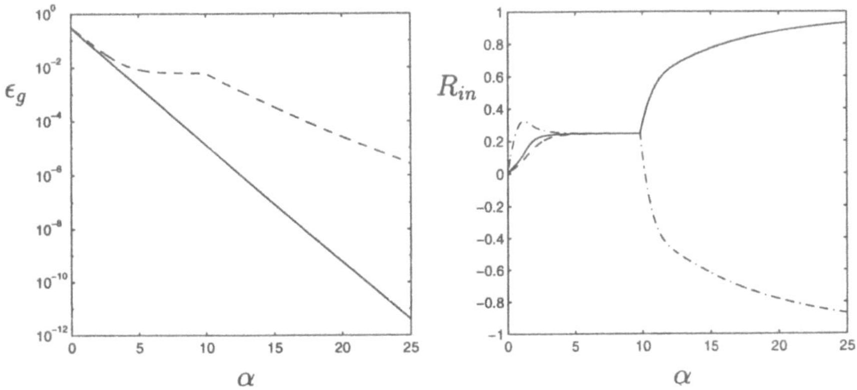

Figure 4: A three node soft committee machine student learns from an analogue perceptron teacher. The figure on the left shows a log plot of the generalization error for the globally optimal (solid line) and locally optimal (dashed line) algorithms. The figure on the right shows the student-teacher overlaps for the locally optimal rule, which exhibit a symmetric plateau before specialization occurs.

an upper bound on the performance one could expect from a real algorithm and may be instrumental in designing practical training algorithms.

6 Natural Gradient Descent

Natural gradient descent (NGD) was recently proposed by Amari as a more principled alternative to standard on-line gradient descent [16]. When learning to emulate a stochastic rule with some probabilistic model this learning algorithm has the desirable properties of asymptotic optimality, given a sufficiently rich model which is differentiable with respect to its parameters, and invariance to reparameterizations of our model distribution. This latter property is achieved by viewing the parameter space of the model as a Reimannian space in which local distance is defined by the KL-divergence [17, 18]. The Fisher information matrix then plays the role of a Reimannian metric in this space. The natural gradient learning rule is obtained by pre-multiplying the standard gradient with the inverse of this matrix. In practice, we require knowledge of the input distribution in order to determine the Fisher information matrix. Yang and Amari discuss methods of pre-processing training examples to obtain a whitened Gaussian process for the inputs [18]. If this is possible, then when the input dimension N is large compared to the number of hidden units K, inversion of the Fisher information for two-layer feed-forward networks requires only $O(N^2)$ operations, providing an efficient and practical algorithm.

In this section we quantify the benefits of NGD using the statistical mechanics description of the learning process and this allows us to compare performance with standard gradient descent in both the transient and asymptotic phases, and to obtain generic results in terms of task complexity and non-linearity. We show that trapping time in the symmetric phase is significantly reduced by using NGD and exhibits a slower power law increase as task complexity grows. We also find that asymptotic performance is greatly improved.

Consider a mapping of the type discussed previously for a SCM, where one defines a Gaussian noise model for output $\zeta_{\rm m}$ given input $\boldsymbol{\xi}$ which is parameterized by $\mathbf{J}$,

$$p_{\mathbf{J}}(\zeta_{\rm m}|\boldsymbol{\xi}) = \frac{1}{\sqrt{2\pi\sigma_{\rm m}^2}} \exp\left(-\frac{(\zeta_{\rm m} - \sum_{i=1}^K g(x_i))^2}{2\sigma_{\rm m}^2}\right) . \tag{14}$$

The training error at each learning iteration is proportional to the log-likelihood of the current example under our noise model and under standard gradient descent we adapt the student weights in the negative gradient direction of this error at each iteration. However, such an algorithm is not consistent with our probabilistic interpretation of the problem, since it depends on our particular choice of model parameterization. A more principled learning algorithm can be defined by viewing the manifold of models as a Reimannian space in which local distance is defined by the KL-divergence [18]. The Fisher information matrix $\mathbf{G} = [G_{i\alpha,k\beta}]$ (where $1 \leq i,k \leq K$ and $1 \leq \alpha, \beta \leq N$) defines the appropriate metric in this space [16],

$$G_{i\alpha,k\beta} = \left\langle \frac{\partial \log p_{\mathbf{J}}(\zeta_{\rm m}|\boldsymbol{\xi})}{\partial J_{i\alpha}} \, \frac{\partial \log p_{\mathbf{J}}(\zeta_{\rm m}|\boldsymbol{\xi})}{\partial J_{k\beta}} \right\rangle_{\{\zeta_{\rm m},\xi\}} . \tag{15}$$

The brackets denote an average over $\zeta_{\rm m}$, according to equation (14), followed by an average over the input distribution. The natural gradient direction is found by pre-multiplying the training error gradient by the inverse of this matrix.

Amari has determined the Fisher information matrix for a general two-layer network. For our particular choice of activation function and with components of $\boldsymbol{\xi}$ selected independently each iteration from a zero-mean Gaussian distribution with unit variance, we find $\mathbf{G} = \mathbf{A}/\sigma_{\rm m}^2$, where

$$\mathbf{A}_{ik} = \frac{2}{\pi\sqrt{\Delta}}\left[\mathbf{I} - \frac{1}{\Delta}\left((1+Q_{kk})\mathbf{J}_i\mathbf{J}_i^{\rm T} + (1+Q_{ii})\mathbf{J}_k\mathbf{J}_k^{\rm T} - Q_{ik}(\mathbf{J}_i\mathbf{J}_k^{\rm T} + \mathbf{J}_k\mathbf{J}_i^{\rm T})\right)\right] \tag{16}$$

with $Q_{ik} \equiv \mathbf{J}_i^{\rm T}\mathbf{J}_k$ and $\Delta = (1+Q_{ii})(1+Q_{kk}) - Q_{ik}^2$.

We then use the statistical mechanics description of the learning process where training example outputs are generated by a teacher network corrupted by Gaussian noise,

$$p_{\mathbf{B}}(\zeta^\mu|\boldsymbol{\xi}^\mu) = \frac{1}{\sqrt{2\pi\sigma^2}} \exp\left(-\frac{(\zeta^\mu - \sum_{n=1}^M g(y_n))^2}{2\sigma^2}\right) . \tag{17}$$

The weight update at each iteration of NGD is given by,

$$\mathbf{J}_i^{\mu+1} = \mathbf{J}_i^\mu + \frac{\eta}{N}\sum_{j=1}^K \mathbf{A}_{ij}^{-1}\delta_j^\mu\boldsymbol{\xi}^\mu , \tag{18}$$

where δ_i^μ is defined after Eq.(3). Notice that knowledge of the noise variance is not required to execute this algorithm.

The Fisher information matrix can be inverted using the partitioning method described in [18] and each block is some additive combination of the identity matrix and outer products of the student weight vectors. It is then straightforward to derive equations of motion for the set of order parameters as we did for standard gradient descent.

Although our equations of motion are sufficient to describe learning for arbitrary system size, the number of order parameters is $\frac{1}{2}K(K-1)+KM$ so that the numerical integration soon becomes rather cumbersome as K and M grow and analysis becomes difficult. To obtain generic results in terms of system size we therefore exploit symmetries which appear in the dynamics for isotropic tasks and structurally matched student and teacher ($K = M$ and $T = T\delta_{nm}$). In this case we define a four dimensional system via $Q_{ij} = Q\delta_{ij} + C(1-\delta_{ij})$ and $R_{in} = R\delta_{in} + S(1-\delta_{in})$ which can be used to study the dynamics for arbitrary K and T (here, δ_{ij} denotes the Kronecker delta). In [19] we show how the Fisher information matrix can be inverted for this reduced dimensionality system. At the cost of some generality we therefore obtain a much simplified dynamical system which is amenable to analysis.

As for standard gradient descent [2], the dynamics is characterized by two major phases of learning. Initially, the order parameters are trapped in the symmetric unstable fixed point; small perturbations lead to an escape from this phase and convergence towards zero generalization error. If the teacher is corrupted by noise then the learning rate must be annealed at late times in order for the generalization error to decay. The fastest decay for NGD is achieved by setting $\eta = 1/\alpha$ and this leads to a inverse decay law for the generalization error. This choice saturates the Cramer-Rao bound and provides asymptotic performance equalling the best batch algorithm [16]. Numerical solutions of the dynamical equations show that the benefit of NGD over standard gradient descent is two-fold: the symmetric phase is shortened significantly and better asymptotic performance is obtained. In the following we will focus on the results of the transient dynamics as NGD has already been proved to be optimal asymptotically [16]. For a detailed analysis of the asymptotic case we refer the reader to Ref. [19].

Unfortunately, even for standard gradient descent an analytical study of the symmetric phase is only possible for small learning rates, which are far from optimal. Such an approach is not appropriate for realistic learning rates and often gives misleading results. It is also unclear how to proceed for natural gradient learning even in this limit, since the Fisher information is singular at the fixed point considered in [2]. In order to obtain generic results in this case we apply the variational method of section 3 for obtaining globally optimal performance, with respect to the choice of training parameters, for both gradient descent and NGD.

We note that the impact of output noise on the symmetric phase dynamics is not considered explicitly here. For low noise levels there is no noticeable effect on the length of the symmetric phase, or on the order parameters and

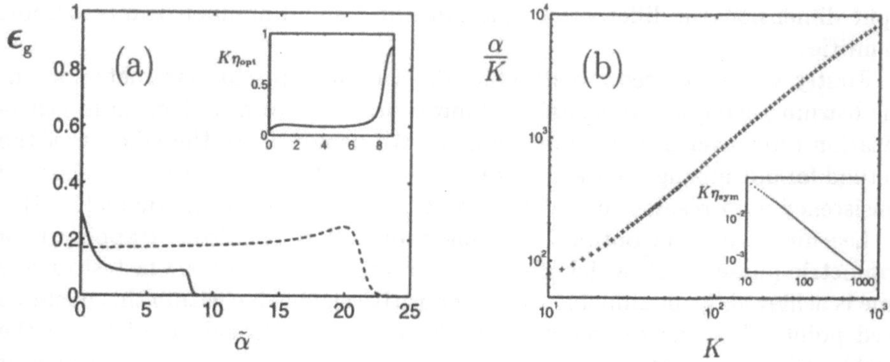

Figure 5: In (a) the generalization error is shown for optimal natural gradient learning (solid line) and optimal gradient descent (dashed line) for $K = 10$, $T = 1$ and zero noise (we define $\tilde{\alpha} = 10^{-2}\alpha$). The inset shows the optimal learning rate for NGD. In (b) the time required for optimal natural gradient learning to reach a generalization error of $10^{-4}K$ is shown as a function of K on a log-log scale. The inset shows the optimal learning rate within the symmetric phase. In both (a) and (b) we used initial conditions $R = 10^{-3}$, $Q = U[0, 0.5]$ and $S = C = 0$.

generalization error within this phase. For larger noise levels the symmetric phase increases in length and the student norms increase, resulting in a larger generalization error. However, we feel that these are secondary effects and that most essential features of this phase are captured by the noiseless dynamics. This is not true for later stages of learning, where the inclusion of noise completely alters qualitative features of the dynamics.

In figure 5(a) we compare optimal performance for $K = 10$ and $T = 1$, which indicates a significant shortening of the symmetric phase for NGD (the inset shows the optimal learning rate). Figure 5(b) shows the time required for NGD to reach a generalization error of $10^{-4}K$ as a function of K (for $T = 1$). The learning time is dominated by the symmetric phase, so that these results provide a scaling law for the length of the symmetric phase in terms of task complexity. We find that the escape time for NGD scales as K^2, while the inset shows that the learning rate within the symmetric phase scales as K^{-2}. Scaling laws for gradient descent were determined in [11], showing a $K^{\frac{8}{3}}$ law for escape time and a learning rate scaling of $K^{-\frac{5}{3}}$ within the symmetric phase. The escape time for the adaptive gradient learning rule studied in [11] scales as $K^{\frac{5}{2}}$, which is also worse than for NGD.

7 Summary and Discussion

In this paper we concentrated on principled methods for optimizing on-line learning. These methods make use of a statistical mechanics model which allows a compact and exact description of the learning process for a SCM with large input dimension via differential equations for a small number of macroscopic quantities.

Firstly, we introduced a method for determining optimal parameters in on-line learning using a variational maximization of the total reduction in generalization error over a fixed time-window. We demonstrate the efficacy of the method for optimizing the learning rate in noisy and noiseless scenarios showing consistency with results obtained via other, more restricted, methods [2, 11].

Learning with the optimal learning rate still suffers from trapping in a symmetric phase [1, 2], which dominates training time, and the fastest escape time is achieved by maximizing the only positive eigenvalue within this unstable fixed point. A separate analytical study of our variational algorithm in the neighbourhood of fixed points [8], shows this to be the expected behaviour of our algorithm and explains the excellent agreement with [11], where the escape eigenvalue within this phase was maximized explicitly. This method also enables one to obtain generic scaling rules for the optimal learning rate and the escape time from the symmetric phase.

Asymptotically, in the absence of noise, we find the same optimal learning rates as obtained in [1, 2], except for one discrepancy: a sharp drop in the optimal learning rate as one approaches the end of the time window. We explained this phenomena which is rooted in a greedy unsustainable rotation of the system's eigenvectors, in a manner that decreases the generalization error for the short run.

Learning from corrupted examples provides a very different picture. After leaving the symmetric phase the optimal learning rate is gradually annealed towards a decay inversely proportional to α. There is a greedy phase, more emphasized than in the noiseless case, towards the end of the optimization time-window, reflecting a change in the decay direction which provides a short-term improvement but is unsustainable for longer times. Our results suggest that there is some danger in annealing the learning rate too early, since losses due to an initially low learning rate might never be recovered and the learning process could even become trapped within the symmetric phase.

Secondly, we employed the same method for determining the optimal on-line rule for a SCM. We provided a simple example of an over-realizable learning scenario where the rule could be determined in closed form and it was shown how the locally optimal rule performed poorly in comparison to the globally optimal rule. It is expected that more involved systems will show even greater difference in performance between local and global optimization and we are currently applying the method to more general teacher mappings. The main technical difficulty is in computing the expected teacher activation in equation (11) which may require the use of approximate methods.

Thirdly, we have analysed natural gradient learning, a novel method motivated by methods of information geometry [16], under the statistical mechanics framework. We find significant improvements over standard gradient descent in both the transient and asymptotic stages of learning, with improved scaling of learning time against task complexity. The major drawback with using NGD is that the input distribution is required in order to determine the Fisher information matrix exactly, and for non-Gaussian inputs it is unclear whether inversion can be carried out efficiently. Efficient averaging and inversion may be achieved using the matrix momentum algorithm suggested by Orr and Leen [21] and we are currently investigating this approach within the present framework.

AcknowledgementThis work was supported by EPSRC grant GR/L19232.

References

[1] M. Biehl and H. Schwarze, "Learning by Online Gradient Descent", *J. Phys.* A, **28**, 643-656, (1995).

[2] D. Saad and S.A. Solla, "Exact Solution for On-Line Learning in Multilayer Neural Networks", *Phys. Rev. Lett.*, **74**, 4337-4340, (1995) ; D. Saad and S.A. Solla, "On-Line Learning in Soft Committee Machines", *Phys. Rev. E*, **52**, 4225-4243, (1995).

[3] G. Cybenko, "Approximation by Buperpositions of a Sigmoid Function", *Math. Control Signals and Systems*, **2**, 303-314, (1989).

[4] P. Riegler and M. Biehl, "Online Backpropagation in 2-layered Neural Networks", *J. Phys.* A, **28**, L507-513, (1995).

[5] D. Barber, D. Saad and P. Sollich, "Finite Size Effects in On-Line Learning of Multilayer Neural Networks", *Euro. Phys. Lett.*, **34**, 151-156, (1996).

[6] H. Robbins and S. Monro, "A Stochastic Approximation Method" *Ann. Math. Stat.* **22** 400-407, (1951).

[7] T.K. Leen, B. Schottky and D. Saad, "Two Approaches to Optimal Annealing", Advances in Neural Information Processing Systems, NIPS **10** - Proceedings of the 1997 conference, Eds. M.I. Jordan, M.J. Kearns and S.A. Solla, MIT press, Cambridge MA, in press, (1998).

[8] M. Rattray and D. Saad, "An Analysis of On-line Training with Optimal Learning Rates", preprint, (1998).

[9] M. Rattray and D. Saad, "Globally Optimal Rules for On-line Learning in Multilayer Networks", *Jour. Phys. A*, **30**, L771-776, (1997).

[10] D. Saad and M. Rattray, "Globally Optimal Parameters for On-line Learning in Multilayer Networks", *Phys. Rev. Lett.*, **79**, 2578-2581, (1997).

[11] A.H.L. West and D. Saad, "On-line Learning with Adaptive Back-propagation in Two-layer Networks", *Phys. Rev. E*, **56**, 3426-3445, (1997).

[12] H. White, "Learning in Artificial Neural Networks: a Statistical Perspective", *Neural Computation*, **1**, 425-464, (1989).

[13] M. Biehl, P. Reigler and M. Stechert, "Learning from Noisy Data: An Exactly Solvable Model", *Phys. Rev.* E, **52**, R4624-4627, (1995).

[14] O. Kinouchi and N. Caticha, "Optimal Generalization in Perceptrons", *J. Phys.* A, **25**, 6243-6250, (1992).

[15] R. Vicente and N. Caticha, "Functional Optimization of Online Algorithms in Multilayer Neural Networks", *J. Phys.* A, **30**, L599-605, (1997).

[16] S. Amari, "Natural Gradient Works Efficiently in Learning", preprint, (1997).

[17] H. H. Yang and S. Amari, "Efficiency and Robustness of Natural Gradient Descent Learning Rule", Advances in Neural Information Processing Systems, NIPS **10** - Proceedings of the 1997 conference, Eds. M.I. Jordan, M.J. Kearns and S.A. Solla, MIT press, Cambridge MA, in press, (1998).

[18] H. H. Yang, S. Amari "Natural Gradient Descent for Training Multi-Layer Perceptrons" submitted to *IEEE Transactions on Neural Networks* (1998).

[19] M. Rattray, D. Saad, S. Amari "Natural Gradient Descent for On-line Learning", in preparation, (1998).

[20] Y. LeCun, P.Y Simard and B. Pearlmutter, "Automatic Learning Rate Maximization by On-Line Estimation of the Hessian's Eigenvectors", Advances in Neural Information Processing Systems, NIPS **5** - Proceedings of the 1992 conference, Eds. C.L. Giles, S.J. Hanson and J.D. Cowan, Morgan Kaufmann, San Mateo CA, 156-163, (1993).

[21] G.B. Orr and , T.K. Leen, "Using Curvature Information for Fast Stochastic Search", Advances in Neural Information Processing Systems, NIPS **6** - Proceedings of the 1993 conference, Eds. J.D. Cowan, G. Tesauro and J. Alspector, Morgan Kaufmann, San Mateo CA, 477-484, (1994).

Section 3
Review Paper

Combining Neural Networks and Hidden Markov Models for Speech Recognition

Edmondo Trentin
ITC-Irst (Istituto per la Ricerca Scientifica e Tecnologica)
Povo (Trento), Italy
and Università di Firenze, V. S. Marta, 3 - Firenze (Italy)
trentin@irst.itc.it

Marco Gori
Dipartimento di Ingegneria dell'Informazione
Università di Siena, V. Roma, 56 - Siena, Italy
marco@neuron.ing.unisi.it

Abstract

Automatic Speech Recognition (ASR) is a challenging classification task over sequences of acoustic features, extracted from the vocal signal, according to some linguistic model. Hidden Markov models (HMM) are the most common and successful tool for ASR, allowing for high recognition performance in a variety of difficult tasks (speaker independent, large vocabulary, continuous speech). In spite of that, standard HMMs suffer from relevant limitations. Artificial neural networks (ANN) have been proposed as an alternative, nonparametric paradigm for ASR since the late Eighties. ANNs were successfully applied in reduced size tasks (e.g., phoneme or isolated word recognition) but they did not succeed in solving the general ASR problem, due to their lack of capability to model long-term dependencies. Combining ANN and HMM within a unifying framework is a suitable approach to ASR, which takes benefit from both techniques and overcomes the corresponding limitations. This paper reviews the main topics concerning hybrid HMM/ANN systems for ASR, summarizing some major architectures of this kind.

Keywords: automatic speech recognition, hidden Markov model hybrid speech recognizer, global optimization, rescoring.

1 Introduction

Among the major topics of speech processing, Automatic Speech Recognition (ASR) [49, 18, 38, 33, 17] is probably the most investigated, as well as challenging. An informal description of ASR formulates the latter as the problem of automatically *recognizing* (not *understanding*) the words from a given dictionary uttered by a speaker, relying only on the information contained in the uttered *speech signal* and on prior knowledge on the problem domain.

Complexity of the problem is strictly related to increasing dictionary size (e.g. more than 50,000 words), continuous speech recognition versus isolately uttered words, number and vocal characteristics of speakers in Speaker Independent (SI) recognizers (versus simpler Speaker Dependent (SD) systems, capable to recognize only speech uttered by the speaker whose acoustic material was used to train the recognizer itself), robustness to environmental conditions (noise and distortion over the channel, multiple microphones spread throughout the room to allow hands-free dictation, etc.), and so on.

State-of-the-art ASR systems are successfully applied in a variety of scenarios: dictation, that is automatic generation of written text from the speech signal, access to databases, human-machine interface, access to remote automatic services on the telephone line, and control of machines.

More formally, the ASR problem can be formulated as a statistical *classification* problem. Once the *classes* have been defined as sequences W of allowable words from a given dictionary, a parametric representation of the speech signal has been chosen (e.g. a sequence of acoustic feature vectors X), and a *Maximum a Posteriori* (MAP) criterion has been adopted, the classification problem can then be stated as finding the sequence of words $\tilde{W}$ which maximizes the quantity $Pr(W \mid X)$. The latter is usually factorized using Bayes' theorem [19] as:

$$Pr(W \mid X) = \frac{Pr(X \mid W)Pr(W)}{Pr(X)}. \tag{1}$$

Given an acoustic observation sequence X, the efforts on the maximization of $Pr(W \mid X)$ can be moved to the search for the class $\tilde{W}$ which maximizes the numerator of the right-hand side of Equation 1, i.e. $Pr(X \mid W)Pr(W)$. The quantity $Pr(W)$, usually referred to as the *language model* [50] depends on high-level constraints and linguistic knowledge about allowed word strings for the specific task. The quantity $Pr(X \mid W)$ is known as the *acoustic model.* It describes the statistics of sequences of parametrized acoustic observations in the feature space given the corresponding uttered words (e.g. certain phonemes).

The feature space is commonly defined by standard signal processing techniques, operating a preprocessing on a raw digital representation of the acoustic signal (the well-known *waveform*) with the goal to obtain a low-dimensional representation of the signal itself in terms of a sequence of discrete-time *acoustic frames*, i.e. vectors of relevant parameters extracted on short-term partially-overlapping windows (e.g. 20msec.) of the waveform. A popular and effective choice is represented by the so-called Mel Scaled Cepstral Coefficients (MSCC) [16], computed by performing spectral analysis of the time-windows and applying a Mel-scaled triangular filterbank to the spectra. A certain number of MSCC (usually ranging from 8 to 12) is retained, along with the energy of the signal, and arranged into acoustic frames completed with the addition of their first- and second-order time derivatives. An example of the projection along the first three Principal Components of such a parametric representation for the vocal signal collected from a sample speaker is shown in Figure 1.

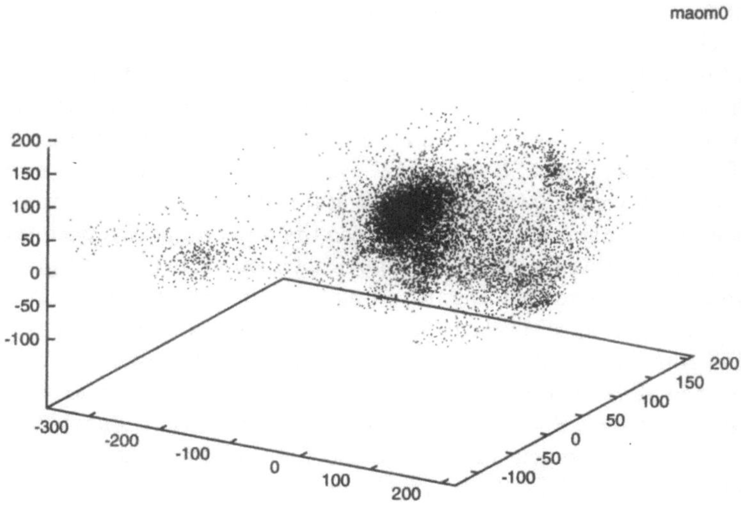

Figure 1: Projection along the first three Principal Components of MSCC representation of the vocal signal collected from a sample speaker.

Hidden Markov models (HMM) [51, 32] are the most popular (parametric) model at the acoustic level. A brief review of HMMs is presented in Section 2. Although HMMs are effective approaches to the problem of acoustic modeling in ASR, allowing for good recognition performance under many circumstances, they also suffer from some limitations. These limitations, discussed at the end of Section 2, are the rationale behind the research for different paradigms.

Section 3 briefly reviews attempts made by researchers between the end of the Eighties and the early Nineties, to use ANNs for ASR. ANNs were successfully applied on reduced size tasks (e.g., phoneme or isolated word recognition) but they did not succeed in solving the general ASR problem, due to their lack of capability to model long-term dependencies. This fact finally lead to the development of hybrid ANN/HMM system, properly combining the basic paradigms within a unifying and effective framework capable to take benefit from both techniques, overcoming their limitations. Section 4 summarizes some major architectures and algorithms for the hybrid approach.

2 Hidden Markov models for ASR and their limitations

An HMM is a pair of stochastic processes: an *hidden* Markov chain and an *observable* process which is a probabilistic function of the states of the former.

This means that observable events in the real world (e.g., acoustic observations) are modeled with (possibly continuous) probability distributions, that are the observable part of the model, associated with individual states of a discrete-time, first-order Markovian process. The semantics of the model (conceptual correspondence with physical phenomena) is usually encapsulated in the hidden part: for instance, in ASR an HMM can be used to model a word in the task-dependent vocabulary, where each state of the hidden part represents a phoneme (or sub-phonetical unit), whereas the observable part accounts for the statistical characteristics of the corresponding acoustic events in a given feature space (e.g. sampled acoustic signal, represented in a proper way, like MSCCs).

More precisely, an HMM is defined by: a set S of Q states, $S = \{S_1, \ldots, S_Q\}$; an *initial state* probability distribution π; a probability distribution that characterizes the allowed transitions between states, that is $\mathbf{a}_{ij} = \{Pr(S_j \text{ at time } t \mid S_i \text{ at time } t-1), S_i \in S, S_j \in S\}$ where the *transition probabilities* $\mathbf{a}_{ij}$ are assumed to be independent of time t; a set of probability distributions (referred to as *emission* probabilities) $\mathbf{b_x} = \{b_i(\mathbf{x}) = Pr(\mathbf{x} \mid S_i), S_i \in S\}$, where $\mathbf{x}$ is an acoustic feature vector. Whenever emission probabilities are modeled by means of continuous probability density functions (e.g. mixtures of Gaussian components) the term *Continuous Density*-HMM (CDHMM) is used.

Proper estimates of π, $\mathbf{a}$ and $\mathbf{b_x}$ from the data are accomplished by using the *forward-backward* (or *Baum-Welch*) [51] and the Viterbi [51] algorithms. Whenever continuous emission probabilities are considered, both of them are based on the general Maximum Likelihood (ML) criterion, i.e. they aim at maximizing the probability of the samples given the model at hand. In particular, the Viterbi algorithm concentrates only on the most alike path throughout all the possible sequences of states in the model. Once trained, HMMs can be used for *decoding* or *recognition*. If the ML criterion is used, the Forward-Backward or the Viterbi algorithms are still suitable for the recognition task.

HMMs have been successfully applied in a variety of tasks, mainly in speech recognition. Unfortunately, standard HMMs in conjunction with the above-mentioned training and decoding algorithms suffer from some major intrinsic limitations. The combination of ANNs with HMMs is intended here as an attempt to overcome some of these limitations (see for instance [10]). Standard CDHMMs, trained with Forward-Backward or Viterbi algorithms, present poor discriminative power among different models, since they are based on the ML criterion, which is itself non-discriminative. The classical HMMs rely on strong assumptions on the statistical properties of the phenomenon at hand. For instance, the stochastic processes involved are modeled by first-order Markov chains, and the parametric form of the probability density functions that represent the emission probabilities associated with all states is heavily constraining. In addition, the number of parameters in HMMs do strongly limit their implementability in hardware. Given these limitations, the use of ANNs with their discriminative training, their capability to perform non-parametric estimation over whole sequences of patterns, and their limited number of parameters (which allows for effective development of neural microchips) appear definitely promising.

3 ANNs applied to ASR and their limitations

Starting from the late Eighties, many researchers began to use Artificial Neural Networks (ANN) for ASR. Neural nets were expected to carry out the recognition task (e.g. classification of phonemes or words) when discriminatively trained on acoustic features. Milestones in this respect are [63, 64, 65, 26, 28, 3, 25, 14, 29, 62, 58, 15, 55, 8, 56], among the others. Lippmann [41] wrote a comprehensive survey of the state of the art in connectionist speech recognition at the end of the Eighties.

Here is a short survey of the fundamental connectionist architectures that were applied to ASR:

1. *Multilayer Perceptron* (MLP): the ability of MLPs to model a linear or non-linear discriminant function for classification tasks, or a regressor in multivariate spaces, makes the feedforward connectionist approach a suitable method for many applications in speech processing. Important examples will be given in the next sections, especially concerning the integration of MLPs in an ASR system based on hidden Markov models. MLPs have been used for recognizing phonemes from portions of spectrograms [5], for grapheme-to-phoneme translation [59, 60], and for mapping acoustic features obtained with a microphone array with the same features obtained with a close mouth [13]. In [20] the *self-supervised* version of MLP is applied to the extraction of acoustic features from spectrograms.

2. *Learning Vector Quantization* (LVQ): LVQ [34, 35] yielded good results in speech coding [36]. In [42] LVQ is used for phoneme-level segmentation of speech signal. Excellent examples of the application of LVQ in ASR can be found in [43, 44], where multi-category, shift-tolerant phoneme recognition experiments are reported.

3. *Time-Delay Neural Network* (TDNN): also known as *tapped delay lines*, TDNNs represent an effective attempt to train a static MLP for time-sequence processing, by converting the temporal sequence into a spatial sequence over corresponding units. The idea is very simple, but useful in a variety of ASR applications, mostly for phoneme recognition [63, 64, 10]. Lang and Hinton [37] obtained a 7.8% error rate in multi talker classification of the isolated letters "B, D, E, V", using acoustic material collected among 100 male speakers. Waibel *et al.* [65] were able to recognize isolated consonants uttered by a Japanese speaker with a low error rate (4.1%), using a combination of specialized TDNNs. A significant 1.4% error rate in vowel recognition was obtained in the same experiments.

4. *Recurrent neural network* (RNN): it has been applied in a variety of speech processing tasks. Particularly remarkable are the results obtained for phoneme recognition by Robinson et al. [54, 58, 56], where RNNs are used as state-space machines, capable to compute an output and the next state, given the input and the current state. This results in a non-linear

extension of linear control theory. The approach gave the best recognition performance on phoneme recognition with the TIMIT corpus and was also used for phone probability estimation. Recently, Robinson *et al.* [57] have integrated RNNs into a hybrid HMM/ANN speech recognizer, where the networks are used to compute posterior state probabilities (see section 4.1). The system had slightly lower performance than some HMM based system in the *Wall Street Journal* dictation task, but it required far less model parameters to train. A recurrent network for plosive recognition was proposed in [6, 21].

In spite of their ability to classify short-time acoustic-phonetic units, such as individual phonemes, ANNs failed as a general framework for ASR, especially with long sequences of acoustic observations like those required in order to represent words from a dictionary or whole sentences. This is mainly due to the lack of ability to model long-term dependencies in ANNs, even when recurrent architectures are considered. The theoretical motivations underlying this problem were well analyzed by [2]. In the early Nineties this fact led to the idea of combining HMMs and ANNs within a single, novel model, broadly known as *hybrid HMM/ANN* [24, 39, 11, 45, 48, 27, 62, 7].

4 Combining ANNs and HMMs within hybrid paradigms

A variety of different architectural and algorithmic solutions have been proposed in the literature. Some ANN/HMM hybrids assume that the output of an ANN is sent to an HMM for ASR [39, 24, 27, 62]. In [5, 7], the ANN is used to transform a vector of acoustic features into more effective observations for HMMs, according to a "global optimization" of the parameters of the combined system.

Earlier approaches (between the late Eighties and the beginning of the Nineties) relied on ANN architectures that attempted to emulate HMMs [12, 48]. Some of them incorporated the Dynamic Programming (DP) algorithm [5] within the ANN itself [27, 40]. The architectures proposed by Bourlard *et al.* rely on a probabilistic interpretation of the ANN outputs [10, 46]. In other cases [5, 27, 62], the ANN outputs are not interpreted as probabilities, but are meant as scores which are used within a DP algorithm to perform alignment and segmentation. As an alternative, the ANN can be used for re-scoring the N-best hypotheses produced with an HMM [66].

In this paper we take into consideration three major classes of hybrid systems:

1. hybrid with ANN estimate of state-posterior probabilities of an underlying HMM

2. hybrid with connectionist feature extraction and global optimization of ANN and CDHMM parameters

3. ANNs for rescoring of standard CDHMMs with *N-best* strategy.

4.1 Hybrid with ANN estimate of state-posterior probabilities of an underlying HMM

Bourlard *et al.* [11, 45, 9, 10] proposed HMM/ANN hybrids for continuous ASR in which a MLP was trained to estimate the posterior probabilities of HMM states, with the ultimate objective of maximizing the posterior probability of a given (left-to-right) Markov model M_i given an acoustic observation sequence X. Posterior probabilities can be approximated by writing [10]:

$$Pr(M_i \mid X) \approx \sum_{q_1^L} \{ \prod_{\ell=1}^{L} Pr(q_\ell \mid \mathbf{x}_{\ell-k}, ..., \mathbf{x}_{\ell+k}, q_{\ell-1}) \} Pr(M_i \mid q_1^L) \qquad (2)$$

where the model M_i is supposed to have Q states, the acoustic observation sequence $X = (\mathbf{x}_1, ..., \mathbf{x}_L)$ is assumed to be of length L, and the sums are extended over all possible sequences q_1^L of states.

The quantity $Pr(M_i \mid q_1^L)$ does not depend on the acoustics (observation sequence X), but only on higher-level choices made in the definition of the models and can thus be computed separately.

A feed-forward network is then trained to estimate the state posterior, called *conditional transition probability*, $Pr(q_\ell \mid \mathbf{x}_{\ell-k}, \ldots, \mathbf{x}_{\ell+k}, q_{\ell-1})$ given a fixed number $2k+1$ of acoustic vectors $\mathbf{x}_{\ell-k}, \ldots, \mathbf{x}_{\ell+k}$ (a *window* or *context* of size k centered in the current acoustic observation $\mathbf{x}_\ell$) and the previous state. This is accomplished using the BP algorithm on a MLP, which has an output unit for each state, that represents the estimate of the state posterior probability. In so doing, advantages are taken from ANNs discriminant training and from their capability to estimate Bayesian posterior probabilities when trained by BP on a Mean Squared Error (MSE) criterion. Other attempts were made to use the MLP to estimate the posteriors in different ways, for example as a function of the current observation and of the previous state only, or as a function also of previous MLP outputs (previous states).

In practice, in order to make the system work, a step backward was made from posteriors to likelihoods. This issue was pursued by using a somewhat standard version of the HMM, in conjunction with neural networks. The latter were trained to perform exactly the same probability estimation as in Equation (2), but with their outputs divided by the *a priori* probabilities of the corresponding states, in order to reduce probabilities to scaled likelihoods, normalized by the unconditional likelihood of each observation (using Bayes' theorem). Priors can be computed apart, from the training data or from statistical considerations on the constraints given by the specific task. This solution was effective and allowed the system to reach the recognition performance of state-of-the-art HMM recognizers on large vocabulary continuous speech tasks, but at the expense of the original theoretical framework.

One central point raised with this approach concerns the training procedure. Indeed, networks are trained by BP, which would require knowledge of target values for the outputs in order to compute the gradient of the cost function. With the exception of toy tasks, no supervised labeling of acoustic frames is actually available (labeling by hand in real-sized databases is not feasible). Bourlard *et al.* suggested an iterative training procedure that starts up with an initial segmentation of the acoustic observations, performs training of the networks according to that segmentation, then uses the Viterbi algorithm, in conjunction with the newly trained networks as estimators of the state-posterior probabilities, in order to produce a new and more reliable segmentation of the data, that in turn is used to train again the networks, and so on in an iterative fashion. The initial segmentation may be obtained using a standard HMM, or by dividing the observation sequence into equally-sized segments; each segment is associated to the corresponding state in the correct HMM state-sequence.

The approach by Bourlard *et al.* was successfully adopted in [53] for the DARPA Resource Management (RM) SI, continuous speech task, with a vocabulary size of 991. An HMM with 2 or 3 states left-to-right phone models was used, with parameter tying of the pdfs associated to the states of a given model. Experimental results obtained averaging on the three datasets used for testing, using 12 MSCCs plus energy and the corresponding first-order time-derivatives and a word-pair grammar (that reduced the perplexity to 60), yielded an average error of 8.7% for the context-independent hybrid, a 7.0% for the context-dependent CDHMM and a 5.5% for the interpolation of the two models. Moreover, results on one of the test sets (February91) showed considerably better performance of the context-independent hybrid (5.8%) with respect to the context-independent CDHMM (11.0%).

Variants on Bourlard's approach are those proposed by [61, 22, 22, 23, 56, 31, 30, 52] and others.

4.2 Hybrid with connectionist feature extraction and global optimization of ANN and CDHMM parameters

[7, 4, 5] describes an hybrid model obtained by using a connectionist *feature extractor*, the output of which is fed into a standard CDHMM. The novelty of the approach, originally inspired from the Alpha net [12], relies on a gradient-based global optimization technique for the parameters of the ANN and those of the HMM according to a common, overall criterion.

The ANN is trained with the goal to transform input acoustic representations into compact, but significant, low-dimensional representations that are more suitable to be modeled by the emission probabilities of the HMM than standard acoustic parameters. In so doing, an increased recognition rate is expected.

The ANN maps a raw input sequence $\mathbf{u}_1^L$ onto a low-dimensional observation sequence $\mathbf{x}_1^L$. An assumption is made that the optimization criterion C

used to train the HMM is a continuous and differentiable function of the observations $\mathbf{x}_t, t = 1, \ldots, L$, so that a gradient descent optimization method can be applied. The partial derivatives $\frac{\partial C}{\partial x_t}$ (here x_t denotes a generic component of t-th observation $\mathbf{x}_t$) can then be computed and used (along with the *chain rule*) to train the weights w of the ANN.

The computation of $\frac{\partial C}{\partial x_t}$ turns out to be very simple, since the Forward-Backward algorithm implicitly computes the derivative of the HMM likelihood $Pr(\mathbf{x}_1^L|Y)$ with respect to the emission probabilities $b_{i,t} = Pr(\mathbf{x}_t|q_t = S_i)$ [1].

Beering in mind the definitions and the notation introduced in Section 2 and referring to [51] for details, it can be stated that in the forward pass the Baum-Welch algorithm computes:

$$\alpha_{i,t} = b_{i,t} \sum_j a_{ij}\alpha_{j,t-1}, \tag{3}$$

where $\alpha_{i,t} = Pr(q_t = S_i, \mathbf{x}_1^t)$. It can be shown [51] that the likelihood $Pr(\mathbf{x}_1^L)$ can be obtained from the α's at the last time step, for the set $\mathcal{F}$ of "final" states of the HMM, as follows:

$$Pr(\mathbf{x}_1^L) = \sum_{i \in \mathcal{F}} \alpha_{i,L}. \tag{4}$$

Let us define $\beta_{i,t} = Pr(\mathbf{x}_{t+1}^L|q_t = S_i)$, which can be computed recursively [51] as:

$$\beta_{i,t} = \sum_j a_{ji}\beta_{j,t+1}b_{j,t+1}. \tag{5}$$

Proceeding by induction and applying the chain rule, it is possible to write:

$$\frac{\partial Pr(\mathbf{x}_1^L)}{\partial \alpha_{i,t}} = \beta_{i,t}. \tag{6}$$

Derivatives with respect to the emission probabilities can now be obtained from Equation (3):

$$\frac{\partial Pr(\mathbf{x}_1^L)}{\partial b_{i,t}} = \frac{\alpha_{i,t}\beta_{i,t}}{b_{i,t}}. \tag{7}$$

When considering the case in which the training criterion C is the likelihood of the observations, and using an HMM constrained by the correct word sequence Y, the gradient of the criterion C with respect to the emission probabilities can be immediately obtained from Equation (7). To compute the gradient with respect to the ANN outputs and thus, via chain rule, with respect to the ANN weights, the computation of the partial derivatives of the emission density

[1] For notational convenience, we write $b_{i,t}$ instead of $b_i(\mathbf{x}_t)$ to denote emission probability of state S_i at time t, keeping in mind that the observation sequence $\mathbf{x}_1^L$ is under consideration.

function with respect to the observations is needed. [7, 5] give an example in the case of Gaussians or Gaussian mixtures emission pdfs.

Experimental results showed that training the ANN jointly with the HMM in the proposed hybrid architecture improved recognition performance over the standard CDHMM [7, 5], lowering the error rate on a plosive recognition task from 19% to 14%.

4.3 ANNs for *rescoring* of hypothesis in CDHMMs with *N-best* strategy

The idea of connectionist *rescoring* was proposed in [66] and further developed in [67]. An ANN is applied to the problem of rescoring the hypothesis generated by an HMM which uses an N-best strategy. In this case the network does not compute scores on individual acoustic frames, but on whole *segments* (sub-sequences) of frames, corresponding to phonemes. In this way, correlations between frames that are close in time - and that correspond to the same phoneme - are exploited, thus overcoming the usual limitations following the assumption of independence made in standard HMMs.

The connectionist model used for this purpose is called Segmental Neural Network (SNN) [1]. It takes a given acoustic segment in input, and produces an estimation of the posterior probability of a certain phoneme given that segment. The overall score, which is actually used in the rescoring process of a whole sentence is obtained by multiplying the scores of the segments in which the sentence has been segmented by the HMM. The score obtained this way is then linearly combined with that computed by the HMM, yielding a total score for the sentence given a specific segmentation. The weights for the linear combination are empirically determined by cross-validation on a development set.

Since the input layer of the networks has fixed size, while the segments are made up of a variable number of frames, two alternative techniques of "normalization" of the length of the input window are proposed. The former is a semilinear temporal sampling, which drops some of the original frames if the segment is longer than the input dimensionality, or replicates some of them if the segment is shorter. The alternative is a discrete cosine transform (DCT) applied to the whole segment, retaining as many parameters as it is necessary to fill the input layer.

Training of the system is accomplished using a segmentation performed by the HMM, possibly considering the first N-best hypothesis of segmentation - where segmentations from the 2nd best down to the Nth best are used as negative examples. Networks are trained on a relative entropy criterion; the outputs are normalized in order to fit the probabilistic framework.

Different neural architectures were tried (1-layer Perceptron, MLP, HyperBF) [67] without significant fluctuations in performance. They were also combined altogether to obtain a more robust rescoring process.

The system was evaluated on the DARPA RM, SI, continuous speech task, with a 1,000 words vocabulary and a bigrams grammar (perplexity 60), using also duration modeling, one left context (at the segment level) and a regularization technique to improve performance. The rescoring approach obtained combining CDHMM, SNN based on MLP and HyperBF, improved the WER of the bare CDHMM from 3.4% to 2.7% on the test set RM Feb91, and from 6.0% to 5.5% averaging on the two test sets RM Sep92.

Another application of ANNs for rescoring of a standard HMM with N-best strategy can be found in [47].

5 Conclusion

ASR is a difficult classification task, defined over sequences of acoustic features extracted from speech signal by means of proper signal processing techniques. HMMs allow to model the local statistical properties of acoustic events relying on discrete or continuous-density probability distributions, known as emission probabilities, and to model the long-term dependencies by incorporating them within an hidden first-order Markov chain, the states of which are used to represent acoustic units (e.g. phonemes). HMMs have been successfully applied to ASR throughout the last years, but they suffer from some major limitations. The latters are due both to theoretic as well as to practical drawbacks. Several attempts have been made by researchers, between the end of the Eighties and the beginning of the Nineties, to overcome these limitations with the adoption of a different, non-parametric paradigm for recognition, relying on ANNS. ANNs lad to satisfactory performance on small ASR tasks, namely phoneme and isolated word recognition, but failed in the general ASR problem of large vocabulary, continuous speech recognition. This fact was probably due to the difficulty of modeling long-term dependencies with ANNs. Hybrid ANN/HMM systems were then proposed as a viable attempt to join the basic paradigms, taking advantage from their specific properties and overcoming their limitations. Successful examples of such hybrids, relying on connectionist posterior-probability estimation of HMM states, on global optimization of a connectionist feature extractor for an HMM, or on ANNs for rescoring the N-best hypothesis of a standard HMM, have been illustrated. Experimental results show that combining ANNs with HMMs indeed allow to improve recognition performance, and even to reduce the overall number of parameters of the system whenever ANNs are used for probability estimate.

Acknowledgments

The authors wish to thank Yoshua Bengio and Renato De Mori for their precious suggestions and stimulating discussions. Special thanks to Cesare Furlanello.

References

[1] S. Austin, G. Zavaliagkos, J. Makhoul, and R. Schwartz. Speech recognition using segmental neural nets. In *International Conference on Acoustics, Speech and Signal Processing*, pages I–625–628, San Franscisco, March 1992.

[2] Y. Bengio, , P. Simard, and P. Frasconi. Learning long-term dependencies with gradient descent is difficult. *IEEE Transactions on Neural Networks*, 5(2):157–166, 1994. Special Issue on Recurrent Neural Networks, March 94.

[3] Y. Bengio. Radial basis functions for speech recognition. In *Speech Recognition and Understanding: Recent Advances, Trends and Applications*, pages 293–298. NATO Advanced Study Institute Series F: Computer and Systems Sciences, 1990.

[4] Y. Bengio. A connectionist approach to speech recognition. *International Journal on Pattern Recognition and Artificial Intelligence*, 7(4), 1993.

[5] Y. Bengio. *Neural Networks for Speech and Sequence Recognition*. International Thomson Computer Press, London, UK, 1996.

[6] Y. Bengio, R. De Mori, G. Flammia, and R. Kompe. Phonetically motivated acoustic parameters for continuous speech recognition using artificial neural networks. In *Proceedings of EuroSpeech'91*, 1991.

[7] Y. Bengio, R. De Mori, G. Flammia, and R. Kompe. Global optimization of a neural network-hidden Markov model hybrid. *IEEE Transactions on Neural Networks*, 3(2):252–259, 1992.

[8] Y. Bengio, M. Gori, and R. De Mori. Learning the dynamic nature of speech with back-propagation for sequences. *Pattern Recognition Letters*, 13(5):375–386, 1992. (Special issue on Artificial Neural Networks).

[9] H. Bourlard and N. Morgan. Continuous speech recognition by connectionist statistical methods. *IEEE Trans. on Neural Networks*, 4(6):893–909, November 1993.

[10] H. Bourlard and N. Morgan. *Connectionist Speech Recognition. A Hybrid Approach*, volume 247 of *The Kluwer international series in engineering and computer science*. Kluwer Academic Publishers, Boston, 1994.

[11] H. Bourlard and C. Wellekens. Links between hidden Markov models and multilayer perceptrons. *IEEE Transactions on Pattern Analysis and Machine Intelligence*, 12:1167–1178, 1990.

[12] J.S. Bridle. Alphanets: a recurrent 'neural' network architecture with a hidden Markov model interpretation. *Speech Communication*, 9(1):83–92, 1990.

[13] C. Che, Q. Lin, J. Pearson, B. Devries, and J. Flanagan. Microphone arrays and neural networks for robust speech recognition. In *Proceedings of ARPA HLT Workshop*, pages 342–347, New Jersey, 1994.

[14] P. Cosi, Y. Bengio, and R. De Mori. Phonetically-based multi-layered networks for acoustic property extraction and automatic speech recognition. *Speech Communication*, 9(1):15–30, 1990.

[15] P. Cosi, P. Frasconi, M. Gori, and N. Griggio. Phonetic recognition experiments with recurrent neural networks. In *Proc. of the International Conference on Spoken Language*, pages 1335–1338, Banff (Canada), October 1992.

[16] S. B. Davis and P. Mermelstein. Comparison of parametric representations of monosyllabic word recognition in continuously spoken sentences. 28(4):357–366, 1980.

[17] R. De Mori. *Spoken Dialogues with Computers*. Academic Press, London, UK, 1997. (in press).

[18] J. R. Jr. Deller, J. G. Proakis, and J. H. Hansen. *Discrete Time Processing of Speech Signals*. Macmillan Publishing Company, New York, USA, 1993.

[19] R. O. Duda and P. E. Hart. *Pattern Classification and Scene Analysis*. Wiley, New York, 1973.

[20] J.L. Elman and D. Zipser. Learning the hidden structure of speech. *Journal of the Acoustical Society of America*, 83:1615–1626, 1988.

[21] G. Flammia. Speaker independent consonant recognition in continuous speech with distinctive phonetic features. Master's thesis, McGill University, School of Computer Science, 1991.

[22] H. Franco, M. Cohen, N. Morgan, D. Rumelhart, and V. Abrash. Context-dependent connectionist probability estimation in a hybrid hidden markov model-neural net speech recognition system. *Computer Speech and Language*, 8:211–222, 1994.

[23] H. Franco and V. Digalakis. Temporal correlation modeling in a hybrid neural network/hidden markov model speech recognizer. In *Proc. of EUROSPEECH*, pages 1681–1684, Madrid, 1995.

[24] M.A. Franzini, K.F. Lee, and A. Waibel. Connectionist Viterbi training: a new hybrid method for continuous speech recognition. In *International Conference on Acoustics, Speech and Signal Processing*, pages 425–428, Albuquerque, NM, 1990.

[25] P. Frasconi, M. Gori, and G. Soda. Recurrent networks for continuous speech recognition. In *Computational Intelligence 90*, Milano (Italy), 1990. Elsevier.

[26] M. Gori, Y. Bengio, and R. De Mori. BPS: A learning algorithm for capturing the dynamical nature of speech. In *Proceedings of the International Joint Conference on Neural Networks*, pages 643–644, Washington D.C., 1989. IEEE, New York.

[27] P. Haffner, M. Franzini, and A. Waibel. Integrating time alignment and neural networks for high performance continuous speech recognition. In *International Conference on Acoustics, Speech and Signal Processing*, pages 105–108, Toronto, 1991.

[28] P. Haffner, A. Waibel, and K. Shikano. Fast back-propagation learning methods for large phonemic neural networks. In *Proceedings of Eurospeech'89*, 1989.

[29] John B. Hampshire and Alexander H. Waibel. A novel objective function for improved phoneme recognition using time-delay neural networks. *IEEE Transactions of Neural Networks*, 1(2):216–228, June 1990.

[30] M. M. Hochberg, S. J. Renals, A. J. Robinson, and G. D. Cook. Recent improvements to the abbot large vocabulary csr system. In *International Conference on Acoustics, Speech and Signal Processing*, pages 69–72, Detroit, 1995.

[31] M. M. Hochberg, S. J. Renals, A. J. Robinson, and D. J. Kershaw. Large vocabulary continuous speech recognition using a hybrid connectionist-hmm system. In *Proc. of ICSLP*, pages 1499–1502, Yokohama, 1994.

[32] X. D. Huang, Y. Ariki, and M. Jack. *Hidden Markov Models for Speech Recognition.* Edinburgh University Press, Edinburgh, 1990.

[33] J. C. Junqua and J. P. Haton. *Robustness in Automatic Speech Recognition: Fundamentals and Applications.* Kluwer Academic Publishers, Boston, USA, 1996.

[34] T. Kohonen. *Self-Organization and Associative Memory.* Springer-Verlag, Berlin, 3 edition, 1989.

[35] T. Kohonen. The self-organizing map. *Proceedings of the IEEE*, 78(9):1464–1480, 1990.

[36] T. Kohonen, G. Barna, and R. Chrisley. Statistical pattern recognition with neural networks: Benchmarking studies. In *IEEE International Conference on Neural Networks*, volume 1, pages 61–68, San Diego 1988, 1988. IEEE, New York.

[37] K. J. Lang and G. E. Hinton. The development of the time-delay neural network architecture for speech recognition. Technical Report CMU-CS-88-152, Carnegie-Mellon University, 1988.

[38] C. H. Lee, F. K. Soong, and K. K. Paliwal, editors. *Automatic Speech and Speaker Recognition: Advanced Topics.* Kluwer Academic Publishers, Boston, USA, 1996.

[39] E. Levin. Word recognition using hidden control neural architecture. In *International Conference on Acoustics, Speech and Signal Processing*, pages 433–436, Albuquerque, NM, 1990.

[40] E. Levin, R. Pieraccini, and E. Bocchieri. Time-warping network: a hybrid framework for speech recognition. In J.E. Moody, S.J. Hanson, and R.P. Lipmann, editors, *Advances in Neural Information Processing Systems 4*, pages 151–158, Denver, CO, 1992.

[41] R.P. Lippmann. Review of neural networks for speech recognition. *Neural Computation*, 1:1–38, 1989.

[42] Jyri Mäntysalo, Kari Torkkola, and Teuvo Kohonen. Experiments on the use of LVQ in phoneme-level segmentation of speech. In *Proc. of the Second Workshop on Neural Networks for Speech Processing*, pages 39–52, Firenze (Italy), 1992. LINT.

[43] E. McDermott and S. Katagiri. Shift-invariant, multi-category phoneme recognition using Kohonen's LVQ2. In *International Conference on Acoustics, Speech and Signal Processing*, volume 1, pages 81–84, Glasgow, Scotland, 1989. IEEE.

[44] E. McDermott and S. Katagiri. LVQ-based shift-tolerant phoneme recognition. *IEEE Transactions on Signal Processing*, 39(6):1398–1411, 1991.

[45] N. Morgan and H. Bourlard. Continuous speech recognition using multilayer perceptrons with hidden Markov models. In *International Conference on Acoustics, Speech and Signal Processing*, pages 413–416, Albuquerque, NM, 1990.

[46] N. Morgan, Y. Konig, S.L. Wu, and H. Bourlard. Transition-based statistical training for asr. In *Proceedings of IEEE Automatic Speech Recognition Workshop (Snowbird)*, pages 133–134, 1995.

[47] T. Moudenc, R. Sokol, and G. Mercier. Segmental phonetic features recognition by means of neural-fuzzy networks and integration in an n-best solutions post-processing. In *Proc. of ICSLP*, pages 338–341, Philadelphia, 1996.

[48] L.T. Niles and H.F. Silverman. Combining hidden Markov models and neural network classifiers. In *International Conference on Acoustics, Speech and Signal Processing*, pages 417–420, Albuquerque, NM, 1990.

[49] L. Rabiner and B. H. Juang. *Fundamentals of Speech Recognition.* Prentice Hall, Englewood Cliffs, NJ, USA, 1993.

[50] L. R. Rabiner, B. H. Juang, and C. H. Lee. An overview of automatic speech recognition. In C. H. Lee, F. K. Soong, and K. K. Paliwal, editors, *Automatic Speech and Speaker Recognition: Advanced Topics*, Boston, USA, 1996. Kluwer Academic Publishers.

[51] La. R. Rabiner. A tutorial on hidden Markov models and selected applications in speech recognition. *Proceedings of the IEEE*, 77(2):257–286, 1989.

[52] W. Reichl and G. Ruske. A hybrid rbf-hmm system for continuous speech recognition. In *International Conference on Acoustics, Speech and Signal Processing*, pages 3335–3338, Detroit, 1995.

[53] S. Renals, N. Morgan, H. Bourlard, M. Cohen, and H. Franco. Connectionist probability estimators in hmm speech recognition. *IEEE Trans. on Speech and Audio Processing*, 2(1):161–1174, January 1994.

[54] A.J. Robinson and F. Fallside. Static and dynamic error propagation networks with application to speech coding. In D.Z. Anderson, editor, *Neural Information Processing Systems*, pages 632–641, Denver, CO, 1988. American Institute of Physics, New York.

[55] T. Robinson. A real-time recurrent error propagation network word recognition system. In *International Conference on Acoustics, Speech and Signal Processing*, volume I, pages 617–620, 1992.

[56] T. Robinson. The application of recurrent nets to phone probability estimation. *IEEE Transactions on Neural Networks*, 5(2):298–305, March 1994.

[57] T. Robinson, L. Almeida, J. M. Boite, H. Bourlard, F. Fallside, M. Hochberg, D. Kershaw, P. Kohn, Y. Konig, N. Morgan, J. P. Neto, S. Renals, M. Saerens, and C. Wooters. A neural network based, speaker independent, large vocabulary, continuous speech recognition system: The wernicke project. In *Eurospeech93*, Berlin, Germany, 1993.

[58] T. Robinson and F. Fallside. A recurrent error propagation network speech recognition system. *Computer Speech and Language*, 5(3):259–274, July 1991.

[59] T. J. Sejnowski and C. R. Rosenberg. *NETtalk: A parallel network that learns to read aloud.* Technical report 86-01, Department of Electrical Engineering and Computer Science, Johns Hopkins University, Baltimore, MD., 1986.

[60] T.J. Sejnowski and C.R. Rosenberg. Parallel networks that learn to pronounce english text. *Complex Systems*, 1:145–168, 1987.

[61] E. Singer and R. P. Lippmann. A speech recognizer using radial basis function neural networks in an hmm framework. In *International Conference on Acoustics, Speech and Signal Processing*, volume 1, pages 629–632, San Franscisco, March 1992.

[62] J. Tebelskis, A. Waibel, B. Petek, and O. Schmidbauer. Continuous speech recognition using linked predictive networks. In R. P. Lippman, R. Moody, and D. S. Touretzky, editors, *Advances in Neural Information Processing Systems 3*, pages 199–205, Denver, CO, 1991. Morgan Kaufmann, San Mateo.

[63] A. Waibel. Modular construction of time-delay neural networks for speech recognition. *Neural Computation*, 1:39–46, 1989.

[64] A. Waibel, T. Hanazawa, G. Hinton, K. Shikano, and K. Lang. Phoneme recognition using time-delay neural networks. *IEEE Transactions on Acoustics, Speech, and Signal Processing*, 37:328–339, 1989.

[65] A. Waibel, H Sawai, and K. Shikano. Modularity and scaling in large phonemic neural networks. *IEEE Transactions on Acoustics, Speech, and Signal Processing*, 37:1888–1898, 1989.

[66] G. Zavaliagkos, S. Austin, J. Makhoul, and R. Schwartz. A hybrid continuous speech recognition system using segmental neural nets with hidden Markov models. *Int. Journal of Pattern Recognition and Artificial Intelligence*, pages 305–319, 1993. Special Issue on Applications of Neural Networks to Pattern Recognition (I. Guyon Ed.).

[67] G. Zavaliagkos, Y. Zhao, R. Schwartz, and J. Makhoul. A hybrid segmental neural net/hidden markov model system for continuous speech recognition. *IEEE Trans. on Speech and Audio Processing*, 2(1):151–160, January 1994.

Section 4

Eduardo R. Caianiello Lecture

Cellular Neural Networks: an Approach to Circuit Design and Information Processing

Gianluca Setti
DEIS, University of Bologna
Viale Risorgimento 2, I-40136 Bologna, Italy
e-mail: gsetti@deis.unibo.it

Patrick Thiran
ICA-DI, Swiss Federal Institute of Technology (EPFL)
CH-1015 Lausanne, Switzerland
e-mail: Patrick.Thiran@epfl.ch

Claudio Serpico
DIE, University of Napoli "Federico II"
Via Claudio 21, I-80138 Napoli, Italy
e-mail: serpico@kaos.dis.unina.it

Abstract

Cellular Neural Networks are arrays of simple nonlinear dynamical systems in which each cell is connected to its nearest neighbors only, which makes them particularly well-suited for VLSI implementations. To highlight this key point, the design of a low power 1.5 μm CMOS CNN architecture is reported, which correctly operates within a wide range of power supply voltages. The proposed circuit is useful because it can operate at low voltages whenever allowed by operating speed requirements and at larger voltage (and power consumption) whenever speed constraints prevail. CNNs applications mainly lie in image preprocessing tasks, for which it is essential to understand how the information (an image), introduced as initial condition, propagates in the network. Two different and somehow opposite behaviors are possible, namely local diffusion between neighboring cells only, or global propagation through the entire lattice. Some practical consequences of this mode of operation are also reported from [5, 6]

1 Introduction

Recently, many studies and proposed applications in the Cellular Neural Network (CNN) field deal with networks made of cells having more and more complex dynamics and coupled by increasingly complex interactions. Nevertheless, in spite of this useful generalizations, the interest about the "simple" original setting described in the first paper of Chua and Yang [1], has not decreased, both from theoretical and practical point of view. In fact, as far as implementation issues are concerned, the use of first neighboring cells connections only, the architecture simplicity of both processing units and interconnections elements, and the space invariant property, make CNNs particularly well suited for an analog circuit implementation in a VLSI technology, like CMOS. Moreover, it is worthwhile to stress that a complete understanding of the behavior of a CNN, even in the original model is not yet available. Indeed, also such a CNN is a high-dimensional non linear dynamical system, and is therefore very difficult to analyze. Therefore, to take full profit of CNN chips signal processing capabilities, it is important to study and understand as deeply as possible the way the network works, to point out its potentials and limitations to process the information (mainly images).

Aim of this contribution is twofold. On one hand, we show CNNs implementation easiness, Section 2 reporting on the design, in a 1.5 μm CMOS technology, of a low power CNN architecture. Although several CNN implementations, both in the voltage-mode [4] and current-mode [2]-[3] domains, have been proposed so far, it is worthwhile to note that all these implementations and designs refer to fixed power supply voltages, typically 5V. The key property of the network proposed in the paper is the ability of correctly operating in a wide range of supply voltages, thus exhibiting a high power supply rejection ratio (PSRR). This is a particularly attractive feature, since *i*) the CNN is characterized by a greater versatility, because it can work with low power supply voltages if the speed requirement can be relaxed for some applications on the chip or with high value if speed is a key objective of the application, and *ii*) it allows a significant power saving during standby operations, similarly to what happens in MOS static memories.

On the other hand, to have a glance on how the information, introduced as initial condition, propagates in the network, in Section 3 we show that two different behaviors are possible, namely *local diffusion* [5] between neighboring cells only and *global propagation* through the entire lattice [6] and we report some practical consequences on the kind of admissible steady-state solutions, on the length of transient, on sensitivity to parameters variations, and on possible ways of processing the information.

2 Design of a high PSRR CNN

In this section, through the adoption of a normalized notation, we will show how it is possible to design a CNN in which a change in power supply voltages has no influence on the signal processing operations performed by the network.

2.1 Normalization of network equations

Refer to a 2-dim CNN with $N_1 \times N_2$ cells described by [1]

$$C\frac{dVx_{i,j}}{dt} = -\frac{Vx_{i,j}}{R} + \sum_{C_{k,l}\in\mathcal{N}_{i,j}} a_{i-k,j-l}\, Vy_{k,l} + \sum_{C_{k,l}\in\mathcal{N}_{i,j}} b_{i-k,j-l}\, Vu_{k,l} + \imath \qquad (1)$$

with initial condition $Vx_{i,j}(0)$, time-independent input $Vu_{i,j}$, bias $\imath$, where $Vy_{i,j} = 1/2\left(|Vx_{i,j}+V_s|-|Vx_{i,j}-V_s|\right)$, and where $a_{0,0} > 1/R$ is assumed. Each cell is connected only to those lying inside their neighborhood $\mathcal{N}_{(i,j)}$, through the A- and B-template coefficients $a_{k,l}$ and $b_{k,l}$ which are space-invariant in this paper.

To explain some choices made at design level, and to introduce some notation, let us recast equation (1) as [1]

$$C\frac{dVx_{i,j}}{dt} = -g_N\left(Vx_{i,j}\right) + \sum_{\substack{k,l\neq i,j \\ C_{k,l}\in\mathcal{N}_{i,j}}} a_{i-k,j-l}\, Vy_{k,l} + \sum_{C_{k,l}\in\mathcal{N}_{i,j}} b_{i-k,j-l}\, Vu_{k,l} + \imath \quad (2)$$

where

$$g_N\left(Vx_{i,j}\right) = \frac{Vx_{i,j}}{R} - \frac{a_{0,0}}{2}\left(|Vx_{i,j}+V_s| - |Vx_{i,j}-V_s|\right)$$

can be considered as the current-voltage relationship of a N-shaped resistor with knee currents $\pm I_N^{max}$ given by

$$I_N^{max} = V_s(a_{0,0} - 1/R). \quad (3)$$

By referring to the circuit meaning of Equation (2), one immediately gets that for implementing a CNN cell one needs to design: *a*) an independent current source $\imath$; *b*) an active resistor with N-shaped current-voltage characteristic $g_N\left(Vx_{i,j}\right)$; *c*) voltage controlled current sources, for carrying out the control and feedback template elements.

Observe that, for the A-template elements, instead of *linear* voltage controlled current sources driven by the output voltage $Vy_{k,l}$ (for $C_{k,l} \in \mathcal{N}_{i,j}$ and $(k,l) \neq (i,j)$), it is fully equivalent in terms of network behavior to consider sources with a *piecewise-linear* characteristic depending on the neighboring cell state voltage $Vx_{k,l}$ and with saturation currents $\pm I^{sat}_{a_{i-k,j-l}}$ given by:

$$I^{sat}_{a_{i-k,j-l}} = a_{i-k,j-l}V_s \qquad C_{k,l} \in \mathcal{N}_{i,j},\ (k,l) \neq (i,j). \quad (4)$$

Such a choice allows for a simpler design and appreciable area saving.

Moreover, considering the constraint $|Vu_{i,j}| \leq V_s$ on the input voltage [1], it is immediate to recognize that current sources with similar I/O characteristic and saturation currents $\pm I^{sat}_{b_{i-k,j-l}}$ expressed by

$$I^{sat}_{b_{i-k,j-l}} = b_{i-k,j-l}V_s \qquad C_{k,l} \in \mathcal{N}_{i,j}. \quad (5)$$

can be also used to implement the B-template elements.

It is worthwhile to stress that this choice in the design process allows to obtain *i*) voltage controlled current sources with an (almost) linear I/O characteristic for all input voltages within the maximum allowable input dynamic range $[-V_s, V_s]$, as needed for realizing a circuit corresponding to the ideal model; *ii*) a saturation effect in the generator output current that prevent undesired behaviors when the circuit experiences unexpected large input voltages.

To understand how it is possible to design a CNN able to operate in a wide range of power supply voltages, let us recast the state equations in normalized form as

$$\dot{x}_{i,j} = -x_{i,j} + \sum_{C_{k,l}\in\mathcal{N}_{i,j}} A_{i-k,j-l}\, y_{k,l} + \sum_{C_{k,l}\in\mathcal{N}_{i,j}} B_{i-k,j-l}\, u_{k,l} + I \tag{6}$$

where the normalized state, output and input variables, A- and B-template elements, offset and time are defined by

$$x_{i,j} = \frac{Vx_{i,j}}{V_s} \quad y_{i,j} = \frac{Vy_{i,j}}{V_s} \quad u_{i,j} = \frac{Vu_{i,j}}{V_s} \tag{7}$$

$$A_{i-k,j-l} = R\, a_{i-k,j-l} \quad B_{i-k,j-l} = R\, b_{i-k,j-l} \quad I = \frac{R\,\imath}{V_s} \tag{8}$$

and $t^* = t/RC$, so that $y_{i,j} = 1/2\,(|x_{i,j}+1| - |x_{i,j}-1|)$. If we are able to design a circuit such that $A_{i-k,j-l}$, $B_{i-k,j-l}$, and I are insensitive, as far as possible, to changes in the power supply voltages, the final goal will be immediately achieved. By using (3), the first equation in (8) and (4) for $(k,l) = (i,j)$, the normalized quantities in (8) can be recast in the more useful form for design purpose as

$$A_{0,0} = I^{sat}_{a_{0,0}} \frac{R}{V_s} = \frac{I^{sat}_{a_{0,0}}}{I^{max}_N}(A_{0,0} - 1).$$

Solving for $A_{0,0}$, from the previous expression one obtains

$$A_{0,0} = \frac{I^{sat}_{a_{0,0}}}{I^{sat}_{a_{0,0}} - I^{max}_N}. \tag{9}$$

Similarly, it is immediate to get, for $C_{k,l} \in \mathcal{N}_{i,j}$

$$A_{i-k,j-l} = \frac{I^{sat}_{a_{i-k,j-l}}}{I^{sat}_{a_{0,0}} - I^{max}_N} \quad B_{i-k,j-l} = \frac{I^{sat}_{b_{i-k,j-l}}}{I^{sat}_{a_{0,0}} - I^{max}_N} \quad I = \frac{\imath}{I^{sat}_{a_{0,0}} - I^{max}_N} \tag{10}$$

By changing the power supply values, it is natural to expect that the parameters $I_{a_{i-k,j-l}}$, $I_{b_{i-k,j-l}}$, I^{max}_N and $\imath$ will also vary. We have thus to design the cell circuit elements so that, modifying the power supply values, these quantities will properly scale, to keep the ratios (10) almost constant.

2.2 N-shaped resistor

The CNN basic building blocks were designed with reference to a well-assessed $1.5\mu m$ n-well CMOS technology. Since to realize a network cell, a N-shaped resistor and voltage controlled current sources with piecewise-linear characteristics are needed, CNN implementations can be grounded on CMOS transconductance differential amplifier (OTA) stages, whose I/O sigmoidal characteristic may well approximate the ideal piecewise-linear one. In particular, for realizing the N-shaped resistor we used an OTA with both positive and negative feedback[1] as reported in Figure 1, where the resistors R_n, R_p, and the bias current generator I_{ss} have to be considered for now as ideal elements. More precisely, transistors M2-M6 form the input OTA differential stage, the pair (M5,M10) and the set M6-M9 are two mirror circuits which drive towards the output node currents proportional to the inverting and non-inverting input voltages, while M11 and M12, identical to M10 and M9, simply copy the output current to be fed into resistor R_n. The N-resistor was designed with reference to a CNN cell with $A_{0,0} = 2$, i.e. with $I^{sat}_{a_{0,0}} = 2\ I^{max}_N$.

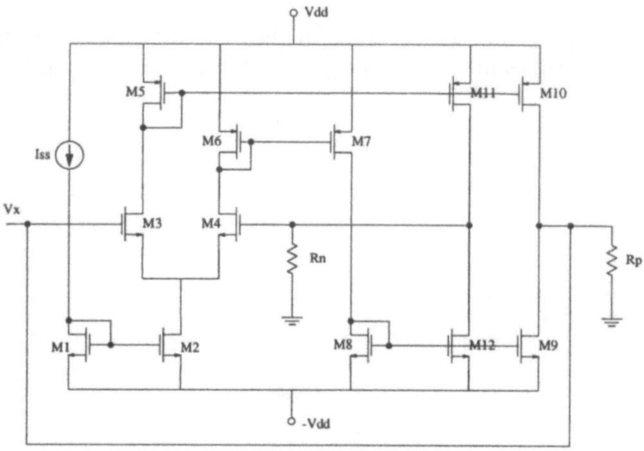

Figure 1: Schematic circuit diagram of the CMOS OTA with both positive (R_p) and negative feedback resistors (R_n) for implementing the N-shaped resistor.

For $V_s \approx 1V$, the values of R_n and R_p needed to satisfy the above constraint on $A_{0,0}$ are reported in Table 1, for $V_{dd} = 5V, 2.5V$, and for several values of I_{ss}. Notice that R_n and R_p have rather large values, and increase for decreasing I_{ss} at constant V_{dd} and for decreasing V_{dd} at constant I_{ss}.

To save chip area, R_n and R_p have been implemented as CMOS active resistors. Although correct CNN cell operations with a nonlinear resistor have been reported in the literature [4], a too large deviation from the ideal case may affect not only the dynamical behavior of the system, but even the final result of the performed processing task, as it will be discussed in Section 3. It is therefore desirable to implement resistors characterized, in addition to large ohmic values, by an almost linear current-voltage relationship for different power supply values.

To this aim, several resistor topologies have been considered. The diagram of the structure which yields to the best results is reported in Figure 2 (a), while Figure 2 (b) shows the simulated $I(V_r)$ characteristic for $2.5V \leq V_{dd} \leq 5V$. Although the achieved current-voltage relationship is almost linear, it can be noticed from Figure 2 (b) that the ohmic values are not large enough.

[1]The positive feedback resistor R_p is an essential component of the N-shaped resistor, while the negative feedback resistor R_n allows to tune the OTA transconductance and the

$V_{dd} = 5V$			
I_{ss} (μA)	1	1.5	2
R_n (kΩ)	250	200	150
R_p (kΩ)	650	550	400
$V_{dd} = 2.5V$			
I_{ss} (μA)	1	1.5	2
R_n (kΩ)	330	250	185
R_p (kΩ)	820	625	465

Table 1: Values of R_n and R_p needed for implementing an N-shaped resistor with $A_{0,0} = 2$, for different values of I_{ss}, and for $V_{dd} = 2.5V$ and $V_{dd} = 5V$.

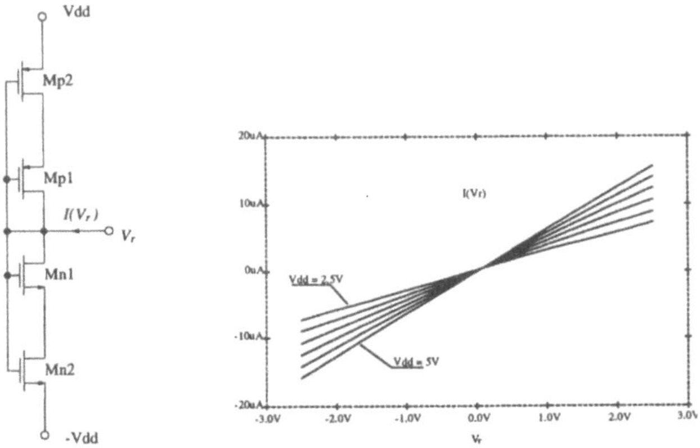

(a) (b)

Figure 2: Schematic circuit diagram of a simple CMOS active resistor (a) and its current-voltage characteristic $I(V_r)$ for several values of $V_{dd} \in [2.5V, 5V]$ (b).

To overcome the impasse, one and two more n-channel and p-channel MOS transistors have been added to the structure for implementing R_n and R_p respectively. In this way, it has been possible to obtain values of 260kΩ and 540kΩ for R_n, and of 680kΩ and 1420kΩ for R_p in the case $V_{dd} = 5V, 2.5V$ respectively, without degrading the linearity of the characteristics. The comparison with the values of R_n and R_p reported in Table 1 highlights that the needed values for obtaining $A_{0,0} = 2$ have been achieved for R_n, but are too large for R_p.

The simulated N-shaped characteristic of the circuit of Figure 1 with R_n and R_p implemented as active resistors, as well as the output current of the OTA, are reported in Figure 3 for a bias current $I_{ss} = 1\mu A$ and for two different values of the power supply. It is worthwhile to notice that the behavior is correct

knee voltage V_s.

V_{dd} (V)	2.5	3	3.5	4	4.5	5
I_{ss} (μA)	0.43	0.52	0.66	0.80	0.94	1.08
I_{DM2} (μA)	0.45	0.56	0.74	0.92	1.10	1.29

Table 2: Values of I_{ss} and I_{DM2} needed for keeping $A_{0,0} = 2$ for different values of V_{dd}.

at $V_{dd} = 5V$ but not at $V_{dd} = 2.5V$; in particular, the value of $A_{0,0}$ computed using (9) with the simulated values of $I_{a_{0,0}}^{sat}$ and I_N^{max} is approximately equal to 2.00 in the first case but grows up to 2.17 in the second one.

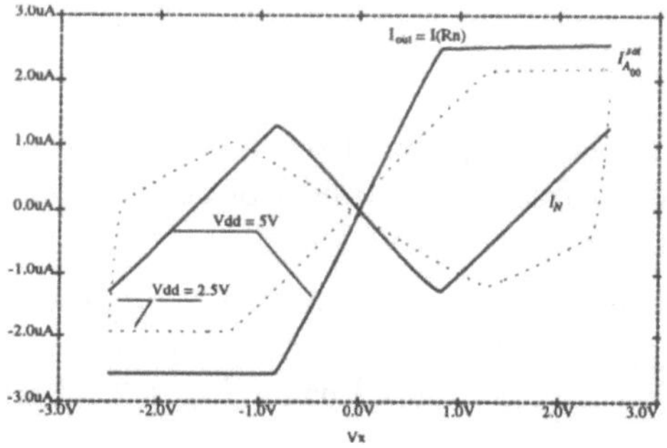

Figure 3: N-shaped resistor and OTA output current characteristic for $V_{dd} = 5V$ (continuous line) and $V_{dd} = 2.5V$ (dashed line), obtained by using the circuit of Figure 1 with R_n and R_p implemented as CMOS active resistors.

This deviation from the ideal behavior at low power supply values can be explained by a too large increase in the resistance values induced in the active resistors by decreasing V_{dd}. To compensate for this effect, the dependence of the resistance on the bias current can be exploited, since, as already mentioned, the needed values of R_n and R_p increase by decreasing I_{ss}. Equivalently, one may refer directly to $A_{0,0}$ which depends on V_{dd} and I_{ss} through $I_{a_{0,0}}^{sat}$ and I_N^{max}. Simulations show that $A_{0,0}$ increases by decreasing V_{dd} at constant I_{ss} and decreases by decreasing I_{ss} at constant V_{dd}. Thus, by balancing the power supply change effect by a matched change in the bias current, $A_{0,0}$ can be kept almost constant.

Table 2 reports, as a function of V_{dd}, the values of I_{ss} needed to maintain $A_{0,0} = 2$, as well as the corresponding current I_{DM2} to be injected into the source-tail pair M3-M4. Therefore, to achieve the goal, one needs to design a tail current generator able to sink the desired current $I_{DM2} = I_{DM2}(V_{dd})$, or, equivalently, to design a nonlinear voltage divider which biases the gate of transistor M2 with a voltage suitably dependent on V_{dd}. This results has been achieved with the ad-hoc designed circuit represented inside the dashed box in Figure 4, which allows to obtain $A_{0,0} \approx 2$ for $2.5V \leq V_{dd} \leq 5V$.

In Figure 5 (a) the ideal characteristic $I_{DM2} - V_{dd}$ (continuous line) is compared with the one obtained from circuit simulations (dashed line). The good

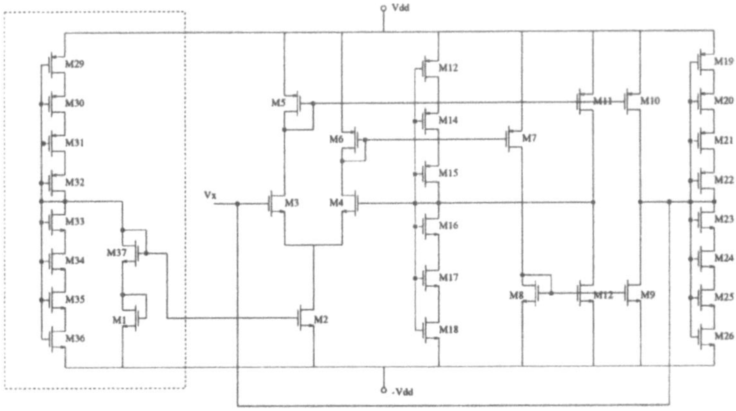

Figure 4: Schematic circuit diagram of the N-shaped resistor with the ad-hoc designed voltage divider (inside the dashed box).

match between real and ideal behavior allows to achieve a good invariance of $A_{0,0}$ with power supply changes. In fact, as reported in Table 3, the maximum relative error on $A_{0,0}$ computed using (9) is 2.5% in the whole interval $2.5V \leq V_{dd} \leq 5V$.

The simulated N-shaped characteristics for the circuit shown in Figure 4 are reported in Figure 5 (b) for several values of V_{dd}. The average power consumption P_d of the whole circuit is about $130\mu W$ for $V_{dd} = 5V$ ($15\mu W$ due to the voltage divider), and decreases to only $15\mu W$ ($1.5\mu W$ due to the divider) for $V_{dd} = 2.5V$. The gate area of the N-resistor is about $5000\mu m^2$ among which $2000\mu m^2$ are due to the voltage divider circuit. Notice however that such a circuit *has to be realized only once in the whole chip.*

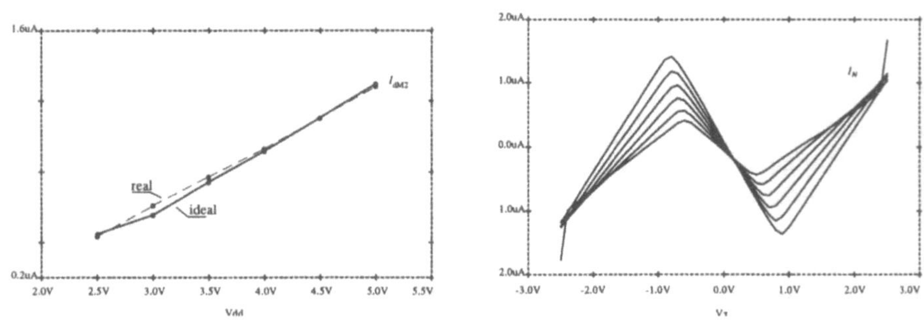

(a) (b)

Figure 5: Ideal (continuous line) and simulated (dashed line) characteristic of I_{DM2} needed for keeping $A_{0,0} = 2$ as function of V_{dd} (a), and N-shaped resistor current-voltage characteristics obtained by simulating the circuit shown Figure 4, for different values of V_{dd}.

V_{dd} (V)	2.5	3	3.5	4	4.5	5
$A_{0,0}$	1.95	1.99	2.00	2.01	2.02	2.03

Table 3: Estimated values of $A_{0,0}$ for the circuit shown in Figure 4 for different values of V_{dd}.

2.3 Feedback and Control Operators

To obtain a correct behavior for the voltage controlled current source needed to realize the A- and B-template elements, it is necessary to keep constant, at different V_{dd} values, the normalized weights $A_{i-k,j-l}$ $((k,l) \neq (i,j))$ and $B_{i-k,j-l}$ given by (10).

In the first case the problem can be easily solved, since a current source with (approximately) piecewise-linear characteristic controlled by $Vx_{i,j}$ can be implemented by mirroring the currents of the N-shaped resistor differential pair with suitable coefficients. In this way, the current injected into the neighbor cell state-voltage node is a scaled copy of the N-shaped resistor OTA's output current, and has therefore the desired characteristic as a function of $Vx_{i,j}$. In particular, the circuits used for implementing positive and negative "lateral" A-template elements are shown in Figures 6 (a) and (b) respectively. Notice that transistors M3 and M4, needed for the sign inversion in the output current, have to be implemented only once for all the "lateral" current generators. Moreover, since the circuit behavior is based on the one of the N-shaped resistor, correct operation at different power supply values is automatically achieved. In fact, as the mirroring coefficient depends only on the shape-factors ratio of MOS transistors, the ratio

$$\frac{I^{sat}_{a_{i-k,j-l}}}{I^{sat}_{a_{0,0}}} = \frac{A_{i-k,j-l}}{A_{0,0}} \qquad C_{k,l} \in \mathcal{N}_{i,j},\ (k,l) \neq (i,j)$$

remains almost constant with power supply. Therefore, one can expect the same range of variation for $A_{i-k,j-l}$ with V_{dd} as for $A_{0,0}$. Indeed, the weights values reported for some typical cases in Table 4 and computed using (10) confirm this conjecture.

As far as the implementation of the B-template coefficients is concerned, we may rely on OTAs with negative feedback similar to those employed in the N-resistor (shown in Figure 7 for the central coefficient), since, as already mentioned, current generators with an (ideally) piecewise-linear I/O characteristic have been assumed. Since this structure can be thought as derived from the N-resistor without the positive feedback, the circuits of Figures 4 and 7 are, in some sense, self-tuned, because the electrical mechanisms which give the knee voltages in their current-voltage characteristic are the same. Moreover, it is possible to obtain correct behavior for different power supply values using the nonlinear voltage divider considered in the previous section, which can therefore be shared among all sub-circuits in the chip.

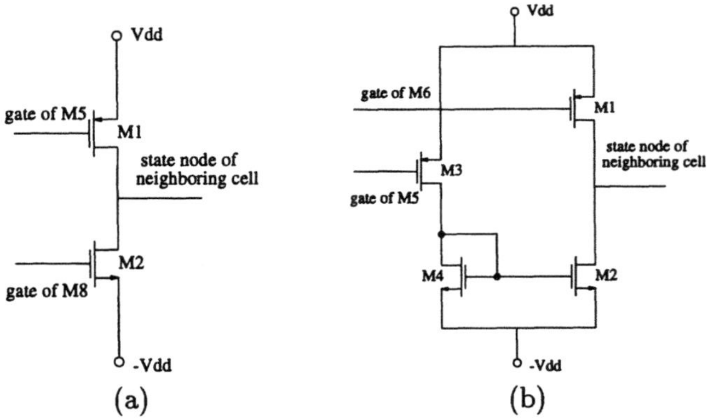

Figure 6: Simple circuit used to implement the positive (a) and negative (b) lateral A and B-template elements.

$V_{dd} = 5V$				
$A_{i-k,j-l}$ (nom.)	2	1	-1	-0.25
$A_{i-k,j-l}$ (sim.)	2.08	1.08	-1.08	-0.27
$V_{dd} = 3.5V$				
$A_{i-k,j-l}$ (nom.)	2	1	-1	-0.25
$A_{i-k,j-l}$ (sim.)	1.98	1.03	-1.03	-0.25
$V_{dd} = 2.5V$				
$A_{i-k,j-l}$ (nom.)	2	1	-1	-0.25
$A_{i-k,j-l}$ (sim.)	1.9	0.95	-0.95	-0.23

Table 4: Comparison between the nominal values of the lateral A-template elements and those computed by using (10) for different values of V_{dd}.

The implementation of the lateral B-template coefficients current generators can be achieved by simply copy the output current of the circuit of Figure 7, through the circuits reported in Figures 6 (a) and (b). As expected, we found that the normalized weights computed using the second of equation of the system (10) have deviations from the ideal values of the same order of magnitude as those found for the A-template elements. Finally, the offset current generator can be simply designed with a few transistors suitably biased with the voltage established by the nonlinear voltage divider.

2.4 Application examples

By using the building blocks described so far, we designed several CNNs with the aim of validating the independence of the performed signal processing tasks on power supply changes for $V_{dd} \in [2.5V, 5V]$. Additionally, the parasitic effects due to overlap and junction capacitances, which depend on transistor dimensions and bias voltages, were accurately considered. The circuits were

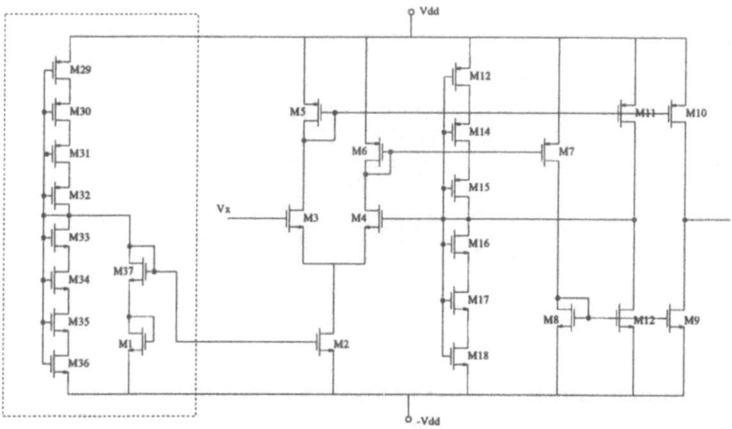

Figure 7: Schematic circuit diagram of the circuit used to implement a central B-template coefficient.

initially designed with a state capacitor of 1pF. However, a careful circuit analysis showed that the total parasitic capacitance associated to each state node of a CNN characterized by an A-template of 5 nonzero elements is of the order of 1.1pF and slightly voltage-dependent. We therefore conducted also simulations without the state capacitor, achieving correct performances in all the cases, with a transient behavior characterized by a settling time approximately reduced by one half. This is very interesting, as shows the feasibility of CNNs with reduced area consumption.

Among the several designed networks, let us mention the 16×1 CNN ($\approx$ 500 transistors) for connected component detector (CCD) operation [7] and the 10×10 CNN ($\approx$ 6000 transistors) for holes filling task [8]. In particular, besides performing a fundamental image-processing operation (e.g. for the realization of an integrated system for written character recognition [9]), a CNN for CCD is well known to be characterized by a particularly complex transient behavior, since, as will be pointed out in Section 3, it operates in a global propagation mode. To validate the capability of the designed circuits to operate in a "heavy" transient conditions (where the state voltages of many cells exhibits several transitions between the two saturation regions before reaching the steady state value) we designed a CNN characterized by $A = [-0.8\ 2\ 0.8]$, $B = 0$, and $I = 0$ and simulated it for several input images. In all cases the final result was correct for any value of V_{dd}. Some more circuit properties can be deduced from Table 5, where τ indicates the maximum transient length for state voltages. In particular, one may remark that increasing the power supply from 2.5V to 5V leads to a power consumption increase by a factor ≈ 8, while τ decreases by a factor ≈ 2.8, so that by reducing V_{dd} from 5V to 2.5V, one can improve the power-delay product per cell by a factor ≈ 3.

For the CNN for holes filler applications, the image to be processed is continuously applied to the input of the network, so that also a nonzero control operator is required. To validate performances of the current controlled voltage

CCD		
V_{dd} (V)	2.5	5
P_d (mW)	0.26	2.1
τ (μs)	85	30
$(P_d \cdot \tau)/N_1$	1.4	3.9
Holes Filler		
V_{dd} (V)	2.5	5
P_d (mW)	3.2	24.8
τ (μs)	60	20
$(P_d \cdot \tau)/(N_1 \times N_2)$	1.9	5.0

Table 5: Maximum transient length (τ), power consumption (P_d) power-delay product per cells ($(P_d \cdot \tau)/N_1$ or $(P_d \cdot \tau)/N_1 \times N_2$) for $V_{dd} = 2.5V, 5V$, for a CDD and holes filler CNN.

generators needed for the B-template elements, we designed a CNN characterized by $A_{0,0} = 2$, $A_{-1,0} = A_{0,-1} = A_{0,1} = A_{1,0} = 1.1$, $B_{0,0} = 3.6$, $I = -0.75$, and simulated it for several input images. In all cases the final result was correct for any value of V_{dd}. As shown in Table 5 the effects of power supply variations on P_d and τ are similar to those of the CCD case.

3 Local Diffusion and Global Propagation of Information

In this section we deal with information propagation in CNNs by studying the phenomena of local diffusion and global propagation in the simplest nontrivial network, namely in a one-dimensional lattice with N cells described by (6) and with the space-invariant A-template $A = [r \;\; p \;\; s]$, with $p > 1$, no B-template and no bias ($I = 0$). Roughly speaking, we will say that a CNN behaves in a *local diffusion* mode when two distant cells do not influence each other if the states of a number r of adjacent cells located between these two cells have reached some value. It behaves in a *global propagation* mode otherwise, i.e. when one of these two cells can always influence the other one, whatever the value of the state of r adjacent cells located in between these two cells. A first attempt of formal definition for in the particular case of a CNN made of an infinite number of cells has been formulated in [5][6] and is as follows.

Consider two adjacent cells C_i and C_{i+1}, for any index i, and consider the three particular solutions $\phi(t)$, $\phi'(t)$ and $\phi''(t)$ of system (6) for $i = 1, \ldots, N$ and for $t \geq t_0$, originating respectively from the three following different initial conditions:

$$\phi(t_0) = [\ldots\ \xi_{i-2}\ \xi_{i-1}\ \xi_i\ \xi_{i+1}\ \xi_{i+2}\ \xi_{i+3}\ \ldots]^T \tag{11}$$

$$\phi'(t_0) = [\ldots\ \xi_{i-2}\ \xi_{i-1}\ \xi_i\ \xi_{i+1}\ \xi'_{i+2}\ \xi'_{i+3}\ \ldots]^T \tag{12}$$

$$\phi''(t_0) = [\ldots\ \xi''_{i-2}\ \xi''_{i-1}\ \xi_i\ \xi_{i+1}\ \xi_{i+2}\ \xi_{i+3}\ \ldots]^T. \tag{13}$$

Note that $\phi_j'(t_0) = \phi_j(t_0) = \xi_j$ for all $j \leq i+1$, and $\phi_k''(t_0) = \phi_k(t_0) = \xi_k$ for all $k \geq i$, whereas $\phi_k'(t_0)$ and $\phi_k(t_0)$ need not being equal for $k > i+1$, nor $\phi_j''(t_0)$ and $\phi_j(t_0)$ for $j < i$.

Definition 1. *This CNN works in a* local diffusion *mode if there exist η_1, η_2 with $|\eta_1| = |\eta_2| = 1$ such that for any initial conditions (11), (12) and (13) satisfying $\xi_i = \eta_1$ and $\xi_{i+1} = \eta_2$, the solutions $\phi(t)$, $\phi'(t)$ and $\phi''(t)$ of (6) are linked by the property that for all $t \geq t_0$*

$$\phi_j'(t) = \phi_j(t) \text{ for all } j < i \tag{14}$$

$$\phi_k''(t) = \phi_k(t) \text{ for all } k > i+1. \tag{15}$$

On the contrary the CNN works in a global propagation *mode if for any η_1, η_2 with $|\eta_1| = |\eta_2| = 1$ there exist particular initial conditions (11), (12) and (13) satisfying $\xi_i = \eta_1$ and $\xi_{i+1} = \eta_2$ such that the solutions $\phi(t)$, $\phi'(t)$ and $\phi''(t)$ of (6) are linked by the property that for some $t > t_0$*

$$\phi_j'(t) \neq \phi_j(t) \qquad \text{for some } j < i \tag{16}$$

or

$$\phi_k''(t) \neq \phi_k(t) \qquad \text{for some } k > i+1. \tag{17}$$

It is worthwhile to notice that this formal definition of local diffusion is stronger than the rough notion that any pair of "distant" cells do not influence each other, because it implies that any pair of cells $\{C_j, C_k\}$ with $j < i$ and $k > i+1$ do not interact at all for $t \geq t_0$, even if they are not so far apart (for instance $j = i-1$ and $k = i+2$). On the other hand, as long as for $0 \leq t < t_0$ no pair of two adjacent cells $\{C_i, C_{i+1}\}$ has reached a saturated value such that (14) and (15) are verified, no precision is provided about the mutual interaction between two distant cells. In fact, we observe that in a network having the local diffusion property defined here, condition $\xi_i = \eta_1$ and $\xi_{i+1} = \eta_2$ is usually rapidly verified, much before distant cells have really had time to influence each other's states. As far as the formal definition of global propagation is concerned, notice that there is global propagation from right to left if (16) is fulfilled, and vice-versa if (17) is fulfilled. This means that any pair of "distant" cells $\{C_j, C_k\}$ with $j < i$ and $k > i+1$ will interact with each other at some time $t > t_0$. Contrary to the diffusion case, the underlying process is now a *traveling wave* that can be maintained, perturbed or even blocked depending on the boundary conditions (fixed Dirichlet) [6].

Although not fully general, Definition 1 has been adopted because we can directly derive the values of the template parameters for which the CNN processes initial conditions in a local diffusion or in a global propagation mode, as formally stated in the following Theorem, proven in [5][6].

Theorem 1. *The CNN has a local diffusion behavior if*

$$p - 1 > |s - r| \tag{18}$$

and a global propagation behavior if

$$p - 1 < |s - r| \tag{19}$$

3.1 Some Practical Consequences

The distinction between local diffusion and global propagation modes of operation in a CNN is not only pure theoretical, but presents practical consequences listed [5][6] and recalled here below.

The admissible *stable steady state solutions* are different in the two cases. In fact a linear or a circular CNN having an even number of cells and operating in a local diffusion mode can be proven to be stable almost everywhere. If it operates in a global propagation mode, this is no longer true, but periodic solutions are then present for some types of boundary conditions.

As expected, *boundary conditions* affect only the cells located on the border of the finite array of a CNN if it works in a local diffusion mode, but induce dramatic changes on the behavior of the whole array if it works in the global propagation mode [6].

In local diffusion mode and in the case of a linear array, the *length of the transients* is not much affected by the number of cells. Indeed, as soon as the condition $\xi_i = \eta_1$ and $\xi_{i+1} = \eta_2$ is satisfied, the CNN can be split into two subsystems that evolve completely independently from each other. Therefore the original CNN of N cells is equivalent to two CNNs of respectively $(i-1)$ and $(N-i-1)$ cells, working in parallel, both in a local diffusion mode. The dynamics of such CNNs can be modeled by splitting them into subsystems with less and less cells, all working in parallel. Therefore, one expects that the length of the transients increases slower than linearly, as a function of the number N of cells. This is indeed what we observe in simulations. Figure 8 (a) shows the average time spent by the network before all cells reach a constant output value, for different values of N, and Neumann boundary conditions. Note that the dependence on N remains similar in the two cases. This is an important property from the implementation point of view, since indicates that a CNN working in the local diffusion mode operates in a really parallel way, thereby keeping their fast processing speed even for large arrays. On the other hand, a global processing of the data will be difficult: it can be based only on local features. Interesting examples from this point of view are CNNs for noise removal or edge detection [10].

In case of global propagation mode of operation, the information must travel from one border of the array to the other one, which indicates a sequential nature of processing. Therefore one expects the *length of the transients* to increase with N. Simulations results for the average transient length shown in Figure 8 confirm this remark. Notice that the shorter transient is obtained for the antisymmetric template ($r = -s$); the more the coefficients become symmetric (and thus the more they approach the limit (19)), the longer the transient. This is due to the period of the underlying periodic solution that would have been present, if the traveling wave had not been interrupted at the end of the array. Note however that the transients increase at least linearly with the number of cells in all cases, and are much longer than in the local diffusion case. This linear dependence of the transient length on the array size might be a handicap if speed is a key factor desired in the considered application.

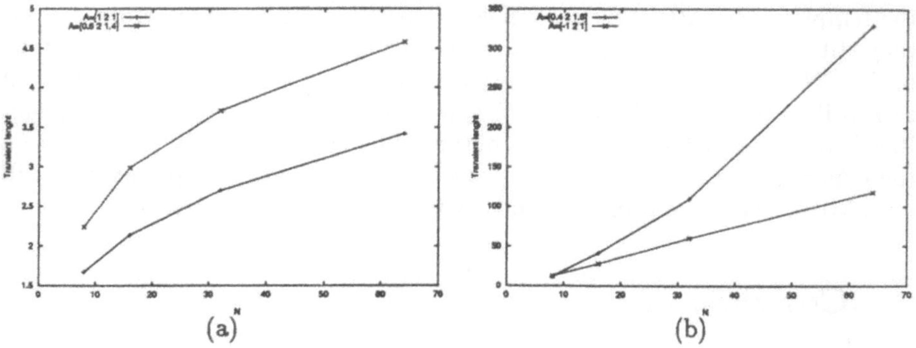

Figure 8: Transient length versus the number N of cells, for two local diffusion templates $A = [1\ \ 2\ \ 1]$ and $A = [0.6\ \ 2\ \ 1.4]$, (a), and for two global propagation templates $A = [-1\ \ \ 2\ \ \ 1]$ and $A = [0.4\ \ \ 2\ \ \ 1.6]$ (b). Neumann boundary conditions have been considered.

On the other hand, due to the influence among distant cells state, the steady state solution depends now on global features present in the initial conditions, thereby allowing a global processing even if the connections are only local. Interesting examples from this point of view are CNNs for CCD or shadow detection [11].

In the local diffusion mode, the essential part of the transient occurs when most cells are operating in the linear region, as most saturated cells simply decouple the array into two smaller sub-arrays of mainly linear cells. Therefore the *design* of templates having a local diffusion behavior should be based more on the spectral approach (by taking a Discrete (Spatial) Fourier Transform of all state variables and template coefficients) [12]. Since all trajectories converge to stable equilibria without having a complicated transient waveform, learning algorithms of the gradient type derived from recurrent backpropagation, or backpropagation-through-time [13] could also be successfully applied.

On the other hand, in the global propagation mode, the essential part of the transient often occurs when most cells are operating in the saturated region. To *design* templates having a global propagation behavior, a purely spatial frequency approach is not sufficient to describe the processing of the initial condition. Learning algorithm are more difficult to apply (at least recurrent back-propagation) because the convergence to fixed-point equilibria is not guaranteed. On the other hand, as most cells are saturated, a design based on a number of rules, allowing or stopping the propagation of the traveling wave if some local condition is (or is not) met, is now possible. The design results then in a set of inequalities linking the different template coefficients [14].

Another likely consequence of the local diffusion property concerns the *sensitivity* to a variation of parameters, which is also of crucial importance for a VLSI implementation. As most of the processing is done in the linear domain, an error in the template coefficients will probably affect the output proportionally to the size of this error. On the contrary, in case of global propagation, if the template parameters are perturbed by small implementation errors, so that they still verify the original system of inequalities mentioned above, the output of the CNN will probably remain very close to the ideal one and the performed task will be the same. However, if the error is such that the system of inequalities is no longer satisfied, then the output is likely to be totally different from the expected one. Therefore, the sensitivity to a variation of parameters will not have a relation of proportionality with the size of this variation.

4 Conclusion

In this work some keys properties of CNNs have been presented and reviewed.

To show the network implementation easiness, the design of a 1.5μm CMOS CNN exhibiting low power dissipation and high power supply rejection ratio has been presented. To achieve the last goal a suitable nonlinear biasing circuit was devised, which, by compensating the effects of changing the power supply voltages, allows to get a system operation in the whole range $2.5V \leq V_{dd} \leq 5V$. It has been shown by device level simulations that the proposed design criteria give, on the whole, small-area, low power circuits offering improved versatility, which may be especially profitable in association with weights programmability.

As far as the network dynamics is concerned, we have distinguished two different possible behaviors, namely local diffusion between neighboring cells and global propagation through the network, giving also the values of the A-template parameters for which the CNN operates in either mode. Some practical consequences have also been considered on the kind of possible steady state solutions, on the length of transient, and on sensitivity to parameters variations.

Acknowledgments

G. Setti wish to thanks Prof. G. Masetti and Prof. S. Graffi for their continuous help and encouragement.

References

[1] L.O. Chua and L. Yang, "Cellular Neural Networks: Theory", *IEEE Trans. on Circuits and Systems*, vol. CAS-35, pp. 1257–1272, 1988.

[2] A. Rodriguez Vazquez, S. Espejo, R. Dominguez-Castro, J. L. Huertas, and E. Sanchez-Sinencio: "Current-mode techniques for the implementation of continuous- and discrete-time cellular neural networks", *IEEE Trans. on Circuits and Systems–II*, vol. CAS-40, pp. 132-146, 1993.

[3] S. Espejo, R. Carmona, R. Dominguez-Castro, and A Rodriguez Vazquez: "A VLSI oriented continous-time CNN model", *Int. Journ. of Circuit Theory and Applications*, vol. 24, pp. 341-356, 1996.

[4] G. F. Dalla Betta, S. Graffi, Z. M. Kovacs, and G. Masetti, "CMOS Implementation of an Analogically Programmable Cellular Neural Network", *IEEE Trans. on Circuits and Systems*, vol. CAS-40, pp. 206-215, 1993.

[5] P. Thiran, G. Setti, and M. Hasler: "An Approach to Information Propagation in 1-dimensional Cellular Neural Networks - Part I: Local Diffusion", *IEEE Trans. on Circuits and Systems–I*, to appear.

[6] G. Setti, P. Thiran and C. Serpico "An Approach to Information Propagation in 1-dimensional Cellular Neural Networks - Part II: Global Propagation". *IEEE Trans. on Circuits and Systems–I*, to appear.

[7] T. Matsumoto, L. O. Chua, and H. Suzuki, "CNN Cloning Template: Connected Component Detector", *IEEE Trans. Circuits and Systems*, vol. CAS-37, pp. 633-635, 1990.

[8] T. Matsumoto, L. O. Chua, and R. Furukawa, "CNN Cloning Template: Hole Filler", *IEEE Trans. Circuits and Systems*, vol. CAS-37, pp. 635-638, 1990.

[9] H. Suzuki, T. Matsumoto and L. O. Chua, "A CNN Handwritten Character Recognizer", *Int. Jour. Circuit Theory and Applications*, vol. 20, pp. 601-612, 1992.

[10] L.O. Chua and L. Yang, "Cellular Neural Networks: Theory", *IEEE Trans. on Circuits and Systems*, vol. CAS-35, pp. 1273–1284, 1988.

[11] T. Matsumoto, L. O. Chua, and H. Suzuki, "CNN Cloning Template: Shadow Detector", *IEEE Trans. Circuits and Systems*, vol. CAS-37, pp. 1070-1073, 1990.

[12] P. Thiran, K. R. Crounse, L.O. Chua and M. Hasler, "Patter n Formation Properties of Autonomous Cellular Neural Networks," *IEEE Trans. on Circuits and Systems – I*, vol. CAS-42, pp. 757–776, 1995.

[13] J. Nossek, "Design and learning with Cellular Neural Networks," *Int. J. Circuit Theory and Applications*, vol. 24, pp. 15–24, 1996.

[14] L.O. Chua and P. Thiran, "An Analytic Method for designing Simple Cellular Neural Networks", *IEEE Trans. on Circuits and Systems*, vol. CAS-38, pp. 1332–1341, 1991.

Section 5
Mathematical Models

Correlation Collapse under Persistent Input Derangement in the Nervous System. Fitting pRAM Models for Thalamo-cortical Circuits

B. Apolloni, D. de Falco, S. Iamundo
Dipartimento di Scienze dell' Informazione, via Comelico 39 - I20135 Milano, Italy
e-mail: apolloni/defalco/iamundo@hermes.mc.dsi.unimi.it

G. Biella, M.L. Sotgiu
Istituto di Neuroscienze e Bioimmagini CNR-Milano and
VI cattedra di Neurologia University of Milan, via Cervi 93 - I20090 Segrate, Italy
e-mail: biella/sotgiu@inb.mi.cnr.it

T. Clarkson, J. G. Taylor
Department of Mathematics, King's College,Strand - London WC2R 2LS, UK
e-mail: udah057@bay.cc.kcl.ac.uk

Keywords: neural networks, pRAM, thalamo-cortical connectivity, sensory processing, noxious derangment.

Abstract

We study the functional connections between neurons of thalamo-cortical circuits of the albino rat brain through their spiking time correlation in normal and neuropathic animals. These correlations are learnt by a special artificial neural network, the pRAM hardware, which virtually maps on the physiological neurons' network, on-line learning conditional probabilities from joint firing records of clusters of neurons. The most relevant probabilities are selected w.r.t. the entropic criterion of efficiently mimicking the actual spiking distribution, and *minimal neural maps* are drawn on the basis of a *strength* parameter defined for the synapses as a measure of the spiking interaction between the connected neurons. Growth of the neural activity, together with a shattering of the minimal maps emerge as a distinguishing feature of persistent nociceptive conditions.

1 Introduction

Inputs (noxious and non-noxious) from the peripheral sources of the body, carried by the peripheral nerves and filtered in the spinal cord and its structural correlates, distribute to several supraspinal stations. The investigation of distribution and processing of these signals in the brain becomes an intractable problem due to the overwhelming number of connections (with many coexisting divergent and convergent neural pathways). Thus some major process must be chosen to adequately represent a small segment of the total system. No true generalization of the entire set of local phenomena can be done, but satisfactory aproximations of the most important circuit dynamics can be attempted. This approach escapes from the absence of a connectivity rule capable of completely describing the synaptic strengths of thalamo-cortical neurons. In addition, it allows for statistical estimates of the overall connectivity for a cluster of a relevant number of neurons. From this view point the thalamus and the cortex are critical sites to analyze.

The thalamus represents one of the major gateways for a number of these pathways and is the most important station of signal modulations before they reach the cortex. The thalamus is connected to the cerebral cortex by several pathways; neurons in the thalamus send widely branched fibers [neurites] to the neurons in the cortex [1].

The cortex is a hypercomplex network, highly inter-connected, where one can distinguish, with appropriate techniques, specific areas involved in the processing of sensory signals, with a precise re-mapping of the *body surface* reproduced on the cortical surface (somatotopy) [2,3]. Backward connections from the cortex to the thalamus are also present. These feed-back pathways are mainly excitatory (as is the excitatory thalamo-cortical 'forward' branch). A special feature in this context is represented by the reticular thalamus, mainly composed of inhibitory neurons. The reticular thalamus receives direct input from the cortex and distributes a diffuse inhibitory message to several thalamic nuclei that symmetrically send excitatory inputs to the reticular cells .

Our general aim is to isolate subsets of neurons chosen from both thalamus and cortex that interact strongly amongst themselves. We assume that spontaneous running or persistent unbalancing conditions, like those present in the neuropathic rats, can change the connectivity rules of the thalamo-cortical circuits in a stable fashion. In particular we are going to propose that in neuropathic rats the reduced possibility of frequency coding, due to a pre-eminent role of spontaneously activated neurons, is primed by a derangement of the overall correlative properties of the network itself [4].

We use the pRAM neural network hardware[1] [5] to identify the inner statistical structure of the traces of the multidimensional boolean variables describing the sequences of spikings and silences of jointly monitored neurons. In particular the network is trained to learn mutual conditional probabilities through the on-chip reward-penalty learning algorithm [5], select the most relevant probabilities through the entropic rule stated in [6] and search for an evidence of the connectivity rules that can soundly represent spontaneous activity states in normal or in neuropathic rats. This representation should yield a kind of graph — we call it a *minimal map* — that could be assumed as a minimal connectivity reqirement for the perceptual outcome.

[1]A pRAM chip was first built in the department of Electrical Engineering of King's College London in 1990. This was followed in 1992 and 1993 by pRAMs with on-chip learning. In April 1998 a PC board supporting up to 5 pRAMs (1280 neurons) will be available.

2 The mathematical model

2.1 The data

We focus on multiple simultaneous signals from the SS-I (somato-sensory primary) cortex. Signals are collected by a couple of multiple microelectrodes, constituted by bundles of 5-7 carbon /tungsten rods. Each rod, whose tip diameter is 2μm, is able to pick up the activity signal of 4-5 neurons. Thus the log of a recording session is constituted by the 10-14 traces, where each track is a cluster of 4-5 neuron signals.The activity profile of each single neuron is accurately separated from the activities of the other neurons merged in the same trace via a dedicated clustering software.
After clustering, the data are composed of a temporal sequence of ν-dimensional (up to 60 components) boolean vectors Y(t) representing the activity of ν neurons during a recording session lasting a few minutes, where $y_i(t) = 1$ if a spike temporally centered in t occurred on the i-th neuron, =0 otherwise. Time is usually discretized in milliseconds, which are then grouped into small intervals called bins. One bin sums the activity of a neuron during an interval of a few milliseconds: we say that a neuron has fired at a given bin t if a spike falls in the corresponding interval. After binning the data are then divided into $\mu \times \nu$-dimensional binary matrices (trials), each representing the activity of the neurons during approximatively one second. With a bin size of 2 ms we have $\mu \approx 500$. Each trial is used as a sample for the pRAM.

2.2 Physiological and computational models

With reference to fig.1, denoting by T and C sets of mammalian thalamus and cortex neurons, respectively, and by RT the reticular nucleus, we schematize the source of the above signal as follows.

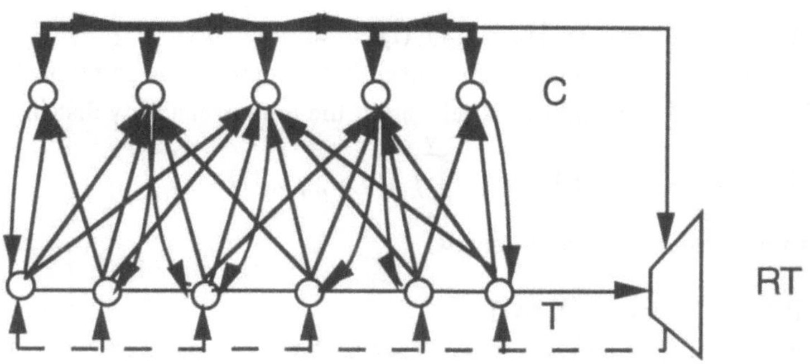

Figure 1. The thalamo-cortical structure

Let us arrange T and C in two layers of neurons and RT as an intermediate layer. We assume that there are no connections spread between the neurons of the first layer, whereas the C layer neurons are fully interconnected.
To reproduce temporal correlations between neurons, we unfolded the elementary loops in the structure in figure 2 mapping the neuron replicas in the layers of the pRAM network [7].

In principle, each neuron in the layer corresponding to time $t=i$ is a pRAM unit receiving input from each unit in layers $t=k$ with $k<i$.

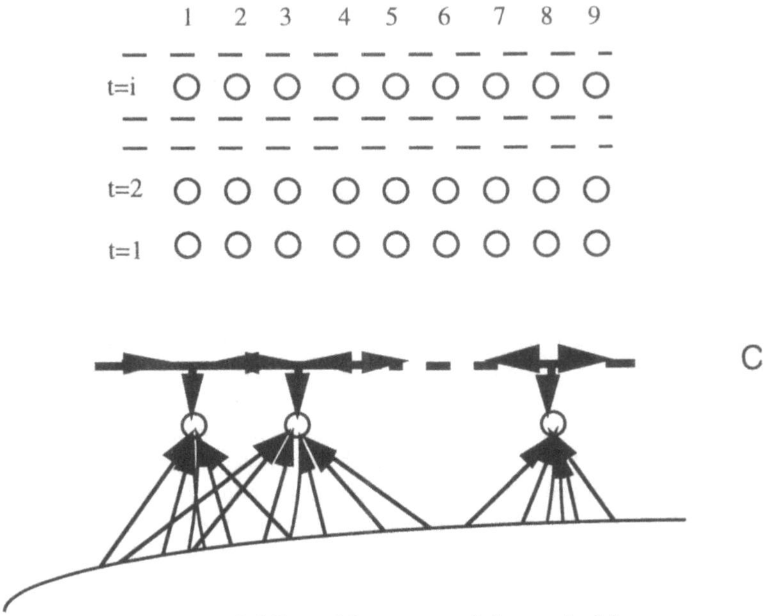

Figure 2 Unfolding of 9 neurons of the cortical layer

2.3 The goal

The goal is to compute a relevant subset of conditional probabilities
$P(Y_i=1|\ Y_{\varphi 1}=y_{\varphi 1},..Y_{\varphi\eta}=y_{\varphi\eta}) = \beta_{y_{j1},\ \ldots,\ y_{jh}}(i)$
with $j_1, \ldots j_k < i$ and $k=$ the maximum connectivity degree allowed by pRAM.
A set of conditional probabilities is relevant if the joint probability distribution law

$$f_{\mathbf{Y}}(\mathbf{Y}) = P(X_1=y_1;\ X_2=y_2;\ldots;\ X_n=y_n) = \prod_{i=1}^{\nu} \beta_{y_{j1},\ \ldots,\ y_{jh}}(i)$$

is close to the actual Y distribution law

$$f_{\mathbf{Y}}(\mathbf{y}) = P(Y_1=y_1;\ Y_2=y_2;\ldots;\ Y_n=y_n) = \qquad (1)$$
$$P(Y_1=y_1)P(Y_2=y2|Y_1=y_1)P(Y_3=y_3|Y_1=y_1,Y_2=y_2)\ldots P(Y_n=y_n|Y_1=y_1;\ldots;Y_{\nu-1}=y_{\nu-1})$$

where X_i is the random bit generated by the i-th pRAM unit u_i.
In case of reduced parametrization, it is easy to show [6] that, whatever the set of the remaining available connections, the model $\beta^{(k)}$, with

- $\beta^{(k)}_{x_{k1(i)},\ldots,\ x_{kN(i)}}(i) = \alpha_{x_{k1(i)},\ldots,\ x_{kN(i)}}(i)$,
- bounded fan-in N,
- $(k_1(i),\ k_2(i),\ldots\ldots,k_N(i)) = k(i)$ the connection pointer of node i indicating the nodes which provide their outputs as inputs to node i and

•$k = (k(1),...,k(\nu))$ the connection matrix,

is the sole model which minimizes the relative entropy $\mathbf{I}(\varphi_\alpha,\varphi_\beta^{(k)})$.

Learning the single conditional probability with the global aim of minimizing the above mentioned relative entropy is a local task incrementally solved as follows.For a given set of pointers $j_1, ...j_k$, the current value of B of β is

$B(t) = (1-\varepsilon)B(t-1) + \varepsilon\, y(t)$, i.e. (2)

$\Delta B= \varepsilon\,(y-B)$

which corresponds to the reward penalty rule

$\Delta B= \rho((y-B)r +\lambda(\overline{y}-B)p)$

with $\rho=\varepsilon$, $r=1$, $p=0$

Highlighting the sole connections affected by these relevant probabilities we obtain a minimal map to explain the observed temporal interactions between the neurons. Suitably processing these probabilities we capture the idea of temporal evolution of the connection strength in relation to some conditioning phenomenon, like "chronic pain" in our application.

We assume the mutual information [8] between units

$$I(u_i;u_j) = H(u_i)-H(u_i|u_j) \quad (3)$$

as the relevance measure for probabilities, $H(u_i)$ and $H(u_i|u_j)$ being the absolute entropy of the i-th unit and the conditional entropy of this unit w.r.t. j-th unit . The general sense of this measure is that the best connections are those which make the output of each neuron depend most on its input. The actual meaning comes from the solution of the constrained optimization problem of minimizing $\mathbf{I}(\varphi_\alpha,\varphi_\beta^{(k)})$. For a given pair of connection matrices k and k', differing for a single alternative connection to neuron u_i, solving this problem consists in selecting k if $\mathbf{I}(\varphi_\alpha,\varphi_\beta^{(k')})>\mathbf{I}(\varphi_\alpha,\varphi_\beta^{(k)})$. This rule translates to our operational rule based on mutual information brought to u_i, by the candidate connections starting from u_j and u_h in k and k' respectively, since:

$$\mathbf{I}(\varphi_\alpha,\varphi_\beta^{(k')})-\mathbf{I}(\varphi_\alpha,\varphi_\beta^{(k)})= H(u_i|u_j)-H(u_i|u_h). \quad (4)$$

From an operational point of view, we compare alternative connections to a given node, and this comparison involves only local computations.

We assume as a probabilistical measure of the connection strength (*pcs*) the quantity

$$\beta_{0_1,..,1_j,..,0_h}(i) - \beta_{0_1,..,0_j,..,0_h}(i) \quad (5)$$

i.e. the difference between the conditional probability of u_i firing when only u_j is firing in its surrounding (solicited response), and the same probability with all surrounding neurons silent (spontaneous firing).

This measure, and its obvious generalization to take into account combined influences of more than one conditioning neurons, results in a different issue of the *efficacy* notion [9] usually measured as $\dfrac{\beta_{0_1,..,1_j,..,0_h}(i)}{\beta(i)}$, with the interesting feature of being 0 when u_i does not react to any activation of u_j.

Neurons have at least two ways to convey information: spike frequency variation and connection strength variation. They are different aspects of the same task. The first is a macroscopic one and can be immediately recognized, the latter is much more subtle to discover. Connection strength variation induces spike frequency modification, but vice-

versa isn't always true, so that spike frequency variation may or may not denote connection strength modification but only a general increase of neurons' activity. In our research we are interested in studying the variations of connection strength among neurons during both basal activity and perceptual tasks, so we need a measure that is independent of spike frequency variation when this doesn't relate to connection strength. Namely, given two neurons u_1 and u_2, let us denote X_i their random states, $p=P(X_2=1|X_1=0)$, i.e. u_2 mean activity when u_1 is silent, and $p=P(X_2=1|X_1=1)$, i.e. the same activity now caused by some u_1 firing. p coincides with the unconditioned u_2 activity in the case of inefficacy of the $\langle u_2,u_1\rangle$ connection, while p', suitably weighted, represents the surplus of activity induced by this connection when an activation starts from u_1 . Thus, p'-p belongs to [0,1] and is insensitive to frequency variations not resulting in connection strength modifications. p'-p = 0 means statistical independence between the two neurons.
A significance level based on the median [10] test is set to discriminate between real statistical dependence and stochastic noise.

3 Numerical results

The general idea is the following. Consider a neural circuit like figure 3.a, possibly representing the activity of three cortical neurons. The circuit can be simulated by a 6-unit pRAM, where the circuit structure should be extracted by the analysis of the difference p'-p of the parameters learned by the pRAM as a function of time (divided in trials) and neurons. In the following we will adopt fan-in=1; thus only the strongest input to neurons will be selected.
Data are summarized in the three-dimensional histograms of figure 4 presenting the activity of one (the head neuron) of a set of simultaneously recorded neurons, matched versus the activities of the other neurons of the same set. ℓ-axis denotes the labels of these neurons. τ-axis reports the sequence of time intervals (trials) dividing the total record. The trials are grouped in blocks of 100 records; each block refers to the same data but with a different temporal shift $\Delta t \in \{1,2,3,4\}$ within the pairs of the activities $\langle y(t),y_\ell(t+\Delta t)\rangle$ sensed from the head and row neuron, respectively. For a given ℓ-neuron, a given trial τ and a given shift Δt, the z-axis gives the strength *pcs* of the connection from head to ℓ-neuron computed on the basis of the joint firing $y(t)\times y_\ell(t+\Delta t)$, with t varying along the trial. For evidence and simplicity sake, for each τ the sole *pcs* of the most relevant connection incoming the head neuron is reported, provided that this *pcs* has been acknowledged significantly different from zero, according to the mentioned median test. All other z are set to zero.
Delays are measured in bins, with a binning of 2ms. A non-zero value of *pcs* at (ℓ,τ) in correspondence of a given Δt means that during trial τ the head neuron has been mostly influenced from neuron ℓ firing with a time delay of Δt bins. If the connection strength is consistently > 0 in a given block, we can assume that there exists a stable connection between the two neurons.
We can clearly distinguish different features for the two kind of rats experimented on.

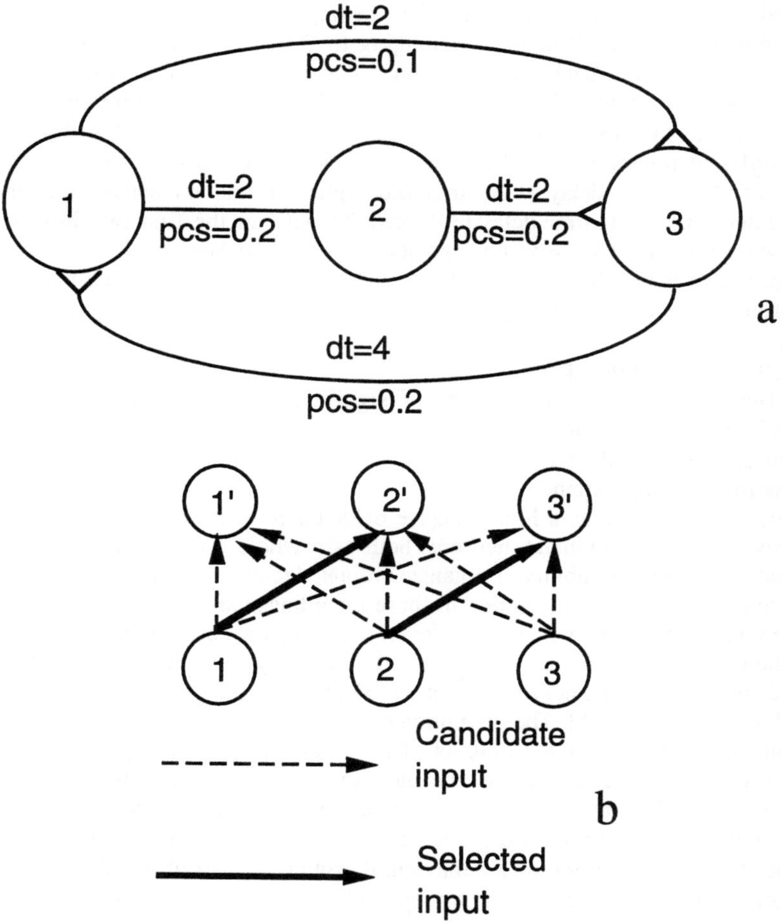

Figure 3: a neural circuit template (a) and pRAM companion (b)

<u>Normal animals</u>

Figure 4.a shows the results of the analysis of real traces recorded from 9 neurons of the cortical region for normal and neuropathic rats. Strong *pcs*'s diffused throughout the whole matrix of coincidences, distributed in all of the delay time windows, are observable. Patterns of tonic (continuous) and bursting (short-lasting episodes of sudden high frequency discharge and abrupt termination) activities of the neuron coupling epoches. These can correspond to analogous discharge patterns of the single cells that simultaneously interact with different rhythmic patterns in the same instant [11,12,13]. Autoconnections *pcs*'s showed recurrent tonic and oscillatory patterns indicating dependence of firing on an efficient temporal distribution of electrochemical events [14,15,16].

In greater detail, a special feature in the pcs peaks with firing delays emerges due to the fact that the analysed traces come from two different tips: neuron 1 to 5 from the first tip, neuron 6 to 9 from the second tip. Simply speaking, neurons in the SSI cortex are organized in microcolumns: neurons that belong to the same microcolumns are usually not interconnected; neurons belonging to different microcolumns are instead widely interconnected. Each microcolumn receives several input fibers from the thalamus, so that same column neurons generally respond to common input. The distance between the tips of the recording microelectrodes is such that each tip records from a different microcolumn. Thus higher *pcs's* appear between same tip recorded neurons at short time delay (1-2 bins) and higher correlation between different tip recorded neurons at long time delay (3-4 bins). Namely, in the first two blocks in the figure we have more peaks between head neuron 5 and the neurons 1 to 5 recorded by the same tip, whilst in the last two blocks we have more peaks between neuron 5 and the neurons 6 to 9 of the other tip.

<u>Animal models of chronic pain</u>

A significant decrease in the number of significant delayed coincidence has been detected in figure 4.b[2]. This happens notwithstanding a higher frequency of discharge of the single units (with twice the original mean), with a consequent increase of the whole neurons ' firing variance.

We interpret this feature as a higher degree of de-coordination of the network activity with respect to normal animals: neurons, both as a whole and on their own, denote a decreased autocorrelative ability of organize ordinated temporal sequences.

The disruption in connectivity found in these recordings can be partially confirmed in the literature by the data obtained in different experimental conditions on the visual and the auditory cortex. Synchrony and delayed synchrony can, in fact, be utilized as coding dimensions by the nervous system, at least for definite functions as vision [17] and auditory processes [18]. The disruption of inner connectivity rules seems to cause derangements in the functional outputs of the neuronal groups exploiting a definite task. No actual data are provided in the literature for the somatosensory system. The available comparable observations on the visual system have shown that animals and humans with divergent squint present a lack of synchrony between neurons driven by the same stimuli and present, on a behavioural ground, a marked inability to integrate the information coming from the two eyes [19]. Similarly studies on the primary auditory cortex have shown that synchrony between single-unit activities is related to coding of auditory signals [18].

4 Conclusions

We have shown that decorrelations of connectivity in pathologic chronic neuropathic states have emerged after analysing data by p-RAM modelling. Previous models of connectivity relied on classical cross-correlograms matching the coincidences between couples of neurons. This method allows for an analysis of all versus all units simultaneously, with large unit sets and using meaningful statistical quantifiers of the interaction between units. The emergent network thus becomes very similar to that 'working' in natural conditions where connectivities are even operating with diverse patterns simultaneously in the diverse areas of the nervous system. The 'propensity' of

---

[2] In absence of chronic implants, the two figures refer to different rats and, obviuosly, different clusters.

the model for natural architectures makes it a most suitable system to model complex natural network representations. More specific features of the connectivity variations in chronic neuropathic conditions will be achieved in future through continuous recording of changing connectivity strengths when chronic implants of recording probes have been set up.

References

[1] E. G. Jones, The Thalamus, Plenum Press, 1985

[2] Abeles, Corticonics, Cambridge Univ. Press, 1990

[3] M.J.Johnson, K.D.Alloway, Sensory modulation of synchronous thalamo-cortical interactions in the somatosensory system of the cat, Exp.Brain Res., 102, 181-197, 1994

[4] Biella, L. Riva and M. L. Sotgiu, Interaction beween neurons in different laminae of the dorsal horn of the spinal cord. A correlation study in normal and neuropathic rats, *European Journal of Neuroscience*, 9:1117-1025, 1997

[5] T.G. Clarkson, D. Gorse, J.G. Taylor, C. K. Ng, Learning probabilistic RAM nets using VLSI structures, IEEE Transactions on computers, vol. 41, pp. 1552-1561, 1992

[6] B. Apolloni, D. De Falco, J.G. Taylor, pRAM layout optimization, Neural Networks 1997, to appear

[7] pRAM-256 data sheet, http://crg.eee.kcl.ac.uk/clarkson.htm

[8] T. Cover, J. Thomas, Elements of information Theory, John Wiley, N.Y, 1991

[9] W. R. Levick, B. G. Cleland and M. W. Dubin, Lateral geniculate neurons of cat: retinal inputs and physiology, Invest. Ophthalmology 11: 302-311, 1997

[10]S. S. Wilks, Mathematical statistics, John Wiley, N.Y, 1962

[11]D.Contreras, M.Steriade, State-dependent fluctuations of low frequency rhythms in cortico-thalamic networks, Neuroscience, 76, 2538, 1997

[12]D.Contreras, A.Destexhe, T.J. Sejnowsky, M.Steriade, Control of spatio-temporal coherence of a thalamic oscillation by corticothalamic feedback, Science, 274, 771-4, 1996

[13]M.Steriade, E. G. Jons, R. R. Llinas, Thalamic oscillations and signalling, Wiley, N.Y., 1989

[14]G.Buszaki, The thalamic clock: emergent network properties, Neuroscience, 41, 351-364, 1991

[15]Koenig, A.K. Engel, Correlated firing in sensory-motor systems, Curr. Biol., 5, 511-519, 1995

[16]M. H. J. Aertsen, G. L. Gerstein, M. K. Habib, G. Palm, Dynamics of neuronal firing correlation:modulation of "effective connectivity".J.Neurophysiol., 61, 900-917, 1989

[17]Singer, C.M. Gray, Visual feature integration and the temporal correlation hypothesis, Ann. Rev. Neurosci., 18: 555-586, 1995

[18]J. Eggermont, Rate and synchronization measures of periodicity coding in cat primary auditory cortex, Hear. Res., 56: 153-167, 1991

[19]K. Von Noorden, Binocular vision and ocular motility. Theory and management of strabismus. Oxford: Blackwell, 1990

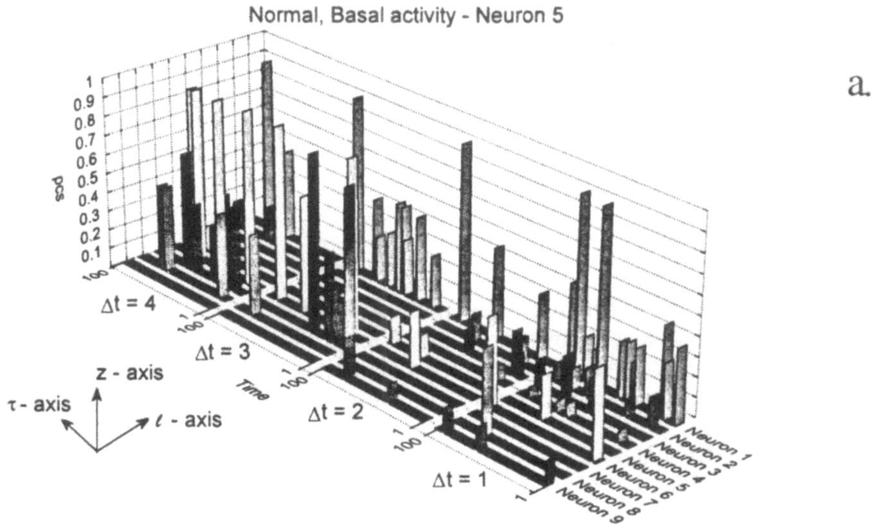

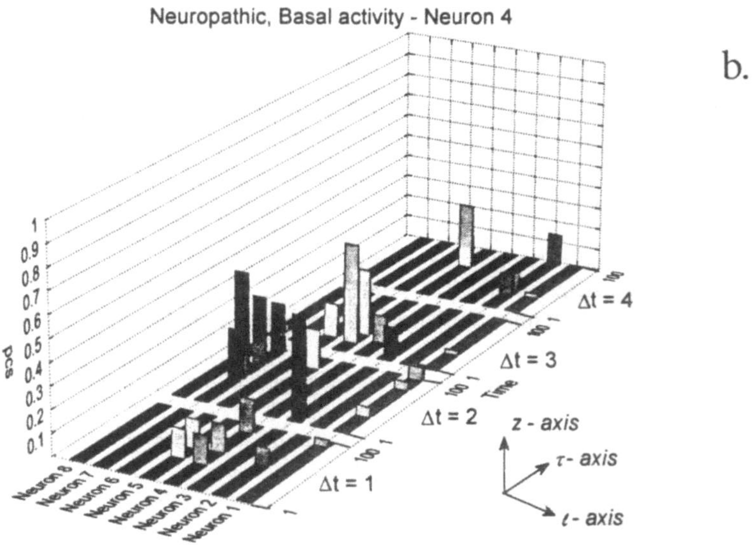

Figure 4. Spread of connection strength between neurons
a. normal animals; b. noxious animals

A New Incremental Strategy for Function Approximation by Radial Basis Function Neural Networks

A. Esposito[a,c],
M. Marinaro[a,b,c],
D.Oricchio[a],
S. Scarpetta[b,c]
[a] International Institute for Advanced Scientific Studies (I.I.A.S.S.) Vietri sul Mare Salerno Italy

[b] Universitá di Salerno, Dipartimento di Fisica Teorica Via S. Allende, Baronissi, I84081, Salerno Italy

[c] INFM unitá di Salerno

Abstract

In this paper we present a new incremental learning algorithm for approximating continuous mapping. The algorithm uses radial basis functions as activation functions. The novelty of the algorithm has different aspects: First, the learning procedure is accomplished modifying only the variances of the activation functions instead of the weights on the synapses; Secondly, the variances of the radial basis functions are trained by a two stage strategy, which includes a local optimization of each new neuron variance. An evolutionary optimization algorithm instead of the usual backpropagation algorithm is used to train the variances. The ability of the net to save training time depends on selectively growing the net structure and on the capability of the algorithm to preserve the locality of the activation functions.

Moreover, reported are comparisons with other incremental algorithm in the literature. Such comparisons show that our net perform better both in terms of net size (few hidden neuron are sufficient to reach very good approximations), and in terms of computational time (the training phase is accomplished on a PC computer and require few minutes).

1 Introduction

Radial Basis function (RBF) network have the ability to approximate any continuous mapping defined on a compact set, given sufficient number of neurons. The universal approximation property has been proved for one hidden layer nets with RBF hidden units (see Hartman et al. [1], Girosi and Poggio [2] and

Park and Sandberg [3]).
These are existence proofs which rely on the availability of an arbitrary large number of hidden units, and they do not offer practical procedure for constructing the networks.
Difficulties lie in deciding the number of hidden units and in learning the adaptable parameters.
We propose a new incremental strategy for selectively growing a RBF [5] network. An error-driven rule is used in order to add the new gaussian node only where it is needed to approximate the desired function. For each new node added, the center of his activation function and the weight connecting it with the output node are settled according to a predetermined rule which make use of both inputs and targets data. Then, the variances of the activation functions are determined by a supervised optimization technique.
The training of the variances consists in two stages: A "local training" where only the variances of the last added node is trained in order to minimize the global error function of the net. A "fine tuning" of the variances of all neurons present at a given instant in the net. In both the stages, the variances are optimized by a genetic algorithm which results more effective than the usual gradient descendent procedure.

2 The Model

The network is designed to rapidly approximate real valued functions $f : \boldsymbol{x} \rightarrow y$ from $k-$dimensional input space, $\boldsymbol{x} \in S_{input}$, to one dimensional output space, $y \in S_{out}$. The net is supposed to learn the function from a finite set of examples $\Gamma_p = \{\boldsymbol{x}_r, y_r; r = 1, \ldots, p\}$ where $y_r = f(\boldsymbol{x}_r)$.
The net architecture at a given instant is composed by a fixed number of input nodes, a hidden layer containing an increasing number of gaussian prototype hidden nodes, h_j, connected to the output through a weight w_j, and one output node.

At the $i - th$ step, when i hidden neurons are present in the network N_i, the net output to an input $\boldsymbol{x}$ is:

$$net_i(\boldsymbol{x}) = \sum_{j=1}^{i} w_j \exp(-\beta_j (\boldsymbol{x} - \mathbf{c}_j)^2) \tag{1}$$

The training phase The training phase consists of an iterative process which adds a hidden node at each step i according the following procedure:

a) A new node h_i is chosen and his parameters $\mathbf{c}_i$, w_i are fixed.

b) The adaptive parameter, β_i, of the new node is optimized according the genetic procedure, while the old node's parameters remain unchanged.

c) Finally the β_js $(j = 1, \ldots i)$ of all nodes present in the net, are updated, always through the genetic procedure

d) The above procedure is repeated until the training error is below a fixed value.

More in details, the ith step of the training phase is accomplished in the following way:
• The point $(\boldsymbol{x}_s, f_s)$ such that

$$(f_s - net_{i-1})^2 = \max_{r=1,\dots,p} (f_r - net_{i-1})^2$$

is chosen from the training set Γ_p, i.e. the training point where the net approximation is worst.

• To correct this error, a new node is added to the net, The center of such node is set equal to $\boldsymbol{x}_s$:

$$c_i = \boldsymbol{x}_s$$

and the weight connection, w_i, from this new node to the output node is set equal to

$$w_i = [f(\boldsymbol{x}_s) - net_{i-1}(\boldsymbol{x}_s)]$$

i.e., the target of the input $\boldsymbol{x}_s$ minus the answer of the net to this input at the previous step (the net with $(i-1)$ hidden neurons).
• Once fixed c_i and w_i, the variance β_i of the new node is optimized using the the evolutionary strategy (see next section) to minimize the Mean Square Error (MSE) of the net with respect to the β_i parameter:

$$MSE = \frac{1}{p}\sum_{r=1}^{p}(f_r - net_i(\boldsymbol{x}_r))^2 \tag{2}$$

• Once learned β_i, the same evolutionary strategy is applied to accomplish the "fine tuning", that is, to optimize the variances β_j of all the hidden nodes present in the net.
Because of the "local" training, once the the βs parameters are trained, their values are very close to the optimal values, and any optimization procedure for the "fine tuning" probably will converge without severe problems. We have choose The evolutionary strategy has been chosen in this case because it is as fast as simple.
• During the β-training the largest squared error between network response $net(\boldsymbol{x}_r)$ and target f_r is stored with the corresponding input and target value $(\boldsymbol{x}_s, f_s)$. This point $(\boldsymbol{x}_s, f_s)$ is the prototype value of the new node that will be added in the next step.

The incremental algorithm always converges. Experimental results show (see sect. 4.) that, for a given net architecture N_i with a training error ϵ_i, adding a neuron and training the net with the above procedure will produce a network N_{i+1} with a training error ϵ_{i+1}, and always $\epsilon_{i+1} < \epsilon_i$. This is mostly due to the fact that, when a neuron is added, the local learning allows to reduce the global error. Therefore, one can add neurons until the net reach the desired approximation.

3 The evolutionary strategy

The evolutionary strategy applied to optimize the β_i of a new neuron, and then, the βs of all the net, is accomplished according the following:

A population of individuals $\beta_i(t+1)$ starting from $\beta_i = \beta_i(t)$, is generated by applying the following mutation :

$$\beta_i(t+1) = \beta_i(t) + N(0,\sigma)\beta_i(t)$$

where $N(0,\sigma)$ is a random number whose probability distribution is a gaussian function with zero mean and variance σ.

If the MSE (cost function) associated to $\beta_i(t+1)$ is less than the cost function associated to $\beta_i(t)$, $\beta_i(t+1)$ becomes the new β_i . Otherwise, we kept the old $\beta_i(t)$.

At beginning, σ is initialized to a chosen value. Then, this value is modified by the following rule:

$$\sigma^{t+1} = \begin{cases} c_1\sigma(t) & \text{if the cost function after the last mutation is increased;} \\ c_2\sigma(t) & \text{if the cost function is reduced;} \\ \sigma(t) & \text{if the cost function is equal to the old one} \end{cases}$$

where c_1, c_2 are constant chosen between $[0, 1.5]$. The β value is always non zero, thanks to a particular condition on the algorithm. The algorithm works as a gradient descent function if $c_1 = 1/c_2$, $c_2 \in [0,1]$. If the individual's position is close to a local minimum of the cost function, then σ is increased (σ is increased until a maximum value, that is fixed). The cycle of minimum-maximum value for σ is repeated a fixed number of times (3 times).

The general idea behind this procedure is to make a non-local search of minimum cost function, with a gradient-descent-like algorithm (local search) when the value of σ is small, and a random-like search when the σ value is high (global search). The advantage of this procedure, with respect to the more popular gradient descendent method, is that there is no need to compute the gradient of the energy cost function and there is no need to choose appropriate values for the learning rate η, and eventually, for the momentum term μ. The same strategy is then used to tune the βs of all the nodes in the net.

4 The net performance

The algorithm has been tested on single value functions, some examples are reported below:

$$y = \sin x + \sin 3x + \sin 6x \qquad x \in [-.35, 3.5] \tag{3}$$

$$y = 0.5\exp(-x)\sin(6x) \qquad x \in [-1, 1] \tag{4}$$

$$y = x \qquad x \in [0, 1] \tag{5}$$

Figures 1(a-c), 2, 3, show the approximation results for the above function using a training and testing set of 100 points. Test data and training data are generated randomly in the input domain. Figures 1a, 1b, 1c report the net output during the growing process (with 1, 2 and 7 neurons respectively) for the first function approximation problem (3).

Each new node has a localized response in order to take care of the data close to his prototype value.

With 7 hidden gaussian nodes, the mean square error is 0.0009. The net performance on the first function are compared with the results reported by Vinod

and Ghose [6] on the same function. Vinod and Ghose used a growing not-uniform networks, with sigmoidal activation functions. Their net was trained through the gradient descent algorithm (backpropagation).
Our net performance is comparable with the one reached by Vinod and Ghose's network on the same function, but our strategy have a less intensive computational cost.
Figure 2 reports the performance of our net for the second function 4. This is the same function on which Yong and Chow [7] have tested their Wavelet Neural Network (WNN), a net for function approximation where the activation function of each node is a combination of Wavelet functions. The WNN networks was tested on an uniformly sampled set of 100 points. The WNN net, in which 25 parameters are adjusted, gives a total training error of 0.0112 when a backpropagation strategy is used; and a total training error of 0.000079 using strategy that combine a backpropagation strategy and QR decomposition. In comparison our net uses only 5 nodes to deliver a total training error of 0.00026. With only seven nodes we obtain a training error of 0.000077.
Figure 3 reports our net approximation for the function 5. The learning process was stopped at 6 hidden gaussian nodes. The corresponding MSE is 0.000023. Finally, Figures 4a, 4b, 4c report our net MSE (eq.2) over the training, Err_{min}, and the testing data, Err_{gen}, as a function of the number of neurons (a maximum of 10 neurons), for each of the above functions (3,4,5).

5 Conclusion

An incremental algorithm for growing RBF and a new learning strategy is proposed. The results show that the net trained with this new strategy is able to obtain good performance for continuous functions. Further theoretical studies and comparative tests (also test for discontinuous function) need to be performed in order to asses the performance of the net.

References

[1] E. Hartman, J. Keeler, J.Kowalsky. *Neural Computation* **2** 1990

[2] F. Girosi and T. Poggio, Networks and the best approximation property. *Biological Cybernetics* **63** 1990

[3] J. Park and I. Sandberg, *Neural Computation* **3** 1991, **5** 1993

[4] J. Moody and C. Darken, Fast Learning in networks of locally-tuned processing units. *Neural Computation* **1** 1989

[5] C.Bishop, *Neural Networks for Pattern Recognition* Clarendon Press Oxford, 1995

[6] V. Vinod and S. Ghose, Growing nonuniform feedforward networks for continuous mappings. *Neurocomputing* **10** 1996

[7] F. Yong and T. Chow, Neural Network Adaptive Wavelets for Function Approximation, Department of Electronic Engineering, City University, Hong Kong, 1996.

Fig. 1a: The dotted lines correspond to the net output. The solid lines to the function to be approximated, which is indicated on the ordinates. Figure refers to a net structure with 1 hidden neuron. On the top of the figure is reported the MSE at this stage, the number of neurons, and the size of the training and testing set (100 points).

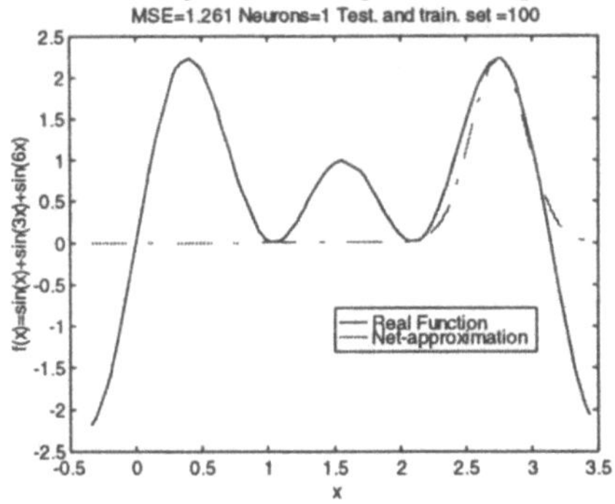

Fig. 1b: The dotted lines correspond to the net output. The solid lines to the function to be approximated, which is indicated on the ordinates. Figure refers to a net structure with 2 hidden neurons. On the top of the figure is reported the MSE at this stage, the number of neurons, and the size of the training and testing set (100 points).

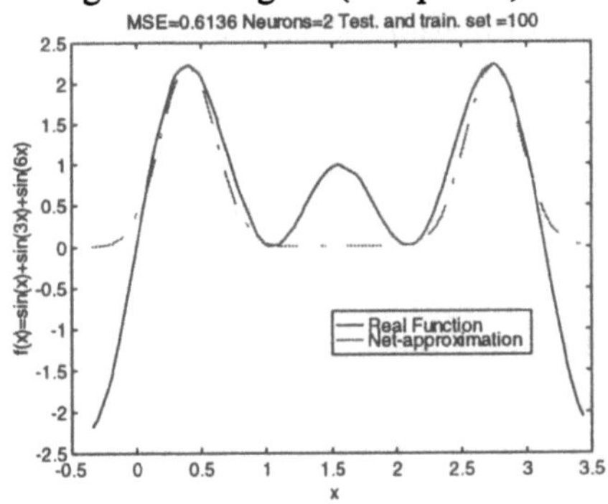

Fig. 1c: The dotted lines correspond to the net output. The solid lines to the function to be approximated, which is indicated on the ordinates. Figure refers to a net structure with 7 hidden neurons. At this stage the performance of the net are comparable with the results reported in the Vinod and Ghose's paper (see text for discussion).

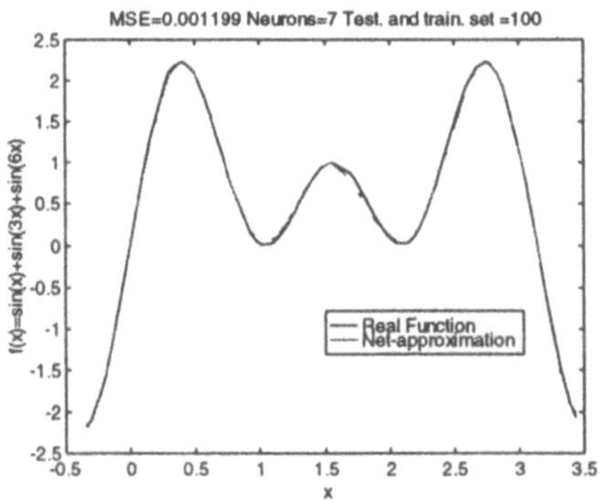

Fig. 2: The dotted lines correspond to the net output. The solid lines to the function to be approximated, which is indicated on the ordinates. Figure refers to a net structure with 7 hidden neurons. At this stage the performance of the net are better than the results reported in the Yong and Chow's paper(see text for discussion).

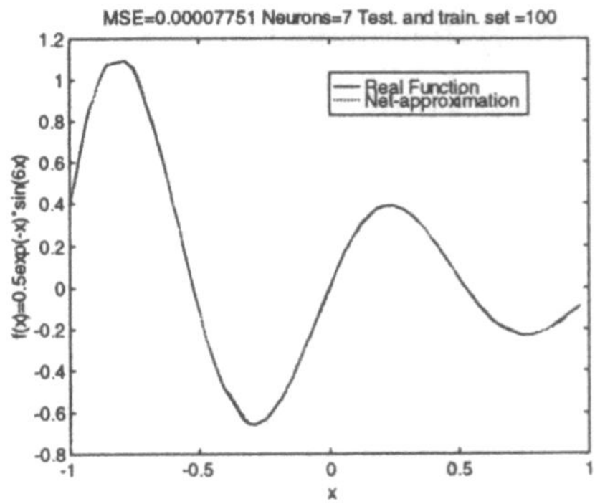

Fig. 3: The dotted lines correspond to the net output. The solid lines to the function to be approximated, which is indicated on the ordinates. Figure refers to a net structure with 6 hidden neurons. On the top of the figure is reported the MSE at this stage, the number of neurons, and the size of the training and testing set (100 points).

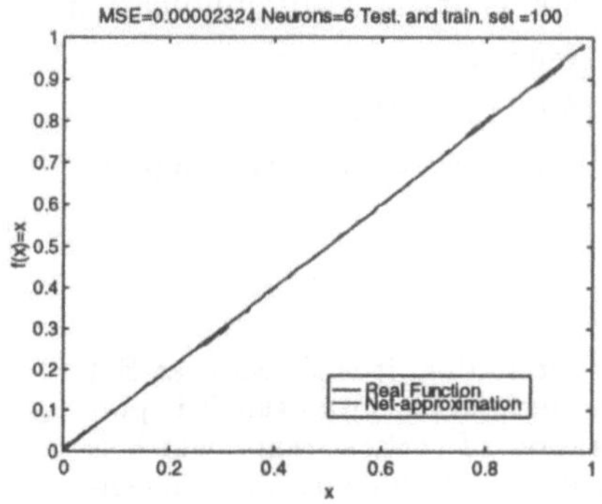

Fig. 4a: The training (solid lines) and the testing (dotted lines) error as a function of the number of neurons in the net (a maximum of 10 neurons is reported). The figure refers to the net trained on the function *f(x)=sin(x)+sen(3x)+sen(6x).*

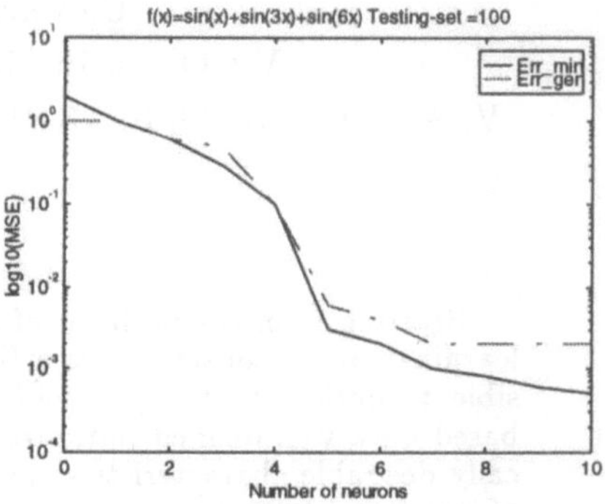

Fig. 4b: The training (solid lines) and the testing (dotted lines) error as a function of the number of neurons in the net (a maximum of 10 neurons is reported). The figure refers to the net trained on the function:*f(x)=0.5exp(x)sin(6x)*

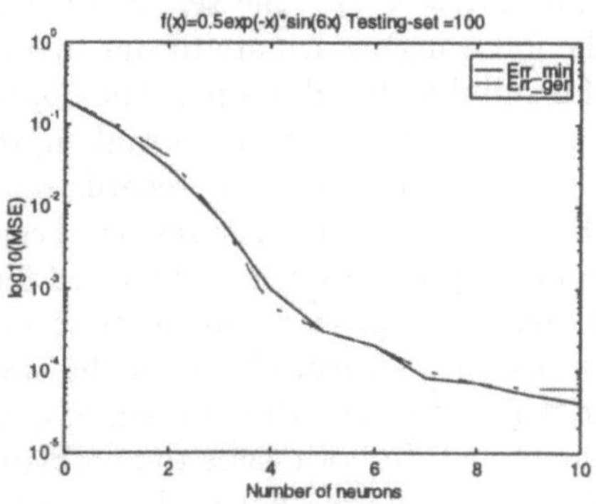

Fig. 4c: The training (solid lines) and the testing (dotted lines) error as a function of the number of neurons in the net (a maximum of 10 neurons is reported). The figure refers to the net trained on the function:*f(x)=x.*

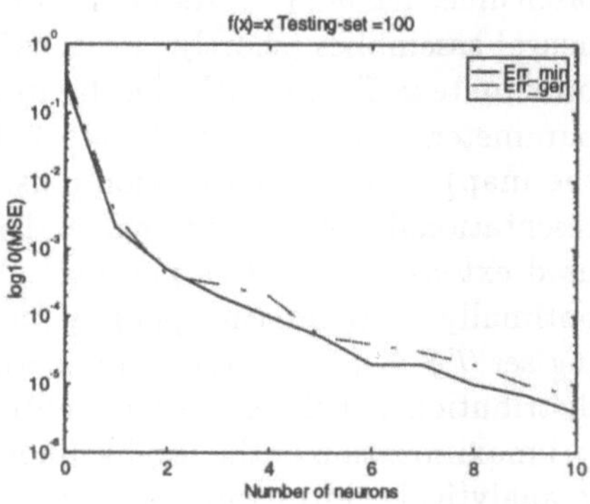

Soft-competitive versus EM learning in cortical map modeling

Francesco Frisone   Pietro G. Morasso
DIST - Department of Informatics, Systems and Telecommunication
University of Genova
Via Opera Pia 13, 16145 Genova, Italy
V: +39 10 3532749; F: +39 10 3532154; E:friso@dist.dist.unige.it

Abstract

Starting from the problem of density estimation, it is shown that EM learning can be considered as a Hebbian mechanism. From this it is possible to outline a theory of self-organization of cortical maps which is based on a well defined optimization process and still preserves biologically desirable characteristics: local computation and uniform treatment of input and lateral connections.

1 Density estimation

Among the multi-faceted aspects of self-organization, density estimation is perhaps the most general one. It allows, in particular, to formulate questions of biological plausibility of cortical map models in a well founded, principled way. Consider a generic sensorimotor vector $\vec{x}$ in d-dimensional space, feeding a cortical area via a thalamic relay nucleus or the association fibers of a transcortical fasciculus; let $p(\vec{x})$ be its unknown pdf and $\mathcal{CM} = \{\backslash_i\}_i^N$ the set of cortical neural assemblies (shortly "neurons", in the following) which are meant to approximate $p(\vec{x})$ implicitly, i.e. by means of a suitable distribution of "neuronal parameters" Θ (center and size of the receptive fields and relative ranking in the map). The approximation must have good characteristics as regards representational power, learning, and modularity. Gaussian mixtures has been used extensively for this purpose[1]. Two main approaches can be followed for optimally approximating $p(\vec{x})$ by means of $gm(\vec{x} \mid \Theta)$, given an adequate training set $\mathcal{TS} = \{\vec{x}_\mu\}_1^P$: (i) minimization of a distance measure between the two distributions, such as the cross-entropy or Küllbak information measure, and (ii) maximization of the log-likelihood of the model. In both cases the problem is analytically intractable and only iterative solutions are known. In particular, the former approach leads to *soft competitive learning* [1] and the latter to *EM-learning* [4].

[1] A Gaussian mixture can be defined as $gm(\vec{x} \mid \Theta) = \sum_i \alpha_i g(\vec{x} \mid \vec{w}_i, R_i)$, where $g(\ )$ is a generic Gaussian $g(\vec{x} \mid \vec{w}_i, R_i) = 1/[(2\pi)^{d/2}||R_i||^{1/2}]e^{-(\vec{x}-\vec{w}_i)^T R_i^{-1}(\vec{x}-\vec{w}_i)}$ and the parameters of the mixture $\Theta = \{\vec{w}_i, R_i, \alpha_i\}_1^N$ are, respectively, the center vectors, covariance matrices and mixing proportions of the Gaussian elements. The mixing proportions (or relative ranking coefficients) must satisfy the constraint $\sum_i \alpha_i = 1$.

1.1 Soft competitive learning

The minimization of the cross-entropy $K(p, gm) = \int p(\vec{x}) log(p(\vec{x})/gm(x|\Theta))dx$ can be solved by means of a stochastic gradient-descent method which yields a simple iterative formula only in the case of a mixture of equal Gaussians ($R_i = \sigma^2 I$) with uniform mixing proportions ($\alpha_i = 1/P$):

$$\Delta\vec{w}_i = \eta(\vec{x}_\mu - \vec{w}_i)u_i(\vec{x}_\mu) \tag{1}$$

where η is the speed or gain factor of the gradient-descent mechanism and the activity level u_i of each neuron is determined by the so called *softmax* function:

$$u_i(\vec{x}) = g(\vec{x} \mid \vec{w}_i, \sigma) / \sum_j g(\vec{x} \mid \vec{w}_j, \sigma) \tag{2}$$

The method has critical parameters which are difficult to adjust: the speed of the gradient-descent η, which must decay to zero with a suitable "freezing curve", and the choice of the common variance σ. The learning equation (1) is found in many different models, with different interpretations and choices of the activation function. In many respect it is has the rank of "the self-organizing equation". The particular version summarized above is perhaps the least heuristic one.

1.2 EM-learning.

The likelihood of the model Θ is the probability it has of generating the observed $\mathcal{TS}$ and, under the hypothesis of independent measures, such probability can be expressed as $\mathcal{L}(\Theta) = \prod_\mu \sum_i \alpha_i g(\vec{x}_\mu \mid \vec{w}_i, R_i)$. Minimizing the logarithm of this function is only analytically tractable in the case of a single Gaussian, yielding well known estimates (the sample mean and sample covariance, respectively). Otherwise, it is necessary to use iterative mechanisms and EM (Expectation Minimization) appears to be simple and very effective. It does not use gradient-descent but rather seeks the fixed point of a transformation which maps the current estimate of Θ into the next one by iterating two steps[2]:

1. A *soft-clustering* step (or E-step) which attributes a *score* $s_i(\vec{x}_\mu)$ (shortly, $s_{i,\mu}$) to each neuron-sample pair, according to the posterior probability that neuron i has generated sample μ. Such score can also be considered the *responsibility* taken by neuron i as regards the sample μ:

$$s_i(\vec{x}_\mu) = \frac{\alpha_i g(\vec{x}_\mu \mid \vec{w}_i, \sigma_i)}{\sum_j \alpha_j g(\vec{x}_\mu \mid \vec{w}_j, \sigma_j)} \tag{3}$$

   Implicitly it induces a subdivision of the $\mathcal{TS}$ into as many soft-clusters as the number of neurons.

[2] Without any loss of generality, we shall assume that the elements of the Gaussian mixture are radially symmetric, i.e. with a diagonal covariance matrix but with different and adaptive variances.

2. A *maximization* step (or M-step) which optimally adapts each neuron to its own cluster, simply computing a weighted sample mean and weighted sample covariance matrix and setting the mixing proportion to the average score of the neuron:

$$\vec{w}_i = \sum_\mu \vec{x}_\mu \frac{s_{i,\mu}}{\sum_\nu s_{i,\nu}} \qquad \sigma_i = \sum_\mu ||\vec{x}_\mu - \vec{w}_i||^2 \frac{s_{i,\mu}}{\sum_\nu s_{i,\nu}} \qquad \alpha_i = \frac{1}{P} \sum_\mu s_{i,\mu} \quad (4)$$

Clearly, this is not a gradient-descent algorithm (rather, it has been shown to be equivalent to a Newton method) [12] and so there is no critical freezing parameter. The fixed point of the transformation is found rather rapidly if the input population is stationary. If transients occur in the input distribution the mechanism has the power to automatically detect the non-stationarity (or novelty) and modulate accordingly the model parameters. An empiric form of this approach has been previously investigated and described under the name of *capture effect* [5].

2 Hebbian learning and topology preservation

The idea that a presynaptic neuron i increases the strength of its synaptic link to a postsynaptic neuron j if both activities are correlated was first formulated by Hebb [6] and a number of recent experimental findings at the neurobiological level seem to confirm its validity[3]. However, Hebbian learning without competition has very limited computational power and the notion of a competitive form of Hebbian learning has been popularized by the SOM (self organized map) model of Kohonen [7] in the framework of optimal vector quantization. If we consider a codebook of prototype vectors $\mathcal{C} = \{\vec{w}_i\}_1^N$ and minimize the quantization error $E_q = \int p(\vec{x})||\vec{x} - \vec{w}_{fb}||^2 d\vec{x}$ by means of a gradient-descent strategy[4], an approximate iterative solution is given by an equation equivalent to (1), with the difference that the type of activation function u_i is "winner-take-all" not "soft-max". The Hebbian nature of equation (1) is readily recognized if we separate the two terms of the equation: the former one ($\vec{x}_\mu u_i$) is clearly Hebbian and the latter ($-\vec{w}_i u_i$) is a decay term with a normalization role. The strong sensitivity to initial conditions of such hard competition approach and its tendency to get stuck in "bad" local minima was a motivation for Kohonen to moderate the crispness of the winner-take-all activation by means of the smoothing action provided by topographically organized recurrent connections,

[3] Various mechanisms of LTP (long term potentiation) of synaptic strength have been studied in detail [3]. The triggering event is an increased production of Ca_{++} ions in the postsynaptic neuron, which is typically mediated by the glutamate neurotransmitter and NMDA channels, although alternative mechanisms are also known to play a role. LTP requires that both neurons are simultaneously depolarized (the time window is around 100 ms). There are global modulating signals but the mechanism is typically local.

[4] $\vec{w}_{fb}$ is the receptive field center of the "first best" neuron, i.e. the neuron for which $||\vec{x} - \vec{w}_i||$ is minimum.

arranged according to a Mexican-hat pattern[5]. Learning is again described by equation (1) (which is applied at steady-state) but with a "bubble-function" activation. *Topology-preservation* has been shown to characterize this heuristic learning procedure, in the sense that the prototype vectors learned by laterally connected neurons tend to be similar. However, this is only a side-effect of the model. The main effect, which allows the network to effectively lower the quantization error to good sub-optimal values, is a sort of simulated annealing mechanism in which the "temperature" of the Monte Carlo simulation is substituted by the size of the bubble-function and the η coefficient: both must be chosen with initial high values and progressively "frozen" during learning. There is a clear link also with soft-competitive learning which requires as well a suitable "freezing" procedure, but only of the gradient-descent coefficient: in fact, the softmax function automatically adapts its degree of sharpness as learning proceeds, different from the bubble-function which requires an explicit modulation.

The EM-learning technique does not have such problems but originally was not inspired by biological concepts and typically is not considered to be Hebbian; we are not aware of attempts to interpret the E-step and/or M-step in biological terms. Nonetheless, the main point of this paper is that EM-learning provides the founding element of a principled approach to a theory of the self-organization of cortical maps, consistent with the Hebbian paradigm. First of all, it is easy to recognize that the scoring function (3) of the E-step is essentially the same as the softmax function, i.e. a soft-competition among the neurons of the map for each input pattern. As regards the M-step, the first one of equations (4) can be obviously interpreted as the batch version of a Hebb rule: $\vec{x}_\mu$ is the synaptic input vector, common to all the neurons of the map, and $s_{i,\mu}$ is the output of the generic neuron. The sum over the training set can be generalized to an integral over a suitable time window (a short-term memory). The normalization effect, which is carried out by the weight decay in (1), here is related to the average score in the same time window. The other two equations in (1) exploit the same time-averaging mechanism which adapts, at the same time, the size and center of the receptive fields of the neurons and their relative ranking. The clear advantage of EM-learning over soft-competitive or SOM learning is that it does not require any freezing procedure since is not gradient-descent based. Rather it self-adapts to a stationary input in a principled way, while remaining sensitive to novelty or drifts, and still preserves the full status of Hebbian mechanism.

In both the soft-competitive and EM learning paradigms there is no element of topology preservation because there are no lateral connections: only input thalamo-cortical connections. Lateral connections are present in related models

[5]The SOM dynamic equation is: $\frac{du_i}{dt} + \gamma u_i = \sum_j c_{ij} u_j + g(\vec{x}_\mu \mid \vec{w}_i, \sigma)$, where c_{ij} are the (symmetric) recurrent or lateral connections, arranged according to a topographic (typically planar) arrangement. The dynamic attractor of such a network, which exists for the symmetry of the lateral connections according to the theory of continuous Hopfield networks, is known as *bubble function*, i.e. a pattern of activity centered around the winning neuron and rapidly decreasing as a function of the network parameters.

like the Neural Gas [8] and the Topology Representing Network [9]. However, such connections do not have a direct role in the self-organization process but rather are limited to the utilization phase (classification, navigation etc.). On the other hand, in the SOM model, the main role of the lateral connections is to support the formation of the bubble function, thus avoiding the global computation implied by the softmax function. The limit is that the connectivity is fixed a priori and is only effective if it matches the number of latent variables, hidden in the $\mathcal{TS}$. The same comment can also by applied to a hybrid model, called GTM [2], which integrates EM learning with an underlying SOM-like grid. In all these cases, lateral connections are treated as different objects from the input connections and such disparity has very little support.

3 Population codes

There is a clear analogy between the softmax function of the soft-competitive model, the bubble function of the SOM model, and the scoring function of the EM-learning model. In all cases they map an input vector into a small cluster of active neurons, i.e. a population code. A probabilistic interpretation of population codes has been proposed by Sanger [10] as the maximum a posteriori estimate of the input vector given the population firing pattern which is fully consistent with the probabilistic model outlined above. However, while in the SOM model it emerges as the attractor of the intrinsic dynamics of the cortical map (this is biologically plausible), in the sof-competitive and standard EM learning paradigm it must be computed explicitly according to a global algorithm (this is *not* biologically plausible). Thus, if we take the best of the different schemes we may define a computational framework which is based on the EM learning strategy but with the following modification: in the the E-step, the explicit function (3) is substituted by a cortical dynamic process which, similar to the SOM model, is driven by a topographic network of recurrent lateral connections but, different from that model, allows such connections to be changed during learning in the same way as the other input connections[6]. In this way, self-organization becomes a principled learning process because inherits the optimal characteristics of the EM framework and, at the same time, is sensitive to the intrinsic dimensionality of the input distribution.

The EM framework is also useful for facing an additional problem: the distributed representation of the input vector $\vec{x}$. In most self-organized models, it is implicitly assumed that such vector is not represented in a distributed manner but with smoothly changing "coordinates", related to the components of $\vec{x}$. In this case it easy to compute the vector difference $\vec{x}-\vec{w}_i$ for each neuron which is the basic element of soft-competitive learning. But such difference becomes fuzzy if the vectors are in fact represented in a distributed way, with varying scale, noise, resolution. On the contrary, the M-step in the EM algorithm,

[6] In [11] we describe a cortical dynamic equation which differs from the SOM model because it does not require recurrent inhibition, typical of the Mexican hat receptive field, but only gating competition: this mechanism does not require inhibitory synapses and is consistent with the excitatory nature of the great majority of synapses on the cortex.

which is based on time averaging, is naturally more robust from this point of view and allows the uniform treatment of lateral and input connections.

Simulations of the model are under way and will be shown at the conference.

References

[1] M. Benaim. On Functional Approximation with Normalized Gaussian Units. *Neural Computation*, 6:319–333, 1994.

[2] C.M. Bishop, M. Svensén, and K.I. Williams. GTM: a principled alternative to the self-organized map. In C. con der Malsburg, W. von Seelen, J.C. Vorbruggen, and B. Sendhoff, editors, *Artificial Neural Networks*, volume LCS-1112, pages 165–170, Berlin, 1996. Springer Verlag.

[3] T.V.P. Bliss and G.L. Collinridge. A synaptic model of memory: long-term potentiation in the hippocampus. *Nature*, 361:31–39, 1993.

[4] A.P. Dempster, N.M. Laird, and D.B. (1977) Rubin. Maximum likelihood from incomplete data via EM algorithm. *J. Royal Stat. Soc. B*, 39:1–38, 1977.

[5] F. Firenze and P. Morasso. The capture effect: a new self-organizing network for adaptive resolution clustering in changing environments. In *IEEE World Congress on Computational Intelligence*, volume 2, pages 653–658. IEEE Service Center, Piscataway, NJ, 1994.

[6] D.O. Hebb. *The organization of Behavior.* J. Wiley, New York, 1949.

[7] T. Kohonen. *Self organization and associative memory.* Springer-Verlag, Berlin, 1988.

[8] T. Martinetz and K. Schulten. A 'Neural-Gas' network learns topologies. In T. Kohonen, K. Makisara, O. Simula, and J. Kangas, editors, *Artificial Neural Networks*, pages 397–402, Amsterdam, 1991. North-Holland.

[9] T. Martinetz and K. Schulten. Topology representing networks. *Neural Networks*, 7:507–522, 1994.

[10] T. D. Sanger. A probability interpretation of neural population coding for movement. In P. Morasso and V. Sanguineti, editors, *Self-organization, Computational Maps, and Motor Control*, pages 75–116. Elsevier Science Publishers, Amsterdam, 1997.

[11] V. Sanguineti, P. Morasso, and F. Frisone. Cortical maps of sensorimotor spaces. In P. Morasso and V. Sanguineti, editors, *Self-organization, Computational Maps, and Motor Control*, pages 1–36. Elsevier Science Publishers, Amsterdam, 1997.

[12] L. Xu and M. I. Jordan. On convergence properties of the EM Algorithm for Gaussian mixtures. *Neural Computation*, 8:129–151, 1996.

Spatio-temporal cortical activity patterns in cognitive tasks using fMRI

F. Frisone, P. G. Morasso, G. Iannò, M. Marongiu
DIST - Department of Informatics, Systems and Telecommunicatio
University of Genova
Via Opera Pia 13, 16145 Genova, Italy
V: +39 10 3532749; F: +39 10 3532154; E:friso@dist.dist.unige.it

P. Vitali, G. Rodriguez
Institute of Neurophysiopathology, DISM - Department of Motor Scie nces
University of Genova
Ospedale S. Martino, L.go R. Bensi 10, 16132 Genova, Italy

Abstract

In this study we investigate the spatio-temporal organization of the human cortical activity in complex tasks. Such complex interactions are crucial, for instance, in the case of voluntary actions, memory and are probably impaired in neurological disorders (e.g. epilepsy) or psychiatric disorders (e.g. schizophrenia).

The aim of the performed fMRI experiments was to evaluate multiple cortical areas involved in verbal output and input during two different covert language tasks, namely "*verbal fluency*" and "*verbal understanding*". We discuss the relevance of such data for a theory of cortex dynamics which emphasizes the role of long-range lateral connections. The future experimental investigation will be aimed at the analysis of the spatio-temporal structure of cortical activity in pathological conditions, such as epilepsy.

1 Introduction

In recent years, there has been an explosive growth in the field of human brain imaging. FMRI has become one of the most promising tools in cerebral neurophysiology. In particular, the BOLD technique (*Blood oxygen level dependent*) is sensitive to increased blood oxygenation related to neural activation. FMRI is a modern, non-invasive tool for brain mapping which adds good temporal resolution to excellent spatial resolution even in deep structures. This new technique is sensitive to increased capillary and venous blood oxygenation related to neural activation: the subsequent decrease in deoxihemoglobin concentration can be measured as an increase in the local $T2^*$-weighted magnetic resonance signal. FMRI has dramatically increased our understanding of localization of the cerebral functions in the last years: previous reports described brain mapping in motor and perceptual (somatosensory, visual, auditory, olfactory) paradigms, and in cognitive tasks;

The goal of this study is to investigate the organization of the human cortical activity in complex tasks like verbal understanding, taking into account some recent experimental results in the fields of neuroanatomy/neurophysiology obtained applying the fMRI system. Language is the main human cognitive process, and its neurophysiology typically reflects the dichotomous categorization of disturbances in linguistic output and input (i.e. expressive/receptive, fluent/non fluent, motor/sensory), which is related to two classical areas located in the inferior frontal gyrus and in the posterior superior temporal gyrus, respectively the Broca's and Wernicke's areas in the dominant hemisphere. Recent studies by Position Emission Tomography and FMRI (review in [6]) have provided strong evidence for more diffuse integration of multiples areas in language tasks: Friston described that as "functional connectivity" [4].

In our fMRI experiments we attempted to evaluate multiple cortical areas dynamically involved during the tasks. For this purpose it is necessary to use two similar stimuli which are presented to the subject in alternating phases, called activation and control phases, differing only for the investigated process.

We discuss the relevance of such data to investigate biologically realistic computational models of cortical maps, which emphasizes the role of long-range lateral connections somehow contradicting the conventional view of the cortex as a continuous computational medium characterized by a localization of tasks and an associated locality of lateral connections. In fact,topologically, the human cerebral cortex is a highly folded sheet and it has a modular organization, characterized by a large number of columns communicating by means of lateral connections, parallel to the cortical surface [7], prompting the (wrong) view of a 2-D *lattice*. However, the observed patterns obtained applying the fMRI technique appears to support the opposite view.

2 Experimental results

Let us now define more precisely the two covert language tasks, called ***verbal fluency*** and ***verbal understanding*** which are used for evaluating multiple cortical areas involved in verbal output and input. ***Verbal fluency***: during the activation phase the subject thinks as many words as possible beginning with a letter said by the examiner. During the control phase the subject silently repeats nonsense word ("bla"). ***Verbal understanding***: the subject mentally finds the words corresponding to the definitions given by the examiner. During the control phase, the subject silently answers "calm" when the examiner says "you must stay".

Eighteen transaxial anatomic slices (T1 weighted, encompassing the whole brain) were acquired by means of a 1.5 T Siemens scanner equipped with head coil and EPI software. Forty T2* weighted functional sequences were acquired during each task, which was composed by 4 cycles. Each cycle alternated 5 acquisitions in the task activation phase and 5 in the control phase. The 720 (40x18) functional images were statistically evaluated on a Siemens workstation by a special software ("Stimulate") which performed cross-correlation analysis

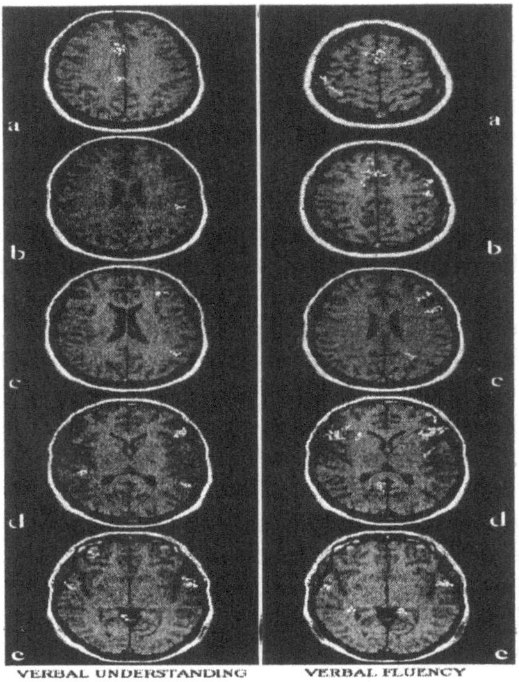

Figure 1: Cortex activity during verbal understanding and verbal fluency tasks. During verbal understanding task we can observe activation in: anterior and posterior cingulus (a), supramarginal gyrus (b), angular gyrus and BA 44 or Broca's area (c), BA 45 of the inferior frontal gyrus and posterior medium temporal gyrus or Wernicke's area (d), bilateral superior temporal gyrus (e). During verbal fluency task we can observe activation in: primary and supplementary motor areas (a), anterior cingulus and dorsolateral prefrontal cortex (b), BA 44 or Broca's area (c), bilateral BA 45 of the inferior frontal gyrus (d), bilateral superior temporal gyrus and hippocampus (e).

and cluster filtering: the significantly activated pixels were superimposed on anatomical images.

As shown by the figure, a widespread activation was present in all subjects in both hemispheres. During verbal fluency the majority of activated pixels was not only in the left inferior frontal gyrus (Broca's area) but also in the left dorsolateral prefrontal cortex, which has been shown [5] to be related to "willed action" tasks; other activation patterns are in premotor and motor areas, in anterior gyrus cinguli, left hippocampus, thalamus, insula and parieto-occipital areas. Premotor and motor areas, seen in other tasks of word production, are probably involved in planning of speech, even if not executed. Limbic areas seem to be related to attentional and memory effort, whereas parieto-occipital areas to visual recall of things corresponding to words [6]. Broca's area (and somewhere the controlateral inferior frontal gyrus) are seen active even during verbal understanding task, with multiple parietotemporal areas, i.e. superior and middle temporal, angular and supramarginal gyrus. Indeed recent acquisitions show primary activations to Wernicke's area in verbal understanding, but this function seems split into semantic (prefrontal) and phonological (parietotemporal) areas. This second task produced an even more diffuse activation with respect to verbal fluency.

3 Discussion

As regards the biological plausibility of cortical map models, the somatotopic layout of many cortical areas has long suggested a kind of topologic organizati on, associated with a dimensionality reduction of the representational space, thus motivating the development of a (large) family of self-organizing network models. From its beginning, however, the effort has been affected by a number of misconceptions, partly due to the over-emphasis in the neurophysiological community on the receptive field properties of cortical neurons. Only recently a new understanding of the cortex is emerging as a continuously adapting dynamical system, shaped by competitive and cooperative lateral connections [13] and such organization is not static but changes with ontogenetic development together with patterns of thalamo-cortical connections. Shortly, it has been suggested that cortical areas can be seen as a massively interconnected set of elementary processing elements, which constitute a computational map [7]. From the modeling point of view, the most common misconceptions about cortical functionality can be reduced to the following three items:

1. flatness of cortical maps (related to the locality of lateral connections);
2. fixed lateral connections (versus plastic thalamo-cortical connections, which determine receptive-field properties);
3. Mexican-hat function of lateral interactions (it implies a significant amount of recurrent inhibition for the formation of localized responses by lateral feedback).

The flatness assumption which characterizes the classic map models [1, 8] is contradicted by the fact that the structure of lateral connections is not genetically determined but depends mostly on electrical activity during development. More precisely, they have been observed to grow exuberantly after birth and reach their full extent within a short period; during the subsequent development, a pruning process takes place so that the mature cortex is characterized by a well defined pattern of connectivity, which includes a large amount of non-local connections: this rules out all the models limited to a purely 2-D circuitry. Moreover, the superficial connections to non-neighboring columns are organized into characteristic patterns: a collateral of a pyramidal axon typically travels a characteristic lateral distance without giving off terminal branches and then it produces tightly terminal clusters (possibly repeating the process several times over a total distance of several millimeters). Such characteristic distance is not a universal cortical parameter and is not distributed in a purely random fashion but is different in different cortical areas [2]. Thus, the development of lateral and afferent connections depends on the cortical activity caused by the external inflow, in such a way to capture and represent the (hidden) correlations in the input channels. Each individual lateral connection is "weak" enough to go virtually unnoticed while mapping the receptive fields of cortical neurons but the total effect on the overall dynamics of cortical maps can be substantial, as is revealed by cross-correlation studies [12]. Lateral connections from superficial pyramids tend to be recurrent (and excitatory) because 80% of synapses are with other pyramids and only 20% with inhibitory interneurons, most of them acting intra-columnarly [10]. Recurrent excitation is likely to be the underlying mechanism which produces the synchronized firing which has been observed in distant columns. The existence (and preponderance) of massive recurrent excitation in the cortex is in contrast with what could be expected, at least in primary sensory areas, considering the ubiquitous presence of peristimulus competition (or "Mexican-hat pattern") which has been observed time ago in many pathways, as the primary somatosensory cortex, and has been confirmed by direct excitation of cortical areas as well as correlation studies; in other words, in the cortex there is a significantly larger amount of long-range inhibition than expected from the density of inhibitory synapses. In general, "recurrent competition" has been assumed to be the same as "recurrent inhibition", for providing an antagonistic organization that sharpens responsiveness to an area smaller than would be predicted from the anatomical funneling of inputs. Thus, an intriguing question is in which manner long-range competition can arise without long-range inhibition and a possible solution is the mechanism of gating inhibition based on a competitive distribution of activation, proposed by [11] and further investigated by [9, 3].

4 Conclusions

According to the illustrated experimental evidence, speech-related tasks involve the simultaneous and concurrent activation not only of the specifically speech-related areas (Broca's and Wernicke's) but also of a number of other cortical and sub-cortical formations. The underlying hypothesis is that the observed patterns of cortex activity, which mix local and non-local areas, have a specific computational function: representing complex tasks onto the cortex. In this framework, different local areas are used for mapping onto the cortex high-dimensional input spaces (e.g. multimodal stimuli) thus setting up a number of inter-related multi-dimensional lattices which cooperate in carrying out complex tasks such as speech. The stability of the mechanism is determined by the delicate balance between the massive recurrent excitation and the non-linear stabilizing effects described above and it is not surprising that such stability can be broken even by small local parametric changes.

The future experimental investigation will be aimed in this direction, i.e. the analysis of the spatio-temporal structure of cortical activity in pathological conditions, such as epilepsy.

References

[1] S. Amari. Dynamics of pattern formation in lateral-inhibition type neural fields. *Biological Cybernetics*, 27:77–87, 1977.

[2] W. Calvin. Cortical columns, modules and hebbian cell assemblies. In M. Arbib, editor, *The handbook of brain theory and neural networks*, pages 269–272. MIT Press, Cambridge, MA, 1995.

[3] F. Frisone and P.G. Morasso. Representing multidimensional stimuli on the cortex. In C. von der Malsburg, W. von Seelen, and J. C. Vorbrüggen, editors, *Artificial Neural Networks - ICANN96*, pages 649–654, Bochum, Germany, July 1996. Springer-Verlag.

[4] K.J. Friston, C.D. Frith, P.F. Liddle, and R.S.J. Frackowiak. Functional connectivity: the principal component analysis of large (pet) data sets. *Journal of cerebral blood flow and metabolism*, 13:5–14, 1993.

[5] C.D. Frith, K.J. Friston, P.F. Liddle, and R.S.J. Frackowiak. Willed action and the prefrontal cortex in man: a study with pet. *Proceedings Royal Society of London*, 244:241–246, 1991.

[6] M. Habib, J.F. Demonet, and R.S.J. Frackowiak. Neuroanatomie cognitive du language: contribution de l'imagerie fonctionelle cerebrale. *Revue Neurologique*, 152:249–260, 1996.

[7] E. I. Knudsen, S. du Lac, and S.D. Esterly. Computational maps in the brain. *Annual Review of Neuroscience*, 10:41–65, 1987.

[8] T. Kohonen. Self organizing formation of topologically correct feature maps. *Biological Cybernetics*, 43:59–69, 1982.

[9] P.G. Morasso and V. Sanguineti. How the brain can discover the existence of external egocentric space. *Neurocomputing*, 12:289–310, 1996.

[10] A. Nicoll and C. Blakemore. Patterns of local connectivity in the neocortex. *Neural Computation*, 5:665–680, 1993.

[11] J.A. Reggia, C.L. D'Autrechy, G.G. Sutton III, and M. Weinrich. A competitive distribution theory of neocortical dynamics. *Neural Computation*, 4:287–317, 1992.

[12] W. Singer. Development and plasticity of cortical processing architectures. *Science*, 270:758–764, 1995.

[13] J. Sirosh, R. Mikkulainen, and Y. Choe. *Lateral interactions in the cortex.* Hypertext Book, 1989. URL:www.cs.utexas.edu/users/nn/webpubs/htmlbook96.

Support Vector Machine for Nonparametric Binary Hypothesis Testing

Davide Mattera Francesco Palmieri

Dipartimento di Ingegneria Elettronica e delle Telecomunicazioni,
Università degli Studi di Napoli Federico II, Italy
Via Claudio, 21, I-80125 Napoli, Italy
e-mail: { mattera , palmieri }@diesun.die.unina.it

Abstract

The Support Vector Machine, introduced in [1] as a practical implementation of the principle of structural risk minimization, constitutes one of the most promising methods for constructing a mathematical model only on the base of a limited amount of measured data. In this paper, we consider the application of this method to the problem of nonparametric binary hypothesis testing (bayesian setting); the main contribution of this paper is the derivation of the Support Vector algorithm in the case of a generic convex approximation of the binary risk function.

1 Introduction

The Support Vectors Machine [6] is a powerful method for constructing a mathematical model from empirical data that has been successfully applied, in different areas, such as the optical character recognition [5]. It is firmly founded on the theory of the Structural Risk Minimization (SRM) [6]. The practical implementation of this principle on the class of the functions

$$\mathcal{F} \triangleq \{f(\boldsymbol{x};\boldsymbol{w}) = \boldsymbol{w}^T \cdot \boldsymbol{\phi}(\boldsymbol{x}) + b \quad , \quad \boldsymbol{x} \in X, \quad \boldsymbol{w} \in \mathcal{R}^d, \quad b \in \mathcal{R}\} \tag{1}$$

does not differ in practice from the regularized radial basis function approach. Both approaches include the norm $||w||^2$ in the cost function as it is directly related to the Vapnik-Chervonenkis dimension of $\mathcal{F}$ and controls the capacity of the class of functions $\mathcal{F}$ as advocated from the principle of structural risk minimization. On the other hand, the radial basis function approach advocates regularization theory to justify the inclusion of the parameter-vector norm $||w||^2$ in the cost function.

Whether or not Vapnik's approach is better motivated from the theoretical point of view, the differences in performances lie, mainly, in the possibility of solving the minimization problem in correspondence of different cost functions.

In this paper, we are concerned with the application of Support Vector Machines to the problem of the nonparametric (or unstructured) binary hypothesis testing. The problem is a classical one: a vector of random variables $\boldsymbol{X}$ is drawn from two different densities $f_1(\boldsymbol{x})$ and $f_2(\boldsymbol{x})$ according to hypotheses H_1 and H_2, respectively. We are looking for a function $f(\boldsymbol{x})$ which minimizes the bayesian cost:

$$p_1 \cdot c_1 \cdot Prob(\overline{f}(\boldsymbol{x}) < 0|H_1) + p_2 \cdot c_2 \cdot Prob(\overline{f}(\boldsymbol{x}) \geq 0|H_2) \quad , \tag{2}$$

where p_i $(i = 1,2)$ and c_i $(i = 1,2)$ are the *a priori* probability and the cost of the error under the hypothesis H_i. In the nonparametric context, no *a priori* information is available about the densities f_1 and f_2, but the only information available is constituted by a set of l realizations of the random vector X under the two hypotheses: $\{(\boldsymbol{x}_1, y_1), \ldots, (\boldsymbol{x}_l, y_l)\}$ where $y_i = 1$ if the ith example has been obtained under H_1 and $y_i = -1$ if the ith example has been obtained under H_2.

A first step for the solution of this problem should be constituted by a supervised feature extraction stage [3]. We suppose in this context that the random vector X is already the output of a preprocessing stage that has achieved dimensionality reduction. We assume here that the dimension of the post-processed vector X is still too high to render ineffective the classical techniques of nonparametric density estimation (see [3]) when only a limited amount of measured data is available.

2 The Support Vector Method

The principle of the structural risk minimization suggests to minimize the mean value of the risk

$$R(\boldsymbol{w}) \triangleq \int C_y(f(\boldsymbol{x};\boldsymbol{w}))dP(\boldsymbol{x},y) \quad , \tag{3}$$

in the class of function $\mathcal{F}$ (1), by solving the following problem

$$\min_{\boldsymbol{w}\in\mathcal{R}^d, b\in\mathcal{R}} \left\{ \frac{1}{2}||\boldsymbol{w}||^2 + C\sum_{i=1}^{l} C_{y_i}(\boldsymbol{w}^T \cdot \boldsymbol{\phi}(\boldsymbol{x}_i) + b) \right\} \quad , \tag{4}$$

where $P(\boldsymbol{x}, y)$ is the distribution function on the space $X \times Y$, $C_y(f(\boldsymbol{x}))$ is the cost of approximating with $sign(f(\boldsymbol{x}))$ the value y which specifies the hypothesis that generated the input point $\boldsymbol{x}$. The value of the constant C can be determined by utilizing cross-validation techniques or SRM bounds [6].

The difficulties encountered in the solution of the problem (4) is mainly due to the form of the cost function for the case of binary test of hypothesis:

$$C_y(\zeta) = c(y) \cdot u(-y\zeta) \tag{5}$$

where $u(\cdot)$ is the usual step function and $c(y)$ is equal to c_1 if $y = 1$ and c_2 if $y = -1$. The computational problem (4) with the cost function (5) is known to be NP complete. For this reason, in practice, only some approximate solutions to the problem can be determined. The first approximation methods, introduced in the literature, considered the following approximation of the step function $u(\zeta)$: $(\zeta+1)^2$; successively, it was noticed that better results can be obtained with the following approximation: $\phi^2(\zeta)$ where $\phi(\cdot)$ is a sigmoidal function such as the logistic function $(1+\exp(-\zeta))^{-1}$ (this approximation was utilized for the derivation of the well-known backpropagation algorithm).

The Support Vector Method [1] is based on approximating the ideal step function $u(\cdot)$ with a convex one $q_y(\cdot)$ to avoid the existence of local minima in the optimization problem[1]; in fact, for any convex optimization problem, every local solution is also global [2]. For suitable choices of the convex function $q_y(\zeta)$, the solution of the problem (4) represents a good approximation of the desired function corresponding to the choice of the ideal function (5).

Problem (4) can be re-written in the form:

$$\min_{\boldsymbol{w}\in\mathcal{R}^d, b\in\mathcal{R}} \left[\frac{1}{2}\|\boldsymbol{w}\|^2 + C\sum_{i=1}^{l} Q_{y_i}(\boldsymbol{w}^T\cdot\boldsymbol{\phi}(\boldsymbol{x}_i)+b)\right] , \tag{6}$$

where $Q_y(\zeta) \triangleq c(y)\cdot q_y(-y\zeta)$. This problem scales in complexity with d; however, one can solve it also if d tends to infinity by noticing that (6) is equivalent to

$$\min_{\boldsymbol{w}\in\mathcal{R}^d, b\in\mathcal{R}, \zeta_i} \left[\frac{1}{2}\|\boldsymbol{w}\|^2 + C\sum_{i=1}^{l} c(y_i) q_{y_i}(\zeta_i)\right] , \tag{7}$$

subject to the following conditions, for each $i\in\{1,\ldots,l\}$,

$$\begin{aligned} &-y_i[\boldsymbol{w}^T\boldsymbol{\phi}(\boldsymbol{x}_i)+b] \le \zeta_i \ , \\ &\zeta_i \ge -a_{y_i} \ , \end{aligned} \tag{8}$$

where $a_y > 0$ is such that $q_y(\zeta) = 0$ for $\zeta \le -a_y$.

By making use of the Kuhn-Tucker theorem (which holds for convex optimization problems), the decision function can be written as

$$\hat{f}(\boldsymbol{x}) = sign(\boldsymbol{w}^T\cdot\boldsymbol{\phi}(\boldsymbol{x})+b) = sign\left(\sum_{i=1}^{l} y_i\alpha_i K(\boldsymbol{x},\boldsymbol{x}_i)+b\right) \tag{9}$$

where $K(\boldsymbol{x}_1,\boldsymbol{x}_2) \triangleq \boldsymbol{\phi}^T(\boldsymbol{x}_1)\cdot\boldsymbol{\phi}(\boldsymbol{x}_2)$ and the coefficients α_i can be determined by solving the following problem:

$$\min_{\zeta_i}\ \max_{\alpha_i,\gamma_i} \left\{ -\frac{1}{2}\sum_{i=1}^{l}\sum_{j=1}^{l} y_i y_j K(\boldsymbol{x}_i,\boldsymbol{x}_j)\alpha_i\alpha_j \right.$$

[1] Also the choice of the class $\mathcal{F}$ of nonlinear functions which depend linearly upon their parameter vector $\boldsymbol{w}$ is, mainly, aimed at avoiding the local minima in the optimization problem.

$$+\sum_{i=1}^{l}[C \cdot c(y_i) \cdot q_{y_i}(\zeta_i) - (\alpha_i + \gamma_i) \cdot \zeta_i - a_{y_i}\gamma_i]\Bigg\} \tag{10}$$

subject to the following conditions, for each $i \in \{1, \ldots, l\}$,

$$\sum_{i=1}^{l} y_i \alpha_i = 0 \ ,$$

$$C \cdot c(y_i) \cdot q'_{y_i}(\zeta_i) - \alpha_i - \gamma_i = 0 \ , \tag{11}$$

$$\alpha_i \geq 0, \qquad \gamma_i \geq 0 \ ,$$

in which q'_y is the first derivative of q_y. The Kuhn-Tucker theorem also implies that the following conditions are satisfied, for each $i \in \{1, \ldots, l\}$,

$$\alpha_i(-bi - \zeta_i) = 0 \tag{12}$$

$$\gamma_i(-\zeta_i - a_{y_i}) = 0 \ , \tag{13}$$

where $b_i \triangleq y_i \cdot \left[\sum_{j=1}^{l} y_j \alpha_j K(\boldsymbol{x}_i, \boldsymbol{x}_j) + b\right]$.

3 Typical convex approximations of the ideal cost function

The simplest convex approximation of the ideal step function $u(\cdot)$ is the following one:

$$q_y(\zeta) \triangleq (\zeta + 1) \cdot u(\zeta + 1) \qquad \forall y \in \{-1, 1\} \ . \tag{14}$$

With this choice for the cost function, the numerical problem (10) reduces to

$$\max_{\alpha_i} \left\{ -\frac{1}{2} \sum_{i=1}^{l} \sum_{j=1}^{l} y_i y_j K(\boldsymbol{x}_i, \boldsymbol{x}_j) \alpha_i \alpha_j + \sum_{i=1}^{l} \alpha_i \right\} \ , \tag{15}$$

subject to the conditions

$$\sum_{i=1}^{l} y_i \cdot \alpha_i = 0 \ , \qquad 0 \leq \alpha_i \leq C \cdot c(y_i) \ . \tag{16}$$

Moreover, from the Kuhn-Tucker conditions (12) and (13), it follows that the following equivalences are valid at the optimal solution:

$$\alpha_i = 0 \quad \Longleftrightarrow \quad b_i \geq 1 \ , \tag{17}$$

$$0 < \alpha_i < C \cdot c(y_i) \quad \Longleftrightarrow \quad b_i = 1 \ , \tag{18}$$

$$\alpha_i = C \cdot c(y_i) \quad \Longleftrightarrow \quad b_i \leq 1 \ . \tag{19}$$

The equivalence (18) allows one to calculate the value of the threshold b. Moreover, the three equivalences show that only a fraction of the examples needs to be considered in the Support Vector expansion, in particular the examples for which $b_i \geq 1$ do not become support vectors. Moreover, all the training examples which are incorrectly classified have $b_i \leq 0$ and, therefore, become support vectors. The algorithm (15) has been introduced in [4]; it reduces to classical one introduced in [6] when the costs c_1 and c_2 have the same value ($c_1 = c_2 = 1$).

Let us now consider the following convex approximation of the ideal step function $u(\cdot)$:

$$q_y(\zeta, \delta_1, \delta_2) \triangleq \delta_1 \cdot (\frac{\zeta}{\delta_2} + 1) \cdot u(\zeta + \delta_2) \qquad \forall y \in \{-1, 1\}\ , \tag{20}$$

with $\delta_1, \delta_2 > 0$. The function (20) reduces to (14) if $\delta_1 = \delta_2 = 1$. Since $q_{y_i}(\zeta, \delta_1, \delta_2) = \delta_1 q_{y_i}(\frac{\zeta}{\delta_2}, 1, 1)$, it can be easily shown that the solution obtained with this loss function is equivalent to that obtained for $\delta_1 = \delta_2 = 1$, provided that the constant C is scaled with the factor $\frac{\delta_1}{\delta_2^2}$.

3.1 Polynomial approximation of the ideal cost function

Let us now introduce the following convex approximation of the ideal step function $u(\cdot)$, $\forall y \in \{-1, 1\}$:

$$q_y(\zeta) \triangleq \begin{cases} 0 & \zeta \leq -1 \\ \dfrac{(\zeta+1)^p}{p} & -1 \leq \zeta \leq \delta \\ \left[(\delta+1)^{p-1}(\zeta+1) - (1-\frac{1}{p})(\delta+1)^p\right] & \zeta \geq \delta \end{cases} \tag{21}$$

where $p \geq 1$. In this case, one simply obtains that the general problem (10) reduces to:

$$\max_{\alpha_i} \left\{ -\frac{1}{2}\sum_{i=1}^{l}\sum_{j=1}^{l} y_i y_j K(\boldsymbol{x}_i, \boldsymbol{x}_j)\alpha_i\alpha_j \right.$$

$$\left. + \sum_{i=1}^{l}\left[\alpha_i - \left(1-\frac{1}{p}\right)\cdot\left(C\cdot c(y_i)\right)^{\frac{-1}{p-1}}\cdot\left(\alpha_i\right)^{\frac{p}{p-1}}\right]\right\}\ , \tag{22}$$

subject to the conditions

$$\sum_{i=1}^{l} y_i \cdot \alpha_i = 0, \qquad 0 \leq \alpha_i \leq C \cdot c(y_i)\cdot(1+\delta)^{p-1}\ . \tag{23}$$

The optimization problems remains a quadratic one only if $p = 2$. Moreover, from the Kuhn-Tucker conditions (12) and (13), it follows that the following equivalences are valid at the optimal solution:

$$\alpha_i = 0 \qquad \Longleftrightarrow \qquad b_i \geq 1 \ , \tag{24}$$

$$0 < \alpha_i < C \cdot c(y_i)(1+\delta)^{p-1} \qquad \Longleftrightarrow \qquad b_i = 1 - \left[\frac{\alpha_i}{C \cdot c(y_i)}\right]^{\dfrac{1}{p-1}} \ , \tag{25}$$

$$\alpha_i = C \cdot c(y_i)(1+\delta)^{p-1} \qquad \Longleftrightarrow \qquad b_i \leq -\delta \ . \tag{26}$$

The equivalences (24)-(26) are analogous to (17)-(19). The optimization problem obtained in correspondence of the choice of other convex cost functions can be easily derived from the general problem (10). Further research is needed to quantify the possible advantages obtainable with a suitable choice of the convex function $q_y(\cdot)$.

4 Conclusions

The support vector algorithm for the case of binary cost function has been generalized to the case of a generic convex approximation of the ideal step function and for different cost of the two types of error. The optimization problem which determines the coefficients remains similar to the classical one [6]. The differences are slight and lie, mainly, in the bounds on the variables of the optimization problem; in particular, the upper bounds are larger for examples generated under the hypothesis for which the error costs more.

References

[1] C. Cortes and V.N. Vapnik, The Support Vector Machines for Pattern Recognition, *Machine Learning*, Vol. 20, n. 5, pp. 273-290, 1995.

[2] R. Fletcher, *Practical Methods of Optimization*, J. Wiley and Sons, II edn., 1987.

[3] K. Fukunaga, *Introduction to Statistical Pattern Recognition*, II edn., Academic Press, 1990.

[4] E. Osuna, R. Freund and F. Girosi, Support Vector Machines: Training and Applications, MIT Technical Report, AI-Memo no. 1602, 1997.

[5] B. Scholkopf, K. Sung, C. Burges, F. Girosi, P. Niyogi, T. Poggio and V. Vapnik, Comparing Support Vector Machines with Gaussian Kernels to Radial Basis Function Classifiers, *IEEE Trans. on Signal Processing*, Vol. 45, pp. 2758-2765, 1997.

[6] V. N. Vapnik, *The Nature of the Statistical Learning Theory,* Springer-Verlag, 1995.

Learning the Amplitude of Activation Functions in Layered Networks

Edmondo Trentin
ITC-Irst (Istituto per la Ricerca Scientifica e Tecnologica)
I-38050 Povo (Trento), Italy
trentin@irst.itc.it

Abstract

This paper introduces a novel algorithm to learn the amplitude of non-linear activation functions (of arbitrary analytical form) in layered networks. The algorithm is based on a steepest gradient-descent technique, and relies on the inductive proof of a theorem that involves the concept of *expansion function* of the activation associated to a given unit of the neural net. Experimental results obtained in a *speaker normalization* task with a mixture of Multilayer Perceptrons show a tangible 12.64% word error rate reduction with respect to the standard Back-Propagation training.

Keywords: speaker normalization, Multilayer Perceptron, backpropagation, activation function, gradient descent, learning.

1 Introduction

Network training algorithms for feedforward and recurrent models have heavily concentrated on the learning of connection weights. The most common example is the well-known *Back-Propagation* (BP) algorithm [6]. No attention has been paid so far to the *amplitude* of the activation functions, that is a measure of the (possibly open) interval in $\mathcal{R}^1$ which defines the range of values that the function can take (the image of the input domain). For instance, the amplitude of the hyperbolic tangent is 2, since its values range from -1 to 1. Once the analytical form of the activation functions is chosen, along with its fixed amplitude, it is commonly assumed that learning the connection weights is sufficient to search the minima of the cost function that should be minimized by the learning process. From a theoretical point of view this assumption is correct: given a certain unit of the network, if an activation $f(x)$ with amplitude λ exists for which a minimum is reached in conjunction with a vector $\mathbf{w}$ of weights for the connections starting from the unit itself, then the minimum is also reached if an activation with amplitude λ' is used in conjunction with weights $\frac{\lambda}{\lambda'}\mathbf{w}$. In practice, it is hardly possible that the BP algorithm is capable to lead to heavily different weights within a given layer, whenever weights are randomly initialized in a uniform manner over a small interval (as it is common practice) and a common learning rate is used for a certain number of training steps. At least, a different learning rate would be needed for each activation function and, in turn, these learning rates should be made trainable.

This paper introduces a novel algorithm to learn the amplitudes of non-linear activation functions in layered networks (see Section 2). No assumption is made on the analytical form of the activations, which may or may not depend upon the trainable amplitude and can possibly vary from unit to unit. The proposed algorithm relies on a stochastic gradient descent [2] technique for the minimization of the least-squares criterion over the training set. Generalization to other cost functions can be accomplished without affecting the major steps of the calculations.

Section 3 illustrates the effects of trainable amplitudes versus standard BP training with fixed amplitudes. *Speaker normalization* [8] in the acoustic feature domain is attempted in order to increase the recognition performance of a pre-trained, continuous speech recognizer. A dramatic decrease in the *Word Error Rate* (WER) with respect to normalization based on a mixture of standard nets was achieved by the networks with trainable amplitudes, on a task with large vocabulary size (10,000 words).

Finally, section 4 draws conclusive remarks.

2 Derivation of a Gradient-based Algorithm to Learn Amplitudes

Let us introduce our notation. A feed-forward network with L layers is considered. Individual layers are denoted by $\mathcal{L}_0, \mathcal{L}_1, \ldots, \mathcal{L}_L$, where $\mathcal{L}_0$ is the input layer (which in not counted, since its units act only as placeholders) and $\mathcal{L}_L$ is the output layer. We write $w_{i,j,l}$ to indicate the weight of the connection holding between the j-th unit in layer $\mathcal{L}_{l-1}$ and the i-th unit in layer $\mathcal{L}_l$. The actual input to the latter unit is denoted by $a_{i,l}$, and the corresponding output is $o_{i,l}$. With this notation we can write

$$a_{i,l} = \sum_{j \in \mathcal{L}_{l-1}} w_{i,j,l} o_{j,l-1} \tag{1}$$

where, for notational convenience, the sum over $j \in \mathcal{L}_l$ is meant to be extended to all the *indexes* of units belonging to layer $\mathcal{L}_l$. Whenever $l = 0$ or $l = L$ these are also the components of the input or output vectors, respectively. An activation function is associated to each unit of the network, the nature of which may be unit-specific, representing the relationship between input and output in the following way:

$$o_{i,l} = f_{i,l}(a_{i,l}). \tag{2}$$

An assumption is made that the activation function associated to the i-th unit in layer $\mathcal{L}_l$ can be either in the form

$$f_{i,l}(a_{i,l}) = \lambda_{i,l} \tilde{f}_{i,l}(a_{i,l}) + \sigma_{i,l} \tag{3}$$

which could be the case of sigmoids or Gaussians with learnable amplitude $\lambda_{i,l}$ and an offset (shift) $\sigma_{i,l}$ along the ordinate axis ($\sigma_{i,l}$ in turn can be a constant or, more generally, a function of $\lambda_{i,l}$); or in the form

$$f_{i,l}(a_{i,l}) = \tilde{f}_{i,l}(a_{i,l}) \tag{4}$$

that could be the case of linear activation, for instance. Anyway, in the following, the symbol $\tilde{f}_{i,l}(a_{i,l})$ will refer to the analytical expression of a function of $a_{i,l}$ which does not explicitly depend on $\lambda_{i,l}$.

Although the notation introduced herein may appear a bit cumbersome, it is indeed useful in order to state clearly the inductive hypothesis that are required to formulate general expressions for the learning rules that are introduced in the next paragraphs.

This paper investigates the case where $\lambda_{i,l} = \lambda$ and $\sigma_{i,l} = \sigma$ for all the units for which equation (3) holds. Further developments (*layer-by-layer* or *unit-specific* amplitudes) are subject of on-going research. Let us consider the training set $\mathcal{T} = (\mathbf{x}, \mathbf{y})$, which specifies the relationship between training input vectors and the corresponding desired (target) output vectors. An on-line learning algorithm is developed in order to minimize the squared error between target output and actual network output, on a sample by sample basis. The criterion function to be minimized at any time a new pattern from the training set is fed into the network is thus in the form:

$$C = \frac{1}{2} \sum_{i \in \mathcal{L}_L} (y_i - o_{i,L})^2 \tag{5}$$

and the stochastic gradient descent prescribes to update the current value of λ, obtaining a new value λ' for the amplitude, according to

$$\lambda' = \lambda + \Delta\lambda \tag{6}$$

where $\Delta\lambda$ is the amount of change and is given by:

$$\Delta\lambda = -\eta \frac{\partial C}{\partial \lambda}. \tag{7}$$

In equation (7) the constant η is the *learning rate*. It is not necessarily the same used to learn the connection weights, if a gradient-descent technique as the BP algorithm is used. By calculating partial derivatives, equation (7) can be rewritten as

$$\Delta\lambda = \eta \sum_{i \in \mathcal{L}_L} (y_i - o_{i,L}) \frac{\partial o_{i,L}}{\partial \lambda}. \tag{8}$$

Let us define the *auxiliary functions*

$$g_{k,l}(a_{k,l}) = \begin{cases} \tilde{f}_{k,l}(a_{k,l}) + \frac{\partial \sigma}{\partial \lambda} & \text{if } f_{k,l}(a_{k,l}) = \lambda \tilde{f}_{k,l}(a_{k,l}) + \sigma \\ 0 & \text{if } f_{k,l}(a_{k,l}) = \tilde{f}_{k,l}(a_{k,l}) \end{cases} \tag{9}$$

and

$$h_{k,l}(a_{k,l}) = \begin{cases} \lambda \tilde{f}'_{k,l}(a_{k,l}) & \text{if } f_{k,l}(a_{k,l}) = \lambda \tilde{f}_{k,l}(a_{k,l}) + \sigma \\ \tilde{f}'_{k,l}(a_{k,l}) & \text{if } f_{k,l}(a_{k,l}) = \tilde{f}_{k,l}(a_{k,l}) \end{cases} \tag{10}$$

We also need to introduce the notion of function *expansion* of unit k of layer $\mathcal{L}_l$, which is defined as:

$$x_{k,l}(a_{k,l}) = \begin{cases} g_{k,l}(a_{k,l}) & \text{if } l = 1 \\ g_{k,l}(a_{k,l}) + h_{k,l}(a_{k,l}) \sum_{n \in \mathcal{L}_{l-1}} w_{k,n,l} x_{n,l-1}(a_{n,l-1}) & \text{otherwise.} \end{cases} \tag{11}$$

The following theorem can now be proved, by induction, to hold true (refer to [7] for details).

Theorem 1: $\frac{\partial o_{k,l}}{\partial \lambda} = x_{k,l}(a_{k,l})$ for each $l = 1, \ldots, L$ and for each $k \in \mathcal{L}_l$.

As a corollary of Theorem 1, equation (8) can now be restated as:

$$\Delta\lambda = \eta \sum_{i \in \mathcal{L}_L} (y_i - o_{i,L}) x_{i,L}(a_{i,L}) \tag{12}$$

that is exactly the formulation we were looking for. Its implementation in the recursive form stated above is straightforward (a recursive function $x()$ can be readily implemented in any computer language), once the auxiliary functions $g()$ and $h()$ have been implemented in terms of the analytical form of the activations for the particular network at hand. Equation (12) is general and does not explicitly depend on the peculiar architecture of the network.

3 An Examples From Speech Processing: Speaker Normalization

Speaker normalization [4, 8] can be formulated as a regression problem between the multidimensional acoustic spaces corresponding to different speakers. The acoustic space of a speaker is represented by means of a set of feature vectors, obtained by acoustic analysis of the speech signals acquired from the speaker himself. The aim is to build a vectorial function able to map the acoustic patterns belonging to a novel speaker, called *new speaker*, onto the corresponding acoustic patterns of another speaker, called *reference speaker*, whose speech data were used to train a speaker-dependent speech recognition system [3]. Application of the estimated transformation on the acoustic data of the new speaker, before sending them to the recognizer, should allow the speaker-dependent (SD) recognition system to perform acceptably, from an applicative point of view.

In the present case the transformation is realized with neural networks. The networks were trained on the time-aligned pairs of feature vectors of the speakers involved.

The feedforward connectionist model that was adopted is a Multilayer Perceptron (MLP), trained with the BP algorithm on a limited amount of acoustic material from each speaker (10 utterances). Stochastic gradient [2] was computed using the usual squared error between target output and actual network output as a criterion function, including a momentum term and an adaptive learning rate. The architecture was a 2-layer MLP, with one hidden layer and one output layer. The hidden layer was built up of 48 units.

Sigmoidal activation functions were used in the hidden layer, with linear output units. More precisely, the chosen sigmoid for hidden unit i of layer l was in the form

$$f_{i,l}(a_{i,l}) = \frac{\lambda}{1 + e^{-\frac{a_{i,l} - \beta_{i,l}}{\theta}}} + \sigma \qquad (13)$$

where $\beta_{i,l}$ is a trainable unit-dependent *bias*, which allows to shift the sigmoid along the abscissa, θ is the smoothness, providing for sharp versus smooth sigmoidal functions, and the meaning of the remaining symbols is the same as that introduced in Section 2.

A mixture of 8 such MLPs (or *experts*) was used to realize a multiple-model regression for speaker normalization. Credit assignment to individual experts was accomplished by partitioning the feature space into 8 phone-class dependent subregions, and assuming a Normal distribution of acoustic patterns within each class. Maximum-likelihood estimates of the parameters of the Gaussian densities were carried out on the training data. Bayesian posterior probability of a certain class given an acoustic observation was used to establish the weight (impact) of the observation on the overall cost function, the minimization of which was pursued by the network associated to the class itself.

Speaker dependent versions of speech recognizer based on Continuous Density Hidden Markov models (CD-HMM) [5] were trained using the Baum-Welch algorithm [5] over the APASCI speech corpus [1]. A set of 34 context independent acoustic-phonetic speech units were adopted. Emission probability densities were modeled with mixtures of 16 Gaussian components, resulting in about 1,500 Gaussians over the whole system.

In order to train the normalization modules and to test the approach, novel speech material from 4 test speakers was collected. The recognition task was the dictation of fragments (chosen at random) of newspaper articles taken from the financial Italian journal *Il Sole 24 Ore*. It is a continuous speech task with large dictionary size (10,000 words). For each speaker, a test set of 30 utterances and an adaptation (training) set of 10 utterances were acquired, on different days, in an office environment.

8 Mel Scaled Cepstral Coefficients (MSCCs) were used as acoustic features, along with the log-energy of the speech signal and their first and second order time-derivatives. The normalization modules were applied just on MSCCs (performing a mapping between 8-dimensional spaces), while derivatives were computed from the transformed parameters and directly fed into the CD-HMMs.

Table 1 reports the results obtained on a database composed of 13 pairs (*new speaker*, *reference speaker*), from the APASCI corpus. Results are expressed in term of WER, which is defined as $WER = 100(Ins + Del + Sub)/\mathcal{N}$, where

Ins is the number of extraneous words inserted by the recognition system into the recognized text, *Del* denotes the number of omitted words, *Sub* represents the number of words substituted with wrong ones, and $\mathcal{N}$ is the overall number of words in the uttered text.

The first row of the Table shows the average WER obtained with the SD recognition systems without any speaker normalization module acting as an acoustic front-end. The WER is definitely high (35.0%) and unfeasible for applicative purposes. This is not surprising: SD systems are not expected to perform well on speech signals acquired from novel speakers. Performance were improved with the adoption of a standard linear regression scheme (see the second row of the Table), intended also as a comparative benchmark for the connectionist approaches, obtaining a 20.7% average WER.

The last two rows of the Table illustrate the effect of speaker normalization accomplished via mixtures of neural networks. In both cases, a significant WER reduction is gained with respect to the linear regression approach. As a first instance, a mixture of 8 MLPs with fixed amplitude of the sigmoids was used, resulting in an average WER of 17.4%. Next, the proposed algorithms for learning the value of the amplitude of the activations were applied during the training of the 8 networks. Row 4 of Table 1 reports a 15.2% average WER for this case, that is a tangible 12.64% WER reduction with respect to the MLPs without trainable amplitude. In the speech community, such an improvement is considered indeed satisfactory. Note that different amplitudes of the sigmoids were learned by the 8 individual experts of the mixture.

Regression model	*Average WER*
Baseline (SD recognizers without normalization)	35.0
Linear multiple regression	20.7
Mixture of 8 MLPs with fixed amplitudes	17.4
Mixture of 8 MLPs with trainable λ	15.2

Table 1: *Average WERs over the 13 pairs (new speaker, refernce speaker), using the corresponding SD recognition systems, obtained with the proposed normalization modules.*

4 Conclusion

This paper introduced a novel algorithm to learn the amplitude of non-linear activation functions (of arbitrary analytical form) in layered networks. Derivation of the algorithm in terms of a steepest gradient-descent technique has been outlined, relying on the inductive proof of a theorem that involves the novel concepts of *auxiliary functions* and *expansion function* of the activation associated to a given unit of the neural net. A compact and general formulation for the learning rule has been obtained, readily implementable starting from the recursive expression that has been given. Experimental results obtained in a *speaker normalization* task with a mixture of MLPs shown a dramatic improvement of performance with respect to the standard BP training. Current research efforts are concentrated on generalization and extension of the present algorithm to *layer-by-layer* and *unit-specific* amplitudes.

References

[1] B. Angelini, F. Brugnara, D. Falavigna, D. Giuliani, R. Gretter, and M. Omologo. Speaker independent continuous speech recognition using an acoustic-phonetic italian corpus. In *Proc. of ICSLP*, pages 1391–1394, Yokohama, 1994.

[2] Y. Bengio. *Neural Networks for Speech and Sequence Recognition*. Internatiuonal Thomson Computer Press, London, 1996.

[3] S. Furui and M. M. Sondhi, editors. *Advances in Speech Signal Processing*. Marcel Dekker and Inc., 1991.

[4] X. D. Huang. Speaker normalization for speech recognition. In *Proc. of ICASSP*, pages I–465–468, San Franscisco, March 1992.

[5] L.R. Rabiner. A tutorial on hidden Markov models and selected applications in speech recognition,. *Proc. of IEEE*, 77(2):267–295, October 1989.

[6] D.E. Rumelhart, G.E. Hinton, and R.J. Williams. Learning internal representations by error propagation. In D.E. Rumelhart and J.L. McClelland, editors, *Parallel Distributed Processing*, volume 1, chapter 8, pages 318–362. MIT Press, Cambridge, 1986.

[7] E. Trentin. Networks with trainable width of activation functions. IRST Technical Report 9703-08, IRST, Povo (TN), ITALY, 1997. Submitted for publication in "Neural Networks".

[8] E. Trentin and D. Giuliani. Speaker adaptation with a mixture of recurrent networks. In *Proc. of ESANN97, European Symposium on Artificial Neural Networks*, Bruges, April 1997.

The BRAIN Learning Algorithm

Salvatore Rampone
Dpt. of Scienze Fisiche "E.R.Caianiello" and INFM
Università di Salerno
Via S.Allende I-84081 Baronissi (Sa) Italy
E-mail rampo@computer.org

Abstract

This paper describes a new learning algorithm (BRAIN), inferring DNF Boolean formulae from examples. The formula terms are computed in an iterative way, by identifying from the training set a relevance coefficient for each attribute. Results on Splice-Junction Gene Sequences and Breast Cancer machine learning data sets are reported.

1 Introduction

The primary task studied in machine learning research has been the acquisition of classification rules from examples. In this task, a *learning algorithm* receives a set of training examples, each labeled as belonging to a particular class. The algorithm's goal is to produce a classification rule (*hypothesis*) for correctly assigning new examples to these classes [5]. This process is called *inductive inference* [9].

In this paper we describe a new learning algorithm, in the following called *BRAIN*, for binary (two class) classification rules. *Restricted hypothesis space* bias and *preference* bias are incorporated in the algorithm, for choosing an appropriate hypothesis.

The former restricts the search of the unknown classification rule in a specific hypothesis space, where the hypotheses are defined in terms of their representations. In the BRAIN algorithm the hypothesis space consists of logical formulae in disjunctive normal form (DNF).

The preference biases attempts to minimize some measure of the hypothesis syntactic complexity. This follows the Occam's razor: the simpler of two competing hypotheses should always be preferred. A theoretical explanation of why Occam's Razor is successful has been presented by Blumer et al. [2]. As preference bias, the BRAIN algorithm tries to minimize the syntactic complexity of the hypothesis by means of a preference value called *relevance*.

The performance of the learning algorithm is measured in terms of the hypothesis classification accuracy on previously unseen examples and of the computational effort required to produce the hypothesis.

While theoretical analysis can provide bounds on the number of examples needed to guarantee a given level of confidence in the accuracy of learning, these bounds are based on worst-case arguments, and hence they tend not to be useful in practice. On the other hand, experimental methods based on the training set / test set methodology can provide good assessments of the accuracy of learned classification rules [16]. This technique has been also widely employed in connectionist systems [1], and this is the technique we apply here.

The computational effort the BRAIN algorithm requires to produce the hypothesis is bounded by $O(n^2t^3)$ where n is the number of variables, and t the number of examples.

Although application of inductive inference to the space of binary sequences of fixed length has undergone a tremendous explosion of interest, reference [5] is an excellent collection of papers on the matter, our algorithm appears to be novel. The general methodology used in developing the learning algorithm is related to the STAR technique of Michalski [9], to the candidate-elimination introduced by Mitchell [11], and to the work of Haussler [6].

This paper can be summarized as follows. In section 2 we explain the learning scenario, in section 3 we introduce the BRAIN algorithm, and in section 4 we show the results obtained on Splice-Junction Gene Sequences and Breast Cancer data sets.

2 Formulae and Properties

Boolean formulae and their properties are known to most of the readers. Anyway, for the sake of clearness, we report some fundamental definitions.

A *Boolean classification rule* (briefly *function*) f_c on n *variables* $x_1, x_2, \ldots, x_n$, each one taking only the values 1 or 0, is a mapping from the n-dimensional *instance space* $\{0,1\}^n$ to $\{0,1\}$

$$f_c : \{0,1\}^n \rightarrow \{0,1\} \tag{1}$$

An *instance* can be viewed as a value assignment to the variables in f_c. An instance is *positive*, and it is indicated as the vector

$$\mathbf{w}_i^+ = w_{i,1}^+, w_{i,2}^+, \ldots, w_{i,n}^+; \qquad w_{i,k}^+ \in \{0,1\} \tag{2}$$

if $f_c(\mathbf{w}_i^+) = 1$, i.e. if $\mathbf{w}_i^+$ belongs to the class c. So p positive instances are indicated as $\mathbf{w}_1^+, \mathbf{w}_2^+, \ldots, \mathbf{w}_p^+$.

An instance is *negative*, and it is indicated as the vector

$$\mathbf{w}_j^- = w_{j,1}^-, w_{j,2}^-, \ldots, w_{j,n}^-; \qquad w_{j,k}^- \in \{0,1\} \tag{3}$$

if $f_c(\mathbf{w}_j^-) = 0$, i.e. if $\mathbf{w}_j^-$ does not belong to the class c. So q negative instances are indicated as $\mathbf{w}_1^-, \mathbf{w}_2^-, \ldots, \mathbf{w}_q^-$.

A Boolean function is *consistent* with a set of instances if and only if it matches every positive instance and no negative instance in the set.

A variable is in *true form*, and it is indicated as the *literal* x_k, if it assumes value 1 when it is assigned to 1, and 0 otherwise. On the contrary, a variable is in *negated form*, and it is indicated as the literal $\overline{x}_k$, if it assumes value 0 when x_k is assigned to 1, and 1 otherwise.

A *term* m is the conjunction of any subset of the n variables $x_1, x_2, \ldots, x_n$, each one in true or negated form. A term is *made true* by an instance if each of its variables is 1 by assigning the corresponding instance values. In the same way a term is *not made true* by an instance if at least one of its variables is 0 by assigning the corresponding instance values.

A Boolean function is expressed as a *DNF* (disjunctive normal form) formula if it is expressed as a disjunction (or) of terms. Given an instance, it assumes value 1 if at least a term of the function is made true by the assignment, and it assumes value 0 otherwise. An L-term DNF function can be represented as

$$f_c = m_1 + m_2 + \ldots + m_L \tag{4}$$

3 BRAIN algorithm

3.1 One-to-One

Let us start finding a DNF formula that is consistent both with a positive instance and a negative one. Given a couple of instances $(\mathbf{w}_i^+, \mathbf{w}_j^-)$, a term m made true by $\mathbf{w}_i^+$ and not made true by $\mathbf{w}_j^-$ must[1] include at least a variable not assigned in the same way in $\mathbf{w}_i^+$ and $\mathbf{w}_j^-$. So given a couple $(\mathbf{w}_i^+, \mathbf{w}_j^-)$, let us consider the set

$$S_{i,j} = \{x_k | w_{i,k}^+ = 1, w_{j,k}^- = 0\} \cup \{\overline{x}_k | w_{i,k}^+ = 0, w_{j,k}^- = 1\} \tag{5}$$

It is immediate to see that, given the set of instances $\{\mathbf{w}_i^+, \mathbf{w}_j^-\}$ then the formula

$$f_c = v_k; \qquad v_k \in S_{i,j} \tag{6}$$

is consistent with the set and of minimum size.

3.2 One-to-Many

Now let us consider to have one positive instance and q negative ones,

$$\mathbf{w}_i^+, \mathbf{w}_1^-, \mathbf{w}_2^-, \ldots \mathbf{w}_q^- \tag{7}$$

In this case different variables can disagree on Different pairs $(\mathbf{w}_i^+, \mathbf{w}_j^-)$. Then a term, to be consistent with the given instances, needs to include at least a variable for each set (5), $j = 1, \ldots, q$. We collect these sets in the set

$$S_i = \{S_1, S_2, \ldots, S_q\} \tag{8}$$

[1]We assume the instances to be self consistent, i.e. an instance can not be positive and negative at the same time

Furthermore, as preference bias, we require a term of the smallest total size. This can be viewed as a set covering problem [4], and while the problem is NP-hard [4], an approximate solution can be found by a generalized greedy method [7], as applied in machine learning by Haussler [6]. In our case, given the $S_{i,j}$ sets, it consists in choosing the variable that is in the most of sets, and to erase such sets, until there are no more sets.

Obviously this greedy choice can be improved by taking into account the size of the $S_{i,j}$ sets (This was also applied by Johnson in the MAX-SAT approximation algorithm improvement [7]). For example if there is only a variable in a $S_{i,j}$ set, that variable will be surely included in the term. We say that variable has a great *relevance*.

Let us formalize this concept. For each $S_{i,j}$ set, we call relevance of v_k in the $S_{i,j}$ set the value

$$R_{i,j}(v_k) = \frac{I_{S_{i,j}}(v_k)}{\#(S_{i,j})} \tag{9}$$

where $I_{S_{i,j}}(v_k)$ is the characteristic function of the $S_{i,j}$ set, that is 1 if $v_k \in S_{i,j}$ and is 0 otherwise. Then we call relevance of a variable v_k in the S_i set the mean value

$$R_i(v_k) = \frac{1}{q}\sum_{j=1}^{q} R_{i,j}(v_k) \tag{10}$$

So, given the S_i set, we repeatedly compute the relevances, choosing the variable with the greatest one, and erasing the $S_{i,j}$ sets including that variable, until there are no more sets. It strictly follows that if s is the minimum variable number of any term including at least a variable for each set, the generalized greedy method is guaranteed to find a consistent term m of size at most

$$s(ln(q+1)) \tag{11}$$

3.3 Many-to-Many: BRAIN algorithm

Now let us consider the case where we have p positive instances and q negative ones,

$$\mathbf{w}_1^+, \mathbf{w}_2^+, \ldots \mathbf{w}_p^+, \mathbf{w}_1^-, \mathbf{w}_2^-, \ldots \mathbf{w}_q^- \tag{12}$$

Not necessarily the formula contains just a term. In the general case a consistent DNF formula will be a disjunction of a set of terms. Each positive instance satisfies at least one term and none of the terms is satisfied by any negative instance. As in the one-to-many case, for each couple $(\mathbf{w}_i^+, \mathbf{w}_j^-)$ a term made true by $\mathbf{w}_i^+$ and not made true by any $\mathbf{w}_j^-$ must include at least a variable not assigned in the same way in $\mathbf{w}_i^+$ and $\mathbf{w}_j^-$ for each couple $\mathbf{w}_i^+, \mathbf{w}_j^-$. Then for each $\mathbf{w}_i^+$ there is a term that contains at least a variable in each $S_{i,j}$ set.

While the problem of finding the minimal consistent DNF formula is NP-hard [8], an approximate solution can also be found by a generalized greedy method as applied by Haussler [6]. Given the instances, it consists in choosing the term that is made true by the greatest number of positive instances and it is not made true by all the negative ones, and to erase such positive instances, until there are no more positive instances. Unfortunately such method is impractical, because the term selection step has high cost.

It can be argued it is possible to use an approximate solution for the term selection, as we will do in this paper, but a general performance evaluation is actually an open problem. What we do is to extend the relevance evaluation to take in account the relative frequency of a variable in the $S_{i,j}$ sets, for $i = 1, \dots p$ and building the function terms in the iterative way of one-to-many case. For each couple $(\mathbf{w}_i^+, \mathbf{w}_j^-)$, let us consider the sets (5) that we collect as

$$S_i = \{S_{i,1}, S_{i,2}, \dots, S_{i,q}\} \tag{13}$$

and the relevances

$$R_i(v_k) = \frac{1}{q}\sum_{j=1}^{q} R_{i,j}(v_k) \tag{14}$$

To take in account the relative frequencies we extend the relevance to all the $S_{i,j}$'s, so having

$$R(v_k) = \frac{1}{p}\sum_{i=1}^{p} R_i(v_k) = \frac{1}{pq}\sum_{i=1}^{p}\sum_{j=1}^{q} R_{i,j}(v_k) \tag{15}$$

By using this extended relevance, the greedy approximation algorithm is applied at the same time both to cover greatest number of positive instances, and to select the less possible number of variables.

The remainder of the procedure is the classical greedy approximation. This procedure can be sketched as follows:

BRAIN Algorithm

Step 1: *Input:* $n=$ variable number,
$G = \{\mathbf{w}_1^+, \mathbf{w}_2^+, \dots \mathbf{w}_p^+, \mathbf{w}_1^-, \mathbf{w}_2^-, \dots \mathbf{w}_q^-\}$ the set of training instances.
Initialization: Set $f_c = \emptyset$.

Step 2: While there are positive instances in G

- 2.1 *$S_{i,j}$-sets:* Build from G the sets $S_{i,j}$ and collect them in S_i's.
- 2.2 *Start a new term:* Set $m = \emptyset$.
- 2.3 *Build the term :* While there are $S_{i,j}$ sets
  - 2.3.1 *Relevances:* Compute the relevances $R(v_k)$
  - 2.3.2 *Add variable:* Select the variable v_k such that $R(v_k)$ is maximum. $m \leftarrow m \cup \{v_k\}$
  - 2.3.3 *Update sets:* Erase the S_i sets not including v_k. Erase the $S_{i,j}$ sets including v_k
- 2.4 *Add the term:* $f_c \leftarrow f_c \cup m$.
- 2.5 *Update instances:* Erase from G the positive instances satisfying m.

Step 3: *Output:* f_c

The resulting f_c is the inferred formula.

It is easy to see the algorithm runs in $O(n^2t^3)$, where n is the variable number on which f_c is defined and t the number of examples seen.

4 Experiments

The BRAIN algorithm has been experimented on several data sets. We report the ones stressing the strenght and the weakness of the algorithm.

4.1 Splice-Junction Gene Sequences data set

A set of experiments has been performed on the StatLog Splice-Junction DNA data set for splice-junctions recognition [12] [14] [15]. Splice junctions are points on a DNA sequence at which 'superfluous' DNA is removed during the process of protein creation in higher organisms. The problem posed is to recognize, given a sequence of DNA, the boundaries between exons (the parts of the DNA sequence retained after splicing) and introns (the parts of the DNA sequence that are spliced out), i.e. given a position in the middle of a window of DNA sequence elements ("nucleotides"), decide if this is an "intron → exon" boundary (IE), "exon → intron" boundary (EI), or neither (N).

There are 3186 examples, (Train: 2000 - Test: 1186, see also Table 1) each labelled to one of 1,2,3. Table 1 shows the example distribution over the classes. Each instance is coded as 240 bit binary vector. We applied the BRAIN algorithm on the training set, building a formula f_c for each class. The results achieved by using a "ten-fold cross-validation" methodology on 1000 examples randomly selected from the complete set of 3190 are reported in Table 2. We should point out that there is very little variance between the results of each experiment, i.e. the algorithm has a very stable behavior. In the same Table there are the error rates produced by various machine learning algorithms (all experiments, except BRAIN, made at the University of Wisconsin).

Table 1: Example distribution.

Class	Train	Test
1	464 (23.20%)	303 (25.55%)
2	485 (24.25%)	280 (23.61%)
3	1051 (52.55%)	603 (50.84%)

4.2 Breast-Cancer Data

The second experiment referring material is the Breast Cancer data set. These data were obtained from the University Medical Center, Institute of Oncology, Ljubljana, Yugoslavia. Thanks go to M.Zwitter and M.Soklic for providing data.

This dataset for evaluating the prognosis of breast cancer has been used by many authors (see for example [10], [13], [3], [16]). It includes 201 instances of one class and 85 instances of another class. The instances are described by 9 attributes, some of which are linear and some are nominal. Each instance has been coded as a 38 bit binary vector. We found several inconsistencies

Table 2: Error rates produced by BRAIN and by various ML algorithms by using a "ten-fold cross-validation" methodology

System	EI(1)	IE(2)	Neither(3)
BRAIN	5.71%	4.08%	4.21%
KBANN	7.56%	8.47%	4.62%
BACKPROP	5.74%	10.75%	5.29%
PEBLS	8.18%	7.55%	6.86%
PERCEPTRON	16.32%	17.41%	3.99%
ID3	10.58%	13.99%	8.84%
COBWEB	15.04%	9.46%	11.80%
Near. Neighbor	11.65%	9.09%	31.11%

between the instances, i.e. some instances labeled to one class are equal to other instances of the other class. Since consistency is a prerequisite for our algorithm, instances 3, 36, 47, 48, 91, and 109 are erased from the dataset. So we have 280 resulting instances, 195 of one class and 85 instances of another class. We created 4 randomly sampled data sets with 70% train and a 30% test partitions, as in Table 3. In this way the results are consistent with the literature [16]. The results are depicted in Table 4. As we can see in this case

Table 3: Example partition.

Class	Train	Test
No-recurrence-events	137(70%)	58(30%)
Recurrence-events	60(70%)	25(30%)
size	**197**	**83**

BRAIN performance is no better than the mean of the known methodologies. In fact this data set stress the algorithm is not noise tolerant.

Table 4: Error percentage produced by BRAIN and by various ML algorithms by using a "four-fold cross validation" methodology.

System	Err-Train	Err-test
BRAIN	0 %	32%
LINEAR	25%	29%
QUADRATIC	25%	34%
Near. Neighbor	0%	35%
Bayes independence	24%	28%
Bayes 2nd order	9%	34%
BACKPROP	24%	29%
PVM rule	23%	23%
ASSISTANT tree	-	28%
CART tree	23%	23%

5 Conclusions

The problem addressed in this paper is to inferring binary classification rules from examples in the binary n-cube. This is due by means of the BRAIN algorithm, described in the paper, a polynomial learning algorithm for DNF formulae. The formula terms are computed in an iterative way, by identifying from the training set a relevance coefficient for each attribute.

The major advantages of the proposed approach are the polynomial computational complexity, and the robust and stable learning. The algorithm weakness is the learning in noisy environment, and this problem will be matter of a future work. Theoretical bounds on the example needed to guarantee a given level of confidence in the accuracy of learning are also necessary.

References

[1] C.M.Bishop, *Neural networks for pattern recognition*, Clarendon, Oxford, 1996.

[2] A.Blumer, A.Ehrenfeucht, D.Haussler, "Occam's razor", *Inf.Proc.Lett.*, **24**, 1987, pp. 377-380.

[3] G.Cestnik, I.Konenenko, I.Bratko, "Assistant-86", *In I.Bratko & N.Lavrac ed.s Progress in Machine Learning*, 31-45, Sigma Press, 1987.

[4] T.H.Cormen, C.H.Leiserson, R.L.Rivest, *Introduction to algorithms*, The MIT Press, Cambridge, 1990.

[5] T.G.Dietterich, J.W.Shaving, ed.s, *Readings in machine learning*, Morgan Kaufmann, San Mateo, 1990.

[6] D.Haussler, "Quantifying Inductive Bias: AI learning algorithms and Valiant's learning framework", *Artificial Intelligence*, **36**, 1988, pp. 177-222.

[7] D.S.Johnson, "Approximation algorithms for combinatorial problems", *J. Comput. Syst. Sci.*, **9** 1974 pp.256-278.

[8] M.Kearns, M.Li, L.Pitt, L.Valiant, "On the Learnability of Boolean formulae", *Proc. 9th Annual ACM Sym. on Theory of computing*, 1987, pp. 285-295.

[9] R.S.Michalski, "A theory and methodology of inductive learning", *Artificial Intelligence* **20** 1983 pp.111-116.

[10] R.S.Michalsky,I.Mozetic,J.Hong,N.Lavrac, "The multi-purpose incremental learning system AQ15" *Proc.5thNCAI*, 1041-1045, PA, Morgan Kaufmann, 1986.

[11] T.M.Mitchell, "Generalization as search", *Art.Int.*, **18**, 1982, pp. 203-226.

[12] M.O.Noordewier, G.G.Towell, J.W.Shavlik, "Training knowledge-based neural networks to recognize genes in DNA sequences". *Advances in Neural Information Processing Systems* **3**, Morgan Kaufmann, 1991.

[13] M.Tan, L.Eshelman, "Using weighted networks to represent classification knowledge in noisy domains", *Proc. 5th Int.Conf.Mac.Le.*, 121-134, Ann Arbor, 1988.

[14] G.G.Towell, J.W.Shavlik, M.W.Craven, "Constructive induction in knowledge-based neural networks", *Proc. 8th Int. Mac. Le. Work.*, Morgan Kaufmann, 1991.

[15] G.G.Towell, J.W.Shavlik, "Interpretation of artificial neural networks", *Advances in Neural Information Processing Systems* **4**, Morgan Kaufmann, 1992.

[16] S.Weiss, I.Kapouless, "An empirical comparison of pattern recognition methods", *Proc. 11th IJCAI*, Morgan Kaufmann, Detroit, 1989.

Training Data Selection for Optimal Generalization with Noise Variance Reduction in Neural Networks

Sethu Vijayakumar
Lab for Information Synthesis, RIKEN Brain Science Institute
Wako, Saitama 351-0106, Japan. Email: sethu@brain.riken.go.jp

Masashi Sugiyama
Department of Computer Science, Tokyo Institute of Technology
Meguro-ku, Tokyo 152, Japan. Email: sugi@cs.titech.ac.jp

Hidemitsu Ogawa
Department of Computer Science, Tokyo Institute of Technology
Meguro-ku, Tokyo 152, Japan. Email: ogawa@cs.titech.ac.jp

Abstract

In this paper, we discuss the problem of active training data selection in the presence of noise. We formalize the learning problem in neural networks as an inverse problem using a functional analytic framework and use the Averaged Projection criterion as our optimization criterion for learning. Based on the above framework, we look at training data selection from two objectives, namely, improving the generalization ability and secondly, reducing the noise variance in order to achieve better learning results. The final result uses the *apriori* correlation information on noise characteristics and the original function ensemble to devise an efficient sampling scheme, which can be used in conjunction with the incremental learning schemes devised in our earlier work to achieve optimal generalization.

1 Introduction

It has been well known that supervised learning in non-asymptotic cases is highly data dependent. The level of generalization, i.e., the ability to correctly respond to novel inputs, achievable using a fixed number of training data is heavily dependent on the quality of the data used[9]. It is also interesting to note that many natural learning systems are not simply passive but make use of at least some form of active learning to examine the problem domain. By *active* learning, we mean any form of learning in which the learning program has some control over the inputs over which it trains.

The problem of "active learning" has been extensively studied in economic theory and statistics [1]. Optimal data selection within the Bayesian framework for interpolation have been studied by Luttrell[4] and MacKay[5]. It has been shown that a smaller training set gathered by an active learner produces generalization performance equal to or better than a much larger data set containing redundant examples [7].Plutowski and White [6] assume that a large amount of data has been collected and work on principles of selecting a subset of that data for efficient training; the entire data sets (inputs and outputs) is consulted at each iteration to decide which example to add, an option that is not permitted in this work.

Here, we look at the learning problem from a functional analytic perspective and define an optimization measure which decides on the usefulness of the training data. Works based on the Shannon entropy and Fisher's information criterion [2] already exist. We use the Averaged Projection criterion described in Section 3.1, a criterion which enforces a trade-off between expanding the approximation space and reducing the noise variance. The training data selection scheme developed here works in two phases. At first, a batch selection of m data to optimize generalization is carried as shown in Section 4.1. Since, selection of the data incorporates additive noise, as a second stage, more data is added incrementally with a view to reduce the effect of noise by exploiting the apriori information on the noise correlation matrix as shown in Section 4.2. The effectiveness of this sampling scheme is demonstrated through a simulation.

2 Functional analytic framework for learning

Let us consider a three-layer feedforward neural network whose number of input, hidden, and output units are L, N, and 1, respectively. It can be easily shown that the input-output relationship of such a network is equivalent to a real valued function of L variables. Based on this interpretation, it follows that the learning in Neural Networks(NNs) is analogous to obtaining an optimal approximation f_m to a desired function f from the set of m training data made up of the inputs $x_i \in R^L$ and the corresponding outputs $y_i \in R$:

$$\{(x_i, y_i) | y_i = f(x_i) + n_i : i = 1, \cdots, m\},$$

where n_i is the additive noise. Let a Hilbert space H, with a reproducing kernel $K(x, x')$, represent the space of all functions to be approximated by the NN. Let D be the domain of the functions to be approximated, which is a subset of the L-dimensional Euclidean space R^L. The reproducing kernel $K(x, x')$ is a bivariate function defined on $D \times D$ which satisfies the following two conditions:

1. For any fixed x' in D , $K(x, x')$ is a function in H.
2. For any function f in H and for any x' in D, it holds that

$$(f(x), K(x, x')) = f(x'), \tag{1}$$

   where the left hand side of eq.(1), represented by the notation $(\cdot, \cdot)$, denotes the inner product in H.

In the theory of Hilbert space, arguments are developed by regarding a function as a point in that space. Thus, things such as 'value of a function at a point' cannot be discussed under the general framework of Hilbert space. However, if the Hilbert space has a reproducing kernel, then it is possible to deal with the value of a function at a point. Indeed, if we define functions $\psi_i(x)$ as

$$\psi_i(x) = K(x, x_i) : 1 \leq i \leq m, \tag{2}$$

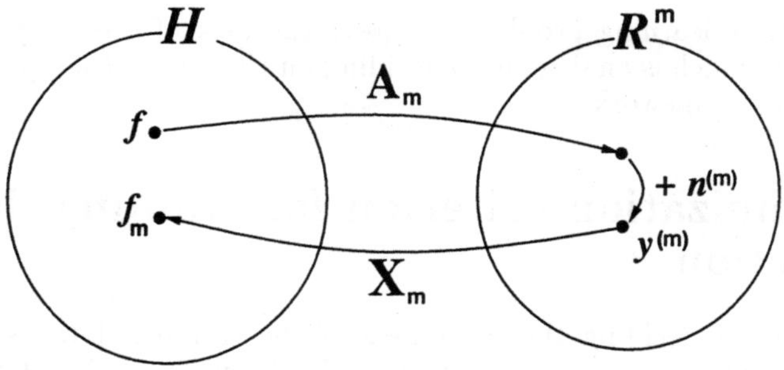

Figure 1: NN learning as an inverse problem

then, the value of f at a sample point x_i is expressed in Hilbert space language as the inner product of f and ψ_i as

$$f(x_i) = (f, \psi_i). \tag{3}$$

Let $\{y_i\}_{i=1}^m$ and $\{n_i\}_{i=1}^m$ form the elements of the m-dimensional vectors $y^{(m)}$ and $n^{(m)}$, respectively. Once the training set $\{x_i\}_{i=1}^m$ is fixed, we can introduce an operator A_m such that

$$y^{(m)} = A_m f + n^{(m)}. \tag{4}$$

The operator A_m, called the *sampling operator*, becomes a linear operator even when we are concerned with nonlinear neural networks. It is expressed by using the Schatten product[1] as

$$A_m = \sum_{i=1}^{m} e_i \otimes \overline{\psi_i}, \tag{5}$$

where $\{e_i\}_{i=1}^m$ is the so-called natural basis [2] in R^m. Now, the learning problem can be reformulated as an *inverse problem* (See Fig.1) of obtaining an operator X_m which provides an optimal approximation f_m of the true function f from the noisy sample values $y^{(m)}$:

$$f_m = X_m y^{(m)}. \tag{6}$$

The generalization ability of the NN, which corresponds to the closeness of the original function f and the approximated function f_m, can be measured using various criterion. In this work, we will restrict ourselves to the Averaged Projection criterion, which will be discussed in the next section. Here, X_m is referred to as the learning operator. Results on obtaining the optimal X_m for a given sampling scheme, both as a batch as well as incremental procedure, will be reviewed in Section 3.2.

[1]The Schatten product denoted by $(. \otimes .)$ is defined by $(e_i \otimes \overline{\psi_i})f = (f, \psi_i)e_i$.

[2]The vector e_i is the m-dimensional vector consisting of zero elements except the element i equal to 1.

In the active learning problem, we have the task of selecting the optimal training data, which is analogous to deciding on the optimal sampling operator A_m under this framework.

3 Optimization criterion for training data selection

As a measure of deciding the usefulness of the training data as well as for obtaining an optimal approximation using the selected training data, we make use of the Averaged Projection criterion, described in the next subsection.

3.1 Averaged Projection criterion

The Averaged Projection optimization corresponds to finding an optimal sampling operator A_m and learning operator X_m such that it minimizes the functional

$$\min_{A_m} J_{AP}^{(0)}[A_m], \tag{7}$$

where

$$J_{AP}^{(0)}[A_m] = \min_{X_m} J_{AP}^{(2)}[X_m] \text{ under the constraint } \min_{X_m} J_{AP}^{(1)}[X_m], \tag{8}$$

$$J_{AP}^{(1)}[X_m] = E_f \|X_m A_m f - f\|^2, \tag{9}$$

$$J_{AP}^{(2)}[X_m] = E_n \|X_m n^{(m)}\|^2. \tag{10}$$

$J_{AP}^{(1)}$ in the functional corresponds to optimizing the generalization ability of the approximated function while $J_{AP}^{(2)}$ corresponds to reducing the effect of noise by reducing the noise variance.

3.2 Optimal learning operator for given sampling scheme

Methods of obtaining the optimal learning operators X_m for a given sampling scheme, i.e. for fixed A_m, have already been obtained based on [3]. This corresponds to minimizing $J_{AP}^{(2)}$ under the constraint of minimizing $J_{AP}^{(1)}$. Here, we provide both the batch and incremental versions of this solution, a result which we will use in optimization of the functional of eq.(7).

3.2.1 Batch solution

Let R represent the correlation operator of the function ensemble, i.e., $R = E_f(f \otimes \overline{f})$ and Q_m represent the correlation matrix of the noise vectors, i.e., $Q_m = E_n(n^{(m)} \otimes \overline{n^{(m)}})$. These are apriori information about the learning problem that we assume to possess. Let $\dagger$ be the Moore-Penrose generalized inverse.

Lemma 3.1 *[3] A learning operator $X_m^{(AP)}$ which satisfies the Averaged Projection criterion is given as*

$$X_m^{(AP)} = R^{\frac{1}{2}} V_m^{\dagger} R^{\frac{1}{2}} A_m^* U_m^{\dagger} + Y_m (I_m - U_m U_m^{\dagger}), \tag{11}$$

where Y_m is an arbitrary operator from R^m to H, and

$$U_m = A_m R A_m^* + Q_m, \tag{12}$$

$$V_m = R^{\frac{1}{2}} A_m^* U_m^{\dagger} A_m R^{\frac{1}{2}}. \tag{13}$$

3.2.2 Incremental solution

The batch solution provided in the previous section can be computed in an incremental manner using the learning results of the previous stage and the newly added training data. The batch and the incremental solutions result in exactly the same optimal learning operator X_m after the m iterations. We define a few notations used for this purpose here.

$$T_m = A_m R^{\frac{1}{2}} \tag{14}$$

$$\Gamma_m = \sum_{n=1}^{m} (e_n^{(m+1)} \otimes \overline{e_n^{(m)}}) \tag{15}$$

$$\phi_{m+1} = R^{\frac{1}{2}} \psi_{m+1} \tag{16}$$

$$\tilde{\phi}_{m+1} = P_{\mathcal{N}(T_m)} \phi_{m+1} \tag{17}$$

$$\xi_{m+1} = \phi_{m+1} - T_m^* U_m^{\dagger} s_{m+1} \tag{18}$$

$$q_{m+1} = E_n(n_{m+1} n^{(m)}) \tag{19}$$

$$s_{m+1} = T_m \phi_{m+1} + q_{m+1} \tag{20}$$

$$\tau_{m+1} = E_n(n_{m+1}^2) \tag{21}$$

$$\alpha_{m+1} = \|\phi_{m+1}\|^2 + \tau_{m+1} - (U_m^{\dagger} s_{m+1}, s_{m+1}) \tag{22}$$

Lemma 3.2 *By using a prior learning operator and new training data, a posterior learning operator satisfying Averaged Projection criterion can be obtained as follows.*

$$X_{m+1}^{(AP)} = \left(X_m^{(AP)} - \zeta_{m+1} \otimes \overline{U_m^{\dagger} s_{m+1} + U_m^{\dagger} T_m V_m^{\dagger} \xi_{m+1}} \right) \Gamma_m^* + \zeta_{m+1} \otimes \overline{e_{m+1}^{(m+1)}}, \tag{23}$$

where ζ_{m+1} is defined as follows.

(a) When $\alpha_{m+1} > 0$ and $\phi_{m+1} \notin \mathcal{R}(T_m^*)$, then

$$\zeta_{m+1} = \frac{R^{\frac{1}{2}} \tilde{\phi}_{m+1}}{\|\tilde{\phi}_{m+1}\|^2}. \tag{24}$$

(b) When $\alpha_{m+1} > 0$ and $\phi_{m+1} \in \mathcal{R}(T_m^*)$, then

$$\zeta_{m+1} = \frac{R^{\frac{1}{2}} V_m^{\dagger} \xi_{m+1}}{\alpha_{m+1} + (\xi_{m+1}, V_m^{\dagger} \xi_{m+1})}. \tag{25}$$

For the remaining conditions, it can be shown that the new training data is redundant from the learning problem perspective and hence, need not be used for training.

4 Active learning

Active learning involves using the apriori information available about the learning problem and devising a sampling scheme to achieve good learning results with limited training data. We will look at the active learning problem from two separate but interrelated objectives. We will first devise a scheme for selecting m training data (batch operation) which will concentrate on improving the generalization ability of the NN by selecting data which extends our projection approximation space in the direction with higher function ensemble probability. Then, in addition to using these m data, we incrementally select the next set of data so as to reduce the noise variance. The second stage of operations is proved to maintain the generalization capability while reducing the effect of additive noise, resulting in a net improvement in the approximation results. Details of each of the stages are provided in the next sections.

4.1 Training data selection for optimal generalization

In this section, we propose a training data selection scheme which selects m data points in a batch operation to optimize the generalization ability, i.e., minimizes the functional

$$\min_{A_m} J_{10}[A_m] = \min_{A_m} E_f \|X_m^{(AP)} A_m f - f\|^2. \tag{26}$$

This corresponds to a modified version of the functional of eq.(7) without including the noise variance minimization criterion. Here, $X_m^{(AP)}$ is the optimal learning operator for a given sampling scheme derived in Section 3.2.1.

4.1.1 Conditions for optimizing generalization capability

Solving the functional of eq.(26),

$$\begin{aligned} J_{10}[A_m] &= \underset{f}{E} \|X_m^{(AP)} A_m f - f\|^2 \\ &= \underset{f}{E}\, tr\{(X_m^{(AP)} A_m - I)(f \otimes \overline{f})(X_m^{(AP)} A_m - I)^*\} \\ &= tr\{(X_m^{(AP)} A_m - I) R (X_m^{(AP)} A_m - I)^*\} \\ &= tr(X_m^{(AP)} A_m R A_m^* X_m^{(AP)*} - X_m^{(AP)} A_m R - R A_m^* X_m^{(AP)*} + R). \end{aligned} \tag{27}$$

For a learning operator $X_m^{(AP)}$ satisfying the Averaged Projection criterion, it can be shown that the following relation holds.

$$X_m^{(AP)} A_m R A_m^* = R A_m^*. \tag{28}$$

Using the relation of eq.(28) in eq.(27), we have

$$J_{10}[A_m] = \{tr(R) - tr(X_m^{(AP)} A_m R)\}. \tag{29}$$

Since, R is a fixed apriori information, we can now consider optimizing an equivalent optimization criterion J_{11}.

$$\min_{A_m} J_{10}[A_m] = \max_{A_m} J_{11}[A_m], \tag{30}$$

where

$$J_{11}[A_m] = tr(X_m^{(AP)} A_m R). \tag{31}$$

Based on [3], it is known that[3]

$$X_m^{(AP)} A_m R^{\frac{1}{2}} = R^{\frac{1}{2}} P_{\mathcal{R}(R^{\frac{1}{2}} A_m^*)} = R^{\frac{1}{2}} P_{\mathcal{R}(T_m^*)}. \tag{32}$$

Using this relation in eq.(31),

$$J_{11}[A_m] = tr(R^{\frac{1}{2}} P_{\mathcal{R}(T_m^*)} R^{\frac{1}{2}}) = (P_{\mathcal{R}(T_m^*)}, R). \tag{33}$$

From now on, we will represent the orthogonal projection onto $\mathcal{R}(T_m^*)$ as P without referring to the subspace of projection. Let $K = dim(\mathcal{R}(T_m^*))$ denote the dimension of the projection space. Using the property of the projection operator P, namely, $PP^* = P$ and $tr(PP^*) = K$, we can convert the variational maximization problem of eq.(33) to a Lagrange maximization problem without constraints :

$$J_{11}[P] = (PR, P) + 2Re(C, P^*P - P) + \lambda[tr(P^*P) - K], \tag{34}$$

where C and λ are the Lagrange multiplier operator and multiplier, respectively. The maximization of the above functional can be done based on the following lemmas.

Lemma 4.1 *[8] The functional represented by eq.(34) is maximized only if*

$$PR = RP. \tag{35}$$

Lemma 4.2 *[8] $PR = RP$ if and only if*

$$R\mathcal{R}(R^{\frac{1}{2}} A_m^*) \subseteq \mathcal{R}(R^{\frac{1}{2}} A_m^*). \tag{36}$$

Lemma 4.3 *[8] $R\mathcal{R}(R^{\frac{1}{2}} A_m^*) \subseteq \mathcal{R}(R^{\frac{1}{2}} A_m^*)$ if and only if $\mathcal{R}(R^{\frac{1}{2}} A_m^*)$ is a Karhunen-Loève subspace of the kernel R.*

Lemmas 4.1, 4.2 and 4.3 mean that when the relation $PR = RP$ holds, the subspace $\mathcal{R}(R^{\frac{1}{2}} A_{m+1}^*)$ is spanned by the eigenfunctions of R. Let λ_n be the n-th eigenvalue of the correlation operator R arranged in decreasing order and φ_n be the corresponding eigenfunction, i.e.,

$$R\varphi_n = \lambda_n \varphi_n \quad (\lambda_1 \geq \lambda_2 \geq \ldots). \tag{37}$$

Then, P can be represented, due to Lemma 4.3, as

$$P = \sum_{n=1}^{K} (\varphi_{m_n} \otimes \overline{\varphi_{m_n}}). \tag{38}$$

where $\{m_n : 1 \leq n \leq K\}$ is a set of indices. Hence, our problem has been reduced to that of obtaining a set of eigenfunctions $\{\varphi_{m_n} : 1 \leq n \leq K\}$ which maximizes eq.(33). Since the functional being maximized, $J_{11}[P]$, is given as

$$J_{11}[P] = (P, R), \tag{39}$$

[3] P_S refers to the orthogonal projection onto a subspace S.

Table 1: Learning conditions and results : sampling for optimal generalization

(1) Correlation operators and their eigen decompositions

$$R = \begin{pmatrix} 9 & 0 & 0 \\ 0 & 4 & 0 \\ 0 & 0 & 1 \end{pmatrix} \quad Q_2 = \begin{pmatrix} 1 & 0 \\ 0 & 1 \end{pmatrix}$$

Eigenvalue	Eigenfunction[3]
$\lambda_1^{(R)} = 9$	$\varphi_1^{(R)} = (1\ 0\ 0)^T$
$\lambda_2^{(R)} = 4$	$\varphi_2^{(R)} = (0\ 1\ 0)^T$
$\lambda_3^{(R)} = 1$	$\varphi_3^{(R)} = (0\ 0\ 1)^T$

(2) Sampling scheme (a) and (b)

Optimal scheme (a)	Non-optimal scheme (b)
$x_1^{(a)} = \frac{\pi}{5},\ \psi_1^{(a)} = (-0.59\ 0\ 0)^T$	$x_1^{(b)} = \frac{\pi}{2},\ \psi_1^{(b)} = (0\ 0\ -1)^T$
$x_2^{(a)} = \frac{\pi}{3},\ \psi_2^{(a)} = (0\ -0.87\ 0)^T$	$x_2^{(b)} = \frac{\pi}{8},\ \psi_2^{(b)} = (0.71\ -0.71\ -0.38)^T$
$\|f - f_2^{(a)}\|^2 = 3.91$	$\|f - f_2^{(b)}\|^2 = 60.74$

from eq.(38), we have,

$$J_{11}[P] = (R, P) = \sum_{n=1}^{K} (R\varphi_{m_n}, \varphi_{m_n}) = \sum_{n=1}^{K} \lambda_{m_n}. \tag{40}$$

Since λ_n are arranged in decreasing order, eq.(33) is maximized if and only if we take

$$\{\lambda_{m_n} : 1 \le n \le K\} = \{\lambda_n : 1 \le n \le K\}. \tag{41}$$

Based on the above analysis, we can write a theorem summarizing the necessary and sufficient condition for selecting the optimal training set, which is analogous to choosing the sampling operator A_m.

Theorem 4.1 *The necessary and sufficient condition for the optimization of the functional for optimal training data selection to maximize generalization capability, represented by eq.(33), is that $\mathcal{R}(R^{\frac{1}{2}} A_m^*)$ is the subspace spanned by $\mathcal{L}\{\varphi_n\}_{n=1}^K$ where $K = dim(\mathcal{R}(R^{\frac{1}{2}} A_m^*))$ and φ_n are the eigenfuctions corresponding to the K largest eigenvalues λ_n of the correlation operator R.*

4.1.2 An illustrative artificial example

Let us consider learning in a Hilbert space H spanned by the functions $\{\sin 6x, \sin 10x, \sin 15x\}$. Let the correlation operator R and the noise correlation matrix Q_2 be given as shown in Table 1-(1). The function to be learned, $f = 9\sin 6x + 4\sin 10x + \sin 15x$, is shown by a solid line in Fig.2. At first we consider selecting two training data from the optimal generalization perspective. For comparison, we look at two sampling schemes (a) and (b) as shown in Table 1-(2). The eigenvalues and eigenfunctions of R in vector representation[4] are

[4] A vector $(a\ b\ c)^T$ denotes a function $a\sin 6x + b\sin 10x + c\sin 15x$.

given in Table 1-(1). In Fig.2, the learning result due to sampling scheme (a) is shown by a dashed line ($f_2^{(a)}$) while the result due to sampling scheme (b) is represented by a dotted line ($f_2^{(b)}$). The comparison of the normed generalization error is shown in the bottom half of Table 1-(2). The space spanned by the sampling functions of scheme (a), i.e., $\mathcal{L}(\psi_1^{(a)}, \psi_2^{(a)})$ forms a K-L subspace of R and it is equivalent to the space spanned by the eigenfunctions corresponding to the two largest eigenvalues of R. On the other hand, the space spanned by the sampling functions of scheme (b), i.e., $\mathcal{L}(\psi_1^{(b)}, \psi_2^{(b)})$, does not form a K-L subspace of R. Based on Theorem 4.1, we predict that sampling scheme (a) will provide a better generalization result, a fact that is supported by the results of the simulation (refer Fig.2 and Table 1-(2)).

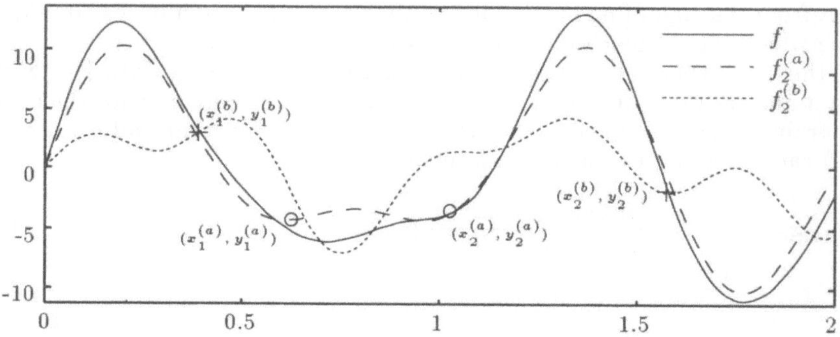

Figure 2: Learning from the generalization perspective: results using optimal (dashed) and non-optimal (dotted) training sets.

4.2 Training data selection for noise variance reduction

In this section, we focus on building upon the results of the previous section and devising a scheme for incrementally finding what training data to add to minimize noise variance of the learned function.

4.2.1 Conditions for maximal noise variance reduction

The criterion to be minimized incrementally for reducing the noise variance at each data selection stage can be written as

$$\min_{x_{m+1}} J_{20}[x_{m+1}], \tag{42}$$

where

$$J_{20} = E_n \|X_{m+1}^{(AP)} n^{(m+1)}\|^2 - E_n \|X_m^{(AP)} n^{(m)}\|^2. \tag{43}$$

By using the incremental solution for $X_{m+1}^{(AP)}$ from Lemma 3.2, J_{20} can be evaluated as follows.

(a) When $\alpha_{m+1} > 0$ and $\phi_{m+1} \notin \mathcal{R}(T_m^*)$, then

$$\begin{aligned} J_{20} &= E_n\left(n_{m+1} - (n^{(m)}, U_m^\dagger T_m V_m^\dagger \xi_{m+1})\right)^2 \frac{\|R^{\frac{1}{2}}\tilde{\phi}_{m+1}\|^2}{\|\tilde{\phi}_{m+1}\|^4} \\ &\geq 0. \end{aligned} \tag{44}$$

(b) When $\alpha_{m+1} > 0$ and $\phi_{m+1} \in \mathcal{R}(T_m^*)$, then

$$\begin{aligned} J_{20} &= -\frac{\|R^{\frac{1}{2}}V_m^\dagger \xi_{m+1}\|^2}{\alpha_{m+1} + (\xi_{m+1}, V_m^\dagger \xi_{m+1})} \\ &\leq 0. \end{aligned} \tag{45}$$

Since the case (a) always results in an increase of the noise variance (as is evident from the sign of eq.(44)), the training data which minimizes eq.(45) is the one which should be chosen for noise variance reduction.

In the case where the noise variance of the new training data is always constant and the noise on the new training data is not correlated with the noise on all training data sampled so far, that is, τ_{m+1} is a constant and $q_{m+1} = 0$, eq.(45) can be reduced to a simple form as

$$J_{20}[x_{m+1}] = -\frac{(B_m u_{m+1}, u_{m+1})}{(u_{m+1}, u_{m+1}) + \tau_{m+1}}, \tag{46}$$

where

$$B_m = Q_m^{\frac{1}{2}} X_m^{(AP)*} X_m^{(AP)} Q_m^{\frac{1}{2}}, \tag{47}$$

$$u_{m+1} = Q_m^{\frac{1}{2}} X_m^{(AP)*} \psi_{m+1}. \tag{48}$$

By using the condition $\tau_{m+1} \geq 0$, we have

$$J_{20}[x_{m+1}] \geq -\frac{(B_m u_{m+1}, u_{m+1})}{(u_{m+1}, u_{m+1})}. \tag{49}$$

This form is known as Rayleigh's quotient. The minimum value of eq.(49) is $-\lambda_1$, where λ_1 is the maximum eigenvalue of B_m. Eq.(49) is minimized if $u_{m+1} = c\ \varphi_1$, where φ_1 is the eigenvector corresponding to λ_1 and c is an non-zero scalar. Thus, when $\tau_{m+1} = 0$, eq.(46) is minimized if ψ_{m+1} satisfies the condition

$$u_{m+1} = Q_m^{\frac{1}{2}} X_m^{(AP)*} \psi_{m+1} = c\ \varphi_1. \tag{50}$$

In this case, the minimum value of eq.(46) is $-\lambda_1$.

4.2.2 Noise variance reduction in the illustrative example

We continue with the artificial learning problem considered in Section 4.1.2 and assume that we have completed training with the two optimal training data $(x_1^{(a)}, y_1^{(a)})$ and $(x_2^{(a)}, y_2^{(a)})$ based on the optimal sampling scheme (a). Now, we consider the problem of selecting an additional training data with the view of reducing noise variance. We can compute $B_2^{(a)}$ corresponding to

Table 2: Learning conditions and results : sampling for noise variance reduction

(1) Eigenvalues and eigenvectors of matrix $B_2^{(a)}$

$$B_2^{(a)} = Q_2^{\frac{1}{2}} X_2^{(a)*} X_2^{(a)} Q_2^{\frac{1}{2}} = \begin{pmatrix} 2.89 & 0 \\ 0 & 1.33 \end{pmatrix}$$

Eigenvalue	Eigenvector
$\lambda_1^{(B)} = 2.89$	$\varphi_1^{(B)} = (-1\ \ 0)^T$
$\lambda_2^{(B)} = 1.33$	$\varphi_2^{(B)} = (0\ \ -1)^T$

(2) Sampling scheme (c) and (d)

Optimal scheme (c)	Non-optimal scheme (d)
$x_3^{(c)} = \frac{\pi}{5}$, $u_3^{(c)} = (1\ \ 0)^T$ $\psi_3^{(c)} = (-0.59\ \ 0\ \ 0)^T$	$x_3^{(d)} = \frac{\pi}{3}$, $u_3^{(d)} = (0\ \ 1)^T$ $\psi_3^{(d)} = (0\ \ -0.87\ \ 0)^T$
noise variance reduction= -2.89	noise variance reduction= -1.33

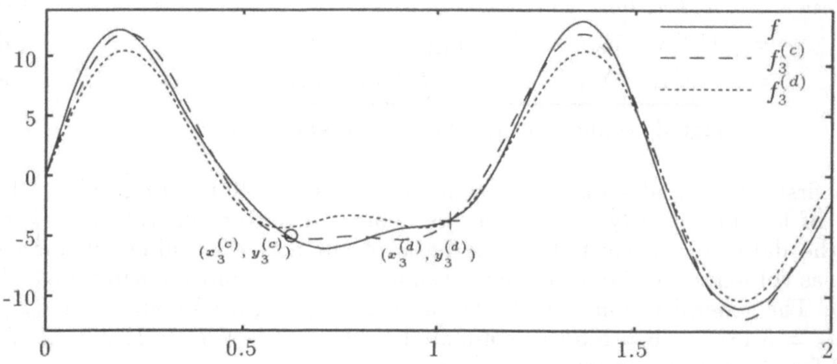

Figure 3: Learning from the noise variance reduction perspective: results using optimal (dashed) and non-optimal (dotted) training sets.

eq.(47) for the sampling scheme (a) as shown in Table 2-(1). We assume that q_3 and r_3, corresponding to eqs.(19) and (21), are both zero and independent of sampling location. The eigenvalues and eigenvectors of $B_2^{(a)}$ are shown in Table 2-(1). As a candidate for the next training data x_3, we consider two locations corresponding to schemes (c) and (d), as shown in Table 2-(2). The value of u_3 corresponding to eq.(48) computed for each sampling scheme is shown alongside. Based on our results from Section 4.2, the sampling scheme (c) should provide a greater noise variance reduction since the vector $u_3^{(c)}$ is a scalar multiple of the eigenvector corresponding to the largest eigenvalue of $B_2^{(a)}$ (refer eq.(50)), while scheme (d) does not satisfy this condition. The plot of the learned results in Fig.3 and the noise variance reduction shown in the bottom half of Table 2-(2) substantiate this prediction.

5 Empirical evaluations

We have demonstrated the mechanism by which the active learning scheme shows selectivity in the training data location through a simple artificial example in the previous section. Here, we demonstrate that the technique scales well with the complexity of the problem by considering a learning problem in high dimensional spaces (infinite dimensional original function space).

We consider learning a function within the band-limited Paley-Wiener space $H = \mathcal{L}(\{\frac{10}{\pi}\text{sinc}(\frac{10}{\pi}x - i)\}_{i=-\infty}^{\infty})$, whose reproducing kernel is written as

$$K(x, x_i) = \psi_i(x) = \frac{10}{\pi}\text{sinc}(\frac{10}{\pi}(x - x_i)). \tag{51}$$

To monitor the learning results, we use a decomposition of the generalization error into the noise and the signal component:

$$\begin{aligned} J_{gen} &= E_f E_n \|f_m - f\|^2 & (52) \\ &= E_f \|X_m A_m f - f\|^2 + E_n \|X_m n^{(m)}\|^2 & (53) \\ &= \underbrace{tr\{R - X_m A_m R\}}_{\text{signal component } J_s} + \underbrace{tr\{X_m Q_m X_m^*\}}_{\text{noise component } J_n} & (54) \end{aligned}$$

At first, we look at a task of learning using a set of 40 training points. The result of learning clearly shows that we achieve a better generalization error with the data points selected on the basis of our active learning scheme (see Fig.5) as compared to the results achieved using passive uniform sampling(refer Fig.4). The generalization error for the active sampling works out to be $J_{gen} = J_s + J_n = 3.18 + 2.36 = \mathbf{5.54}$ as compared to $J_{gen} = J_s + J_n = 12.37 + 2.17 = \mathbf{14.90}$ for the passive uniform sampling.

In the second part of the evaluation, we consider selecting incrementally, an additional 10 data points with an aim of reducing the noise variance while using the 40 optimal training data selected from the first stage. Again, the plots of the learned results, shown in Figs.6 and 7, and analysis of the generalization error ($J_{gen} = J_s + J_n = 3.18 + 1.89 = \mathbf{5.07}$ for the optimal case as against $J_{gen} = J_s + J_n = 3.18 + 2.05 = \mathbf{5.23}$ for the non-optimal case) shows that there is an effective reduction in the noise variance component while maintaining the generalization ability when we use the active sampling scheme devised here.

6 Conclusion

The generalization ability of a learning system in a noisy environment is a delicate balance on how well it can select data to enlarge the approximation space and at the same time, reduce noise variance by redundant sampling. The framework provided here provides an effective mechanism of incorporating apriori information about the function ensemble and the noise correlation matrix to select training data in accordance with the goals of optimal generalization and noise variance reduction.

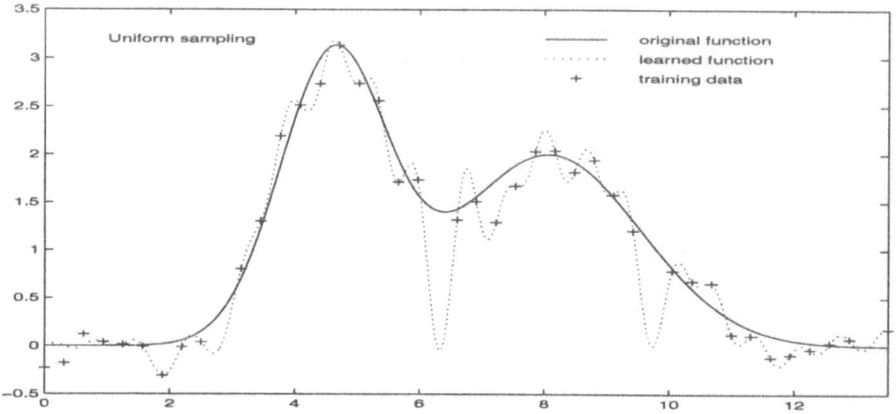

Figure 4: Sampling for improving generalization (Non-optimal)

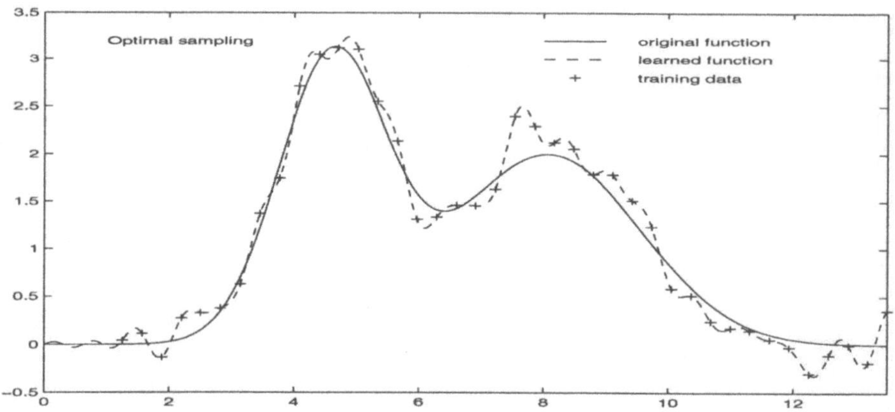

Figure 5: Sampling for improving generalization (Optimal)

References

[1] V.V. Federov. *Theory of optimal experiments.* Academic Press, New York, 1972.

[2] K. Fukumizu. Active learning in multilayer perceptrons. In *Advances in Neural Information Processing Systems*, Vol. 8, pp. 295–301. MIT Press, 1996.

[3] Y. Koide, Y. Yamashita, and H. Ogawa. A unified theory of the family of projection filters for signal and image estimation. *Transactions of the IEICE Japan*, Vol. J-77 D-II, No. 7, pp. 1293–1301, 1994. In Japanese.

[4] S.P. Luttrell. The use of transinformation in the design of data sampling schemes for inverse problems. *Inverse Problems*, Vol. 1, No. 1, pp. 199–218, 1985.

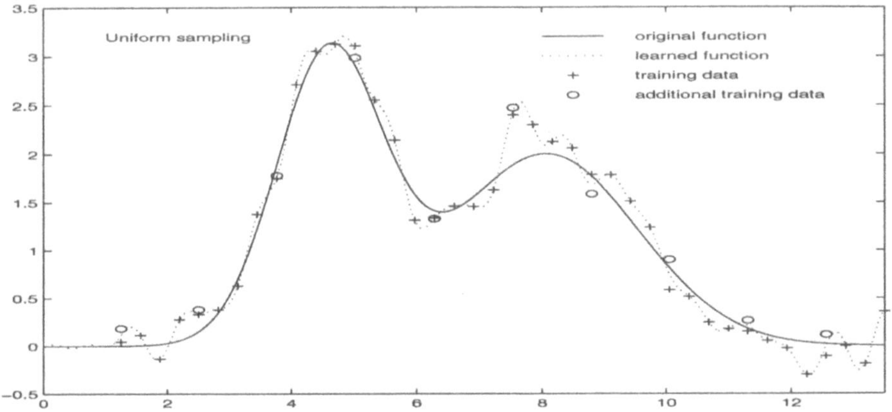

Figure 6: Sampling for noise variance reduction (Non-optimal)

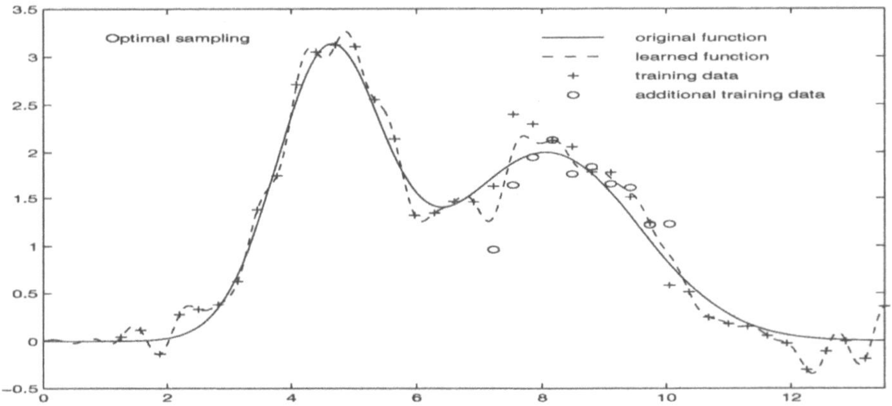

Figure 7: Sampling for noise variance reduction (Optimal)

[5] D. Mackay. Information-based objective functions for active data selection. *Neural Computation*, Vol. 4, No. 4, pp. 590–604, 1992.

[6] M. Plutowski and H. White. Selecting concise training sets from clean data. *IEEE Transactions on Neural Networks*, Vol. 4, No. 2, pp. 305–318, 1993.

[7] P. Sollich and D. Saad. Learning from queries for maximum information gain in imperfectly learnable problems. In *Advances in Neural Information Processing System*, Vol. 7, pp. 287–294. MIT Press, 1995.

[8] S. Vijayakumar. *Computational theory of incremental and active leanrning for optimal generalization*. PhD thesis, Tokyo Institute of Technology, 1998.

[9] M. Wann, T. Hediger, and N.N. Greenbaun. The influence of training sets on generalization in feedforward neural networks. In *Proceedings, International Joint Conf. on Neural Networks*, Vol. 3, pp. 137–142, 1990.

Section 6

Pattern Recognition and Signal Processing

Astronomical Object Recognition by means of Neural Networks

R. Tagliaferri
DMI - Università di Salerno, 84081 Baronissi (Salerno)
and I.N.F.M. Unità di Salerno, 84081 Baronissi (Salerno)
and IIASS "E. R. Caianello", 84019 Vietri s/m (Salerno)
Email: robtag@dia.unisa.it

G. Longo, S. Andreon, S. Zaggia
Osservatorio Astronomico di Capodimonte, via Moiariello 16, 80131 Napoli

N. Capuano, G. Gargiulo
Facoltà di Scienze, Università di Salerno, 84081 Baronissi (Salerno)

Abstract

Astronomical wide field imaging deals with Terabytes data sets and requires new strategies for data reduction and analysis. We discuss here the application of different types of neural nets to the detection and extraction of celestial objects. Preliminary tests show that neural nets are more effective than traditional techniques.

1. Introduction

Wide field imaging and hence sky surveys are used in all fields of observational astronomy and cosmology [1]. Large format photographic plates such as those used for Schmidt telescopes or panoramic (larger than 8kx8k) CCD detectors contain up to 10^6 celestial objects and a typical survey consists of hundreds or thousands of such individual plates or frames. The processing of such an humongous amount of data (in the pixel-space) cannot be handled with the traditional interactive data reduction techniques and calls for A.I. (Artificial Intelligence) based procedures capable to push as far as possible the automatic extraction of useful information from the digitized data. In order to build catalogues of object parameters we need first to detect the objects against the (noisy) background and then to measure some properties of the detected objects such as, for instance, the degree of spatial resolution (point-like or extended objects) geometrical (elongation, position angle of the major axes of the best fitting ellipse, etc.), morphological (type) and photometric (total and isophotal magnitudes, color indexes, etc.). In what follows we shall describe the application of a specifically tailored Neural Net (NN) package to the segmentation of astronomical images, *id est* to the detection of objects against a noisy background. The raw material used for the testing are Schmidt plates obtained at the European Southern Observatory (ESO) Schmidt telescope in the R photometric band. The useful field of each plate covers 5 x 5 sq. deg. of the sky. The plates were digitized using the Perkin & Elmer PDS Microdensitometer at the

Astronomical Observatory of Capodimonte in Naples and calibrated through calibration spots taken simultaneously to the scientific exposition.

2. PCA Neural Nets

Principal Component analysis (PCA) is a widely used technique in data analysis. Mathematically, it is defined as follows: let $\mathbf{C}=E(xx^T)$ be the covariance matrix of L-dimensional zero mean input data vectors x. The i-th principal component of x is defined as x^T $\mathbf{c}$(i), where $\mathbf{c}$(i) is the normalized eigenvector of $\mathbf{C}$ corresponding to the i-th largest eigenvalue λ (i).

The subspace spanned by the principal eigenvectors $\mathbf{c}$(1), ... , $\mathbf{c}$(M), (M<L) is called the PCA subspace (of dimensionality M) [2], [3]. PCA's can be neurally realized in various ways [4], [5], [6], [2], [7], [8]. The PCA network used by us is a one layer feedforward neural network which is able to extract the principal components of the stream of input vectors. Typically, Hebbian type learning rules are used, based on the one unit learning algorithm originally proposed by Oja [6]. Many different versions and extensions of this basic algorithm have been proposed during the recent years (see [9], [10], [3], [8]). The structure of the PCA NN can be summarized as follows: there is one input layer, and one forward layer of neurons totally connected to the inputs; during the learning phase there are feedback links among neurons, that classify the network structure as either hierarchical or symmetric. After the learning phase the network becomes purely feedforward. The hierarchical case leads to the well known GHA algorithm [8], [10]; in the symmetric case we have the Oja's subspace network [6]. PCA neural algorithms can be derived from optimization problems, such as variance maximization and representation error minimization. We can generalize these problems to nonlinear problems, getting nonlinear algorithms (and relative networks). These have the same structure of the linear ones: either hierarchical or symmetric. These learning algorithms can be further classified in: robust PCA algorithms and nonlinear PCA algorithms [9], [10]. We define robust PCA so that the objective function grows less than quadratically. The non linear learning function appears at selected places only. In nonlinear PCA algorithms all the outputs of the neurons are nonlinear function of the responses. We have seen both in preceding experiments [11] and in this paper that the hierarchical robust NN reaches the best performance. Specifically, in the experiments we compared hierarchical robust NN with learning function g(t)= tanh(αx) with linear PCA.

3. Neural nets for image segmentation

For the segmentation we used Hierarchical and Hybrid NNs. The former is a Multilayer Unsupervised NN, while the latter is structured in two layers: the first one being an unsupervised neural net and the second a clustering algorithm. Aim of both setups is to attain a number of elements equal to the number of classes in which we want to segment the input image. In the following sections we shall describe first the neural models and the clustering algorithms and hence the hierarchical and hybrid networks.

3.1 Unsupervised neural nets

Kohonen [12],[13] Self Organizing Maps (SOM) are composed by a neuron layer structured in a rectangular grid. When a pattern x is presented to the net each neuron i receives the components and computes the distance d_i from its weight vector w_i. The unit which has the minimum distance from the input pattern will be the winner. The adaptation step consists in the modification of the weights of the neurons in the following way:

$$w_i^{(t+1)} = w_i^{(t)} + \varepsilon^{(t)} \cdot h_{\sigma^{(t)}}\left(d(i,k)\right) \cdot \left(x - w_i^{(t)}\right)$$

where $\varepsilon^{(t)}$ is a *gain term* $(0 \leq \varepsilon^{(t)} \leq 1)$ decreasing in time, $h_{\sigma^{(t)}}(x)$ is a unimodal function with variance $\sigma^{(t)}$ decreasing with x and $d(i,k)$ is the distance in the grid between the i and the k neurons.

The Neural-Gas NNs have a learning algorithm [14] which works better than the preceding one: in fact, it is quicker and it reaches a lower average distortion value[1]. It uses a soft-max adaptation of the weights and it classifies the neurons in an ordered list $(i_1, i_2, \ldots, i_m)$ following their distance form the input pattern. The weight adaptation depends on the position $rank(i)$ of the i neuron in the list in the following manner:

$$w_i^{(t+1)} = w_i^{(t)} + \varepsilon^{(t)} \cdot h_{\sigma^{(t)}}\left(rank(i)\right) \cdot \left(x - w_i^{(t)}\right) \cdot$$

The algorithm applies the gradient descent technique to the error function:

$$E_{ng} = \frac{1}{2C(\sigma)} \sum_{j=1}^{m} \int P(x) \cdot h_\sigma\left(rank(x)\right) \cdot \left(x - w_j\right)^2 d^n x \quad \text{with} \quad C(\sigma) = \sum_{k=0}^{m-1} h_\sigma(k).$$

The neural net is composed by a linear layer of neurons.

The Growing Cell Structure (GCS) [15] is a NN which is able to change its structure depending on the data set. Aim of the net is to map the pattern space into a two-dimensional discrete structure A in such a way that similar patterns are represented by topological neighbor elements. The structure A is a two-dimensional simplex where the vertices are the neurons and the edges attain the topological information. Every modification of the net always maintains the simplex properties. The learning algorithm starts with a simple three node simplex and tries to obtain an optimal network by a controlled growing process: for each x pattern of the training set the winner and the neighbors weights are adapted as follows:

$$w_k = w_k + \varepsilon_b \cdot (x - w_k); \quad w_i = w_i + \varepsilon_n \cdot (x - w_i) \quad \forall i \text{ connected to } k$$

---

[1] Let $P(x)$ be the pattern probability distribution over the set $V \subseteq \Re^n$ and let $w_{i(x)}$ be the weight vectorof the neuron which classifies the pattern x, therefore we define average distortion as:

$$E = \int P(x)\left(x - w_{i(x)}\right)^2 d^n x$$

where ε_b and ε_n are constants which determine the adaptation strength for the winner and for the neighbors, respectively.
The insertion of a new node is made after a fixed number λ of adaptation steps. The new neuron is inserted between the unit which has win more times than the others and the more distant of its topological neighbors. The algorithm stops when the network reaches a pre-defined number of elements.

A simpler algorithm is the K-means clustering algorithm [16] in its on-line release which applies the gradient descent directly to the average distortion function above defined as follows:

$$w_i^{(t+1)} = w_i^{(t)} + \varepsilon^{(t)} \cdot \left(x - w_i^{(t)}\right).$$

The main limitation of this technique is that the error function presents many local minima which stops the learning before reaching the optimal configuration.

The last unsupervised learning algorithm is the Maximum Entropy [17] which applies the gradient descent with soft-max adaptation of the weights to the error function

$$E_{me} = -\frac{1}{\beta}\int P(x)\ln\sum_{j=1}^{m} e^{-\beta\left(x-w_j\right)^2} d^n x$$

and adaptation step

$$w_i^{(t+1)} = w_i^{(t)} + \varepsilon^{(t)} \cdot \frac{\exp\left(-\beta^{(t)} d_i\right)}{\sum_{k=1}^{m} \exp\left(-\beta^{(t)} d_k\right)} \cdot \left(x - w_i^{(t)}\right)$$

where β is the inverse temperature and takes value increasing in time.

3.2 Hybrid neural nets

Hybrid NNs are composed by a unsupervised single layer NN and a clustering algorithm that uses the information derived by the NN learning algorithm. After the learning of the net, we must apply the clustering algorithm to have a neuron partition in subsets. Their number is equal to the number of the output classes. Furthermore, we want that neurons with similar weight vectors will be in the same class, while neurons with very distant weight vectors will be in different classes. The best strategy is clearly to apply the clustering algorithm directly to the weight vectors of the unsupervised NN after the learning.
A non-neural agglomeration clustering algorithm that divides the pattern set (in this case the weights of the neurons) $W = \{w_1, \dots, w_m\}$ in l cluster $C_1, \dots, C_l$ (with $l<m$) can be briefly summarized as follows:

1. we initially divide W in m clusters $C_1, \dots, C_m$ such that $C_i = \{w_i\}$;
2. we compute the distance matrix D such that $D_{ij} = d(C_i, C_j)$;
3. we find the smallest element D_{ij} of the matrix D and we unify the clusters C_i and C_j in a new one $C_{ij} = C_i \cup C_j$;
4. if the number of clusters is greater than l then go to step 2 else stop.

This is the shared basis of many algorithms appeared in literature [18]. The only difference is the distance function. For example, two different choices can be:

a. $d(C_i,C_j) = \min_{w_{ik} \in C_i \text{ e } w_{jl} \in C_j} \|w_{ik} - w_{jl}\|$ (*nearest neighbor algorithm*);

b. $d(C_i,C_j) = \left\| \frac{1}{|C_i|} \sum_{w_{ik} \in C_i} w_{ik} - \frac{1}{|C_j|} \sum_{w_{jl} \in C_j} w_{jl} \right\|$ (*barycentre method);*

c. $d(C_i,C_j) = \frac{1}{mn} \sum_{1 \le k \le n,\ 1 \le l \le m} \|w_{ik} - w_{jl}\|$ (*average between groups*).

The output of the clustering algorithm will be a labeling of the patterns (in this case neurons) in l different classes.

3.3 Unsupervised hierarchical neural nets

An alternative approach is to use a new unsupervised single layer NN instead of a clustering algorithm. In this way the second layer NN learns from the weights of the first NN and clusters the neurons on the basis of a similarity measure or a distance. If we apply this process several times then we obtain the unsupervised hierarchical NNs. The number of neurons at each layer decreases form the first to the output layer, and, as a consequence, the net takes a pyramidal aspect as illustrated in Figure 1. The net takes as input a pattern x and then the first layer finds the winner neuron. The second layer takes the first layer winner weight vector as input and finds the second layer winner neuron and so on until the top layer. The activation value of the output layer neurons is 1 for the winner unit and 0 for all the others. Briefly, the learning steps of a s layer hierarchical NN with training set X are the following:

1. The first layer is trained on the patterns of X with one of the previous learning algorithms.
2. The second layer is trained by using the same algorithm or one of the other previous ones on the elements of the set X_2 which is composed by the weigh vectors of the first layer winner units.
3. By using the same algorithm or one of the others, we iterate the process to the *i-th* layer NN (*i*>2) on the training set X_i which is composed by the weight vectors of the winner neurons of the *i-1-th* layer when presenting X to the *1-st* layer NN, X_2 to the *2-nd* layer and so on.

By varying the learning algorithms of the layers we obtain different NNs with different properties and abilities. For instance, by using only SOMs we have a Multi-layer SOM (*ML-SOM*) [19] where every layer is a two-dimensional grid. We can easily obtain *ML-Neural-Gas*, *ML-Maximum-Entropy* or *ML-K-means* organized on a hierarchy of linear layers. The ML_GCS has a more complex architecture and has at least 3 units for layer.

We can think to have hierarchical NNs where different layers have different learning algorithms so that we can take advantage from the properties of each model (for example since we cannot have a ML-GCS with 2 output units, then we can use another NN in the output layer).

To solve our basic problem, we need to have a hierarchical NN with a number of output layer neurons equal to the number of the output classes. In this way the labeling becomes a simple problem without reducing the generalization capacity of the net. In fact, the first layer neurons are enough to correctly accomplish the distribution probability density of the input patterns. On the other hand, the number of neurons of a layer cannot rapidly decrease with respect to the number of units of the preceding layer, a hierarchical NN is slower than a single layer NN, and the computing time depends on the layer number. After the learning phase, it is simple to label in a unique way the input neurons depending on the corresponding output units. In this way we use single layer NNs in the computation on the test set.

4. Experimental results

The experiments were performed on a 1° by 1° degree extracted from one of the plates. Each pixel is a 16 bit integer and each plate is a 8 kbytes × 8 kbytes. In order to identify the principal components of the system we used a 5 x 5 running window to feed both a non linear PCA neural network and a traditional linear PCA. In both cases it turned out that 90% of the information is contained in three components only. As it is clear from Figure 2, the non linear PCA neural network outperforms the traditional PCA tool allowing a much better discrimination of the objects against the background at faint light levels. In Figure 3 we also show the advantage of adopting a hyperbolic tangent as activation function of the non linear PCA net. With respect to the linear case, the distance between faint and luminous object is reduced and the contrast between background and objects is greatly increased.

We therefore used both Hierarchical and Hybrid neural nets to segment the image in six classes. The results of the application of the NNs are summarized in Table 1 and Table 2 and Figure 4, which - for the sake of clarity - refers only to a small fraction of the field. The NN dimensions are illustrated in the second column of Table 1. The performance of the NNs has been computed as distortion rate and as percent of correct detected object after the deblending which is applied to all the segmented images in the same way. This third and final step consists in running a simple algorithm capable to resolve partially overlapping objects. The best performing networks are the Hierarchical (3 layers) Neural gas and the GCS + Neural gas (first layer GCS plus two layers of neural gas). In fact, for what concern the average distortion rate, ML_SOM , ML_Neural Gas and GCS + Neural gas reach the same value of the hybrid Neural Gas systems as shown in Table 1. For what concern the performance in the object identification tasks, ML_Neural Gas and GCS + Neural gas reach a similar correctness percent (more than 93%) and the identification of about 720 corret objects, while the only comparable model is the ML_SOM with a 92.33% of correctness but only 699 correct objects. The comparison of GCS + Neural gas (chosen because is faster than ML_Neural Gas with a comparable correctness) with FOCAS is shown in Table 3 and Figure 5, where our system outperforms the standard system.

References

[1] Jarvis J.F., Tyson J.A., FOCAS: Faint Object Classification And Analysis System in The Astronomical Journal, vol. 86, no. 3, pp. 476-495, 1981.

[2] Oja E., Ogawa H., Wangviwattana J., Learning in nonlinear constrained Hebbian network, In T. Kohonen et al. (Eds.), Artificial neural networks, 385-390, Amsterdam: North-Holland, 1991.

[3] Oja E., Karhunen J., Wang L., Vigario R., Principal and independent components in neural networks - recent developments, Seventh Italian Workshop on Neural Networks, Vietri 1995, M.Marinaro & R.Tagliaferri Ed.s, World Scientific Pu. Singapore, 16-35, 1996.

[4] Baldi P., Hornik K., Neural networks for principal component analysis: learning from examples without local minima, Neural Networks, 2, (7), 53-58, 1989.

[5] Jutten C., Herault J., Blind separation of sources, part I: an adaptive algorithm based on neuromimetic architecture, Signal Processing, 24, (1), 1-10, 1991.

[6] Oja E., A simplified neuron model as a principal component analyzer, Journal of Mathematical Biology, 15, 267-273, 1982.

[7] Plumbley M., A Hebbian/anti Hebbian network which optimizes information capacity by orthonormalizing the principal subspace, in Proc. IEE Conf. on Artificial Neural Networks, Brighton, UK, 86-90, 1993.

[8] Sanger T. D., Optimal unsupervised learning in a single-layer linear feedforward network, Neural Networks, 2, 459-473, 1989.

[9] Karhunen J., Joutsensalo J., Representation and separation of signals using nonlinear PCA type learning, Neural Networks, 7, 113-127, 1994.

[10] Karhunen J., Joutsensalo J., Generalization of principal component analysis, optimization problems, and neural networks, Neural Networks, 8, 549-562, 1995.

[11] Rasile M., Milano L., Tagliaferri R., Longo G., Periodicity Analysis of Unevenly Spaced Data by means of Neural Networks, Proceedings of the 9th Italian Workshop on Neural Nets WIRN Vietri '97, M. Marinaro & R. Tagliaferri Ed.s, Springer-Verlag, London (pp. 201-212) (1997).

[12] Kohonen T., Self-organized formation of topologically correct feature maps in Biological Cybernetics, vol. 43, pp. 59-69, 1982.

[13] Kohonen T., Self-organization and associative memory (2nd edition) Springer-Verlag Berlin, 1988.

[14] Martinetz T., Berkovich S., Schulten K., Neural-Gas Network for Vector Quantization and its Application to Time-Series Prediction in IEEE Transactions on Neural Networks, vol. 4, no. 4, pp. 558-568, 1993.

[15] Fritzke B., Growing Cell Structures - A Self-Organizing Network for Unsupervised and Supervised Learning in Neural Networks, vol. 7, no. 9, pp. 1441-1460, 1994.

[16] Lloyd S., Least squares quantization in PCM in IEEE Transaction on Information Theory, vol. IT-28, p. 2, 1982.

[17] Rose, Gurewitz F., Fox G., Statistical mechanics and phase transition in clustering in Physical Review Letters, vol. 65, no. 8, pp.945-948, 1990.

[18] Everitt B., Cluster Analysis Social Science Research Council. Heinemann Educational Books. London, 1977.

[19] Koh J., Suk M., Bhandarkar S., A Multilayer Self-Organizing Feature Map for Range Image Segmentation in Neural Networks, vol. 8, no. 1, pp. 67-86, 1995.

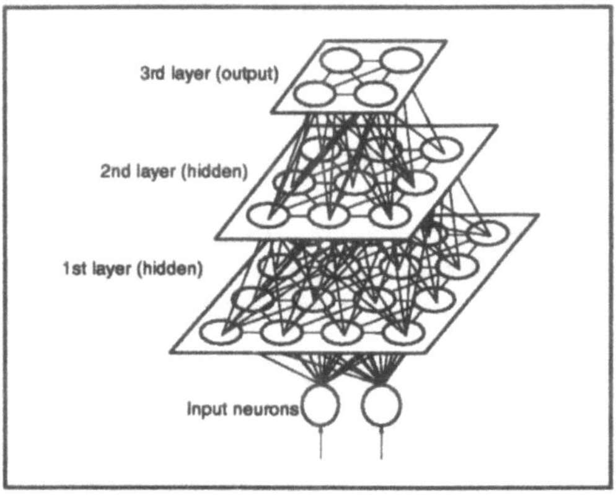

Fig. 1 - A hierarchical NN.

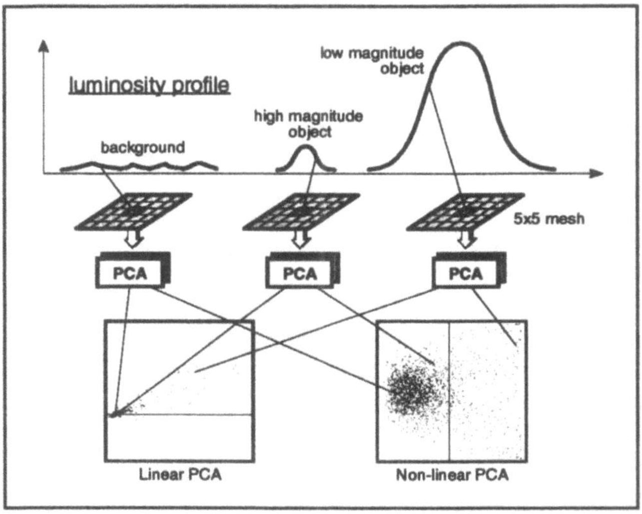

Fig. 2 - Comparison of a Linear PCA and a Non-linear PCA.

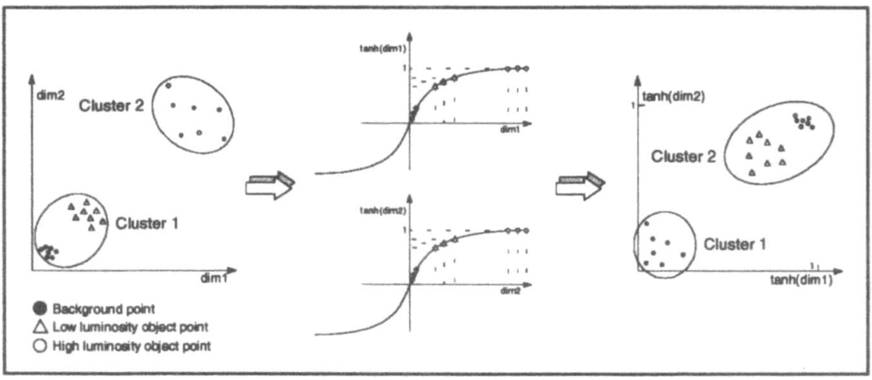

Fig. 3 - Effect of the *tanh* activation function.

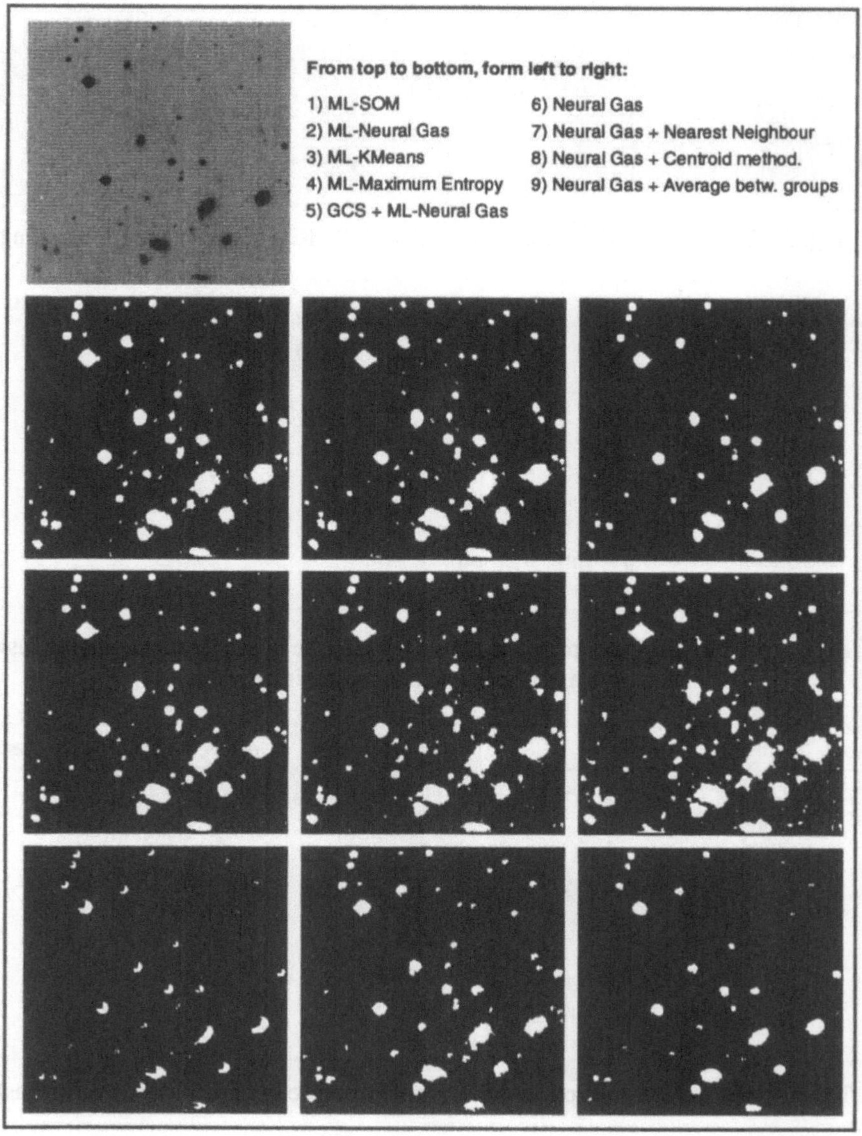

Fig. 4 - Comparison between hierarchical NNs and Hybrid NNs on a real image.

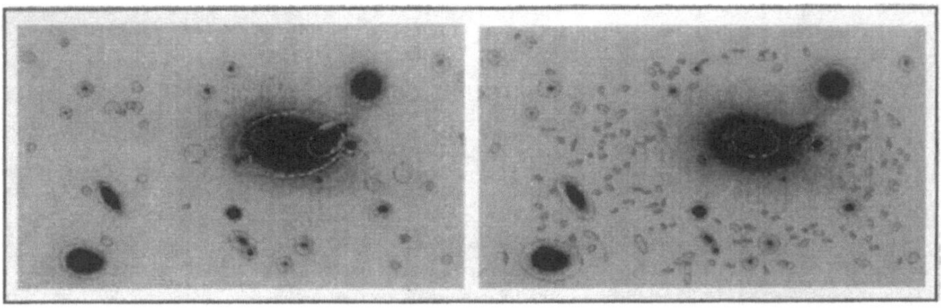

Fig. 5 - Comparison of our method (left) with SKICAT (right) after deblending.

	NN type	neurons	avr. distort.	time (sec.)
1	ML-SOM	50 20 6	0.1188	340
2	ML-Neural Gas	50 20 6	0.1163	302
3	ML-K Means	50 20 6	0.1300	112
4	ML-Maximum Entropy	50 20 6	0.3478	217
5	GCS + ML-Neural Gas	50 20 6	0.1166	235
6	Neural Gas	6	0.2831	103
7	" " + Nearest Neighbour	50	0.1163	256
8	" " + Barycentre method.	50	0.1163	249
9	" " + ave. Between groups	50	0.1163	243

Tab 1 – Hierarchical unsupervised NNs performance in image segmentation tasks: average distortion and computing time.

	NN type	detected obj.	Correct obj.	Corr. Perc.
1	ML-SOM	757	699	92.34
2	ML-Neural Gas	771	720	93.34
3	ML-K Means	426	394	92.49
4	ML-Maximum Entropy	637	607	95.29
5	GCS + ML-Neural Gas	768	719	93.62
6	Neural Gas	1113	720	64.70
7	" " + Nearest Neighbor	384	354	92.19
8	" " + Barycentre Method	438	381	86.99
9	" " + Ave. between Groups	308	297	96.43

Tab 2 – Hierarchical unsupervised NNs performance in object identification tasks.

	FOCAS	Our Method
Total Detected Obiects	2443	1923
Spurius Artifacts	635	120
Right Objects	1808	1803
Correctness percent	74.01	93.76

Tab 3 – Comparison of our method with FOCAS on Coma cluster core.

Learning to balance a pole on a movable cart through RL: what can be gained using Adaptive NN?

N.A. Borghese, C. Biemmi and F. Lo Monaco
Laboratory of Human Motion Study and Virtual Reality
Istituto Neuroscienze e Bioimmagini, CNR, Via f.lli Cervi 83, 20090 Segrate (Milano), Italy
[biemmi, lomonaco, borghese]@inb.mi.cnr.it

Abstract The work of Barto, Sutton and Williams on the ACE/ASE model for Reinforcement Learning is here put in perspective. In their work a state-control (input-output) map, which allows to balance a pole hinged on a moving cart as long as possible, is learned when the only information provided by the environment is the system failure. This work has given rise to a large body of research in the fields of machine learning and artificial intelligence. Its relevance lies in the fact that it can be applied to control all those systems which are only partially known. A critical issue is the exploration of the state space which may require impractical amount of memory and learning time. Adaptive networks, which have been studied in the most recent years, offer a natural solution in the implementation of the learning system allowing an adaptive partitioning of the state space according to the task difficulty experienced in the different regions.

1 Introduction

Balancing a pole hinged on a movable cart by applying forces on the cart's base is a difficult control problem when a) the dynamics of the system is not perfectly known and b) there is no direct knowledge of the state. The only information available to the system is: "success or fail". This problem is of general interest in particular for the community of human motor control as a paradigm to analyse how upright body posture can be learned. In the recent past different non-traditional control techniques have been proposed with some success [1, 2, 3, 4]. All these techniques are based on the "discovery" of a control policy from repeated experimental trials. From the connectionist point of view, these techniques fall under the domain of *Reinforcement Learning* (RL) where the input-output relationship is not given explicitly as in supervised learning, but it is discovered by evaluating the global performance of the controlled system. The key element of RL is a sub-system, the *Critic*, which evaluating the system performance is able to provide to the Controller sub-system an information suitable to improve its performance. We here focus on RL implemented in the ACE/ASE model of Barto et al., [5] which has been the ground model of what has been developed in the last years.

2 The inverse pendulum balancing problem

The system to be controlled is constituted of a cart which is free to translate on a horizontal straight track and of a pendulum which is hinged on the cart and it is free to rotate (Fig. 1). Task of the Controller is to keep the pendulum upright as long as possible by pushing the cart along the track rightwards or leftwards. The system is identified by four variables: the position and velocity of the cart on the track $(x, \dot{x})$

and the angular velocity and orientation of the pole with respect to the vertical $(\theta, \dot{\theta})$. The state of the system, **x**, is therefore a quadri-dimensional vector: $\mathbf{x}(x, \dot{x}, \theta, \dot{\theta})$. In the original approach of Barto, the state space was discretised into N regions which are called boxes following Michie and Chambers [6]. Whenever the a certain box, i, is visited, x_i is set equal to one, otherwise it is set equal to zero.

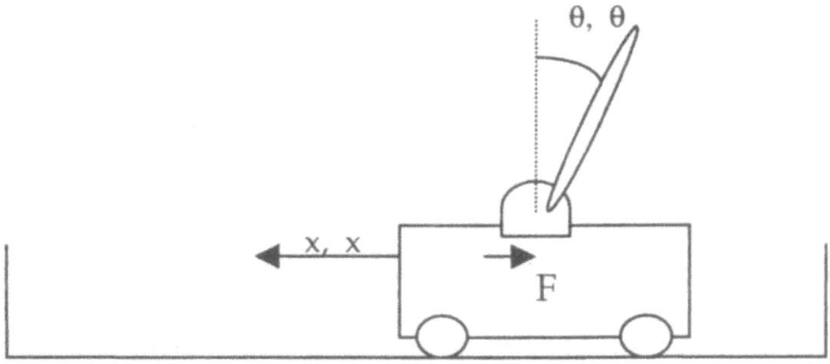

Fig. 1. The test system used. It is constituted of a cart which is free to translate on a track between to end-stops and a pole hinged in the middle of the cart and free to rotate. The Controller applies its force in the cart centre.

3 The ACE-ASE Model

In the RL paradigm, two sub-systems are identified: the Controller and the Critic (cf. Fig. 2). In Barto's model the Controller was implemented in the Adaptive Search Element (ASE) and the Critic in the Adaptive Critic Element (ACE). Both these elements were implemented as Perceptrons. The only information coming from the environment was the Reinforcement signal, r, which is always zero until the Controller fails when it becomes equal to -1. The Controller fails when the cart reaches the end of the track or the pole falls down. A secondary Reinforcement signal, $\hat{r}$, is generated inside the Critic; this is the information used to tune the weights in the ASE.

3.1 The Adaptive Search Element

The word Search (ASE) stresses the fact that the control policy is not known a-priori but it emerges from the learning itself through a search in the space of the possible actions. In fact, at start, the control policy generated by the ASE is random. As learning progresses the weights in the ASE are tuned to build an input-output (state-control) map where those actions which revealed most successful for the task are facilitated. The tuning information is provided by the ACE. It is relevant to stress that the ASE output is not uniquely determined by its set of weights and the input vector, a random component is added which allows to explore different control actions for the same state. The ASE output is given as:

$$y(t) = f\left(\sum_{i=1}^{n} w_i(t)x_i(t) + v_i(t)\right) \tag{1}$$

where f is a threshold function, $\nu(t)$ is a Gaussian random noise and $w_i(t)$ are a set of adjustable weights. $\mathbf{x}(t)$ is the state (four components, n=4) at time t and y(.) corresponds to a push towards right or left whose amplitude is not scalable:

$$y(t) = \begin{cases} +1 & if \ \ f(.) \geq 0 \\ -1 & if \ \ f(.) < 0 \end{cases} \tag{2}$$

The weights w_i can therefore be seen as a bias in the probability to get a push towards one of the two directions.

3.2 The Adaptive Critic Element

Credit assignment is the problem which has to be faced by the Critic. It is the determination of the components in the Controller to which the credit (blame) for increments (decrements) in the overall system performance should be ascribed. A possibility is the association of the most recent reinforcement with the most recent control action: in this case a negative reinforcement, elicited by a Controller failure, would penalise the last action. However, the failure of the system can be ascribed to poor actions taken earlier and the very last action can be even appropriate for the situation. Therefore aim of the Critic is to discover which control actions were responsible for the actual failure. The Critic analyses the entire temporal horizon of the task: without this temporal dimension, the Controller would not succeed in its task. From the analysis, the Critic builds a "risk-map" for each state box; this represents the expected amount of time for which the actual control policy is able to succeed starting from that state. During learning, those actions which keep the system far form the most dangerous boxes are rewarded by increasing the associated weights in the ASE, making these actions more likely to occur. Primary job of the Critic is therefore to learn a correct "risk-map". This is represented as a perceptron in the Barto's model as:

$$p(t) = \sum_{i=1}^{n} v_i(t) x_i(t) \tag{3}$$

where $\{v_i(t)\}$ are a set of adjustable weights (cf. Eq. (1)). p(t) is transformed into an internal reinforcement signal, $\hat{r}$, used by the ASE, by the following updating rule:

$$\hat{r}(t) = r(t) + \gamma p(t) - p(t-1) \tag{4}$$

with $0 < \gamma < 1$. Normally the external reinforcement, r(t), is zero indicating a successful accomplishment of the task. In this case, if p(t) is greater than p(t-1), the system has moved in a less risky state and the actual action can be rewarded, vice-versa if p(t) is less than p(t-1). In case of failure, r = -1, there is no actual state (p(t) = 0), the internal reinforcement is negative and the risk associated to the previous state is increased (cf. Eq. (5a)). If the last state were already labelled as very risky (e.g. p(t-1) = -1), its associated risk was correctly estimated and has not to be updated; it was the risk associated to the previous states that was incorrectly estimated and the undertaken actions will be punished by increasing the risk of failure associated to these states. This learning paradigm has been termed learning with a Critic [7] or with a distant teacher [8, 9] to distinguish it from learning with a teacher as in the supervised learning paradigm where for each state the control action is specified by the environment. The Critic therefore evaluates which control actions lead to an improvement in the overall performance playing therefore the role of a teacher, but it cannot directly specify the control policy.

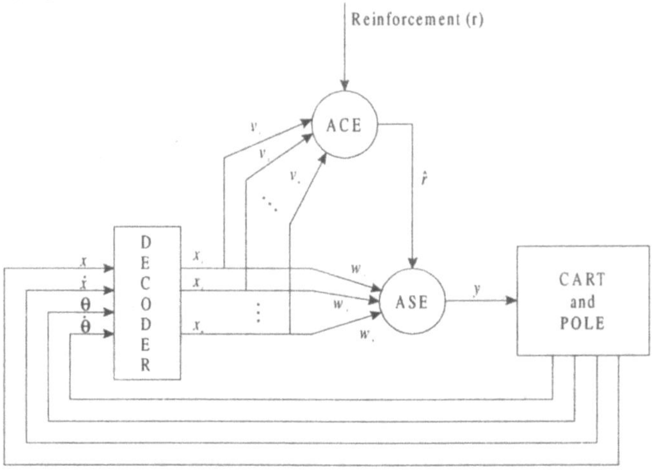

Fig. 2. The overall system. The output of the ASE (y) is the Force F and the output of the ACE ($\hat{r}$) is the internal reinforcement signal which represents the risk estimation for the actual state.

3.3 Learning mechanism

The expected reinforcement or risk is memorised in the p vectors (Eq. (3)). These depend on the weights $\{v_i\}$ which are updated as follows:

$$v(t+1) = v(t) + \beta[\hat{r}(t) + \gamma p(t) - p(t-1)]\bar{x}(t) \quad (5a)$$

$$\bar{x}(t+1) = \lambda\bar{x}(t) + (1-\lambda)x(t) \quad (5b)$$

where $\bar{x}$ is a trace of the decoder output and it represents the frequency with which a certain state has been visited in the recent past. $0 < \lambda < 1$ determines the trace decay. Eq. 5a increases the reliability of a state, if the actual control policy caused a transition to a less risky state (p(t) > p(t-1)). The amount of updating depends on how frequently that state has been visited ($\bar{x}$ (t)). The ACE therefore is able to provide a reinforcement signal also in absence of an external reinforcement (without a Controller failure). This allows to greatly improve the ASE performance by continuously updating its weights. The updating of the ASE weights, $\{w_i\}$, at time t depends on two factors: the eligibility of the related state, $\mathbf{x}_i$, and the internal reinforcement $\hat{r}$, at the same time instant:

$$w_i(t+1) = w_i(t) + a\hat{r}(t)e_i(t) \quad (6a)$$

$$e_i(t+1) = \delta e_i(t) + (1-\delta)y(t)x_i(t) \quad (6b)$$

where $0 < \delta < 1$. α determines the rate of change of the weights. The eligibility associates to each state $\mathbf{x}_i$ a value which expresses the correlation in time between that state and the output: the higher is the correlation, the larger will be the weights updating. A weight is therefore increased if the reinforcement is positive and that state box has been visited; if the reinforcement is negative or the state and the output are uncorrelated the weight is decreased so as to make the Controller more likely do something else.

4 The role played by Adaptive Neural Networks

The Critic is the key element in RL. It estimates the "Cost-To-Go" which is a discounted measure of reward or risk of a given control policy associated to each state. In the Barto's approach the Cost-To-Go is represented by the risk function p(t) (Eqs. 3 and 4). Classical RL models implicitly assume that the state can be represented with as many boxes or elements as needed. This may easily require very large memories which make their use impractical. E.g., in the ASE/ACE model, the boxes number increases exponentially with the number of subdivisions ($N_{boxes} = N^4_{subdivisions}$). Neural Networks models reduce the resources requirement through their generalisation capabilities and have been proposed as constituents of the Critic and Controller sub-systems [2, 10]. However, classical models perform poorly on real problems because of the many local minima produced. Better solutions rely on models which make use of local or semi-local units (e.g. [11, 12, 13, 14, 15, 16]) which partition the state space not rigidly as in the box model, but with some degree of overlapping. To take full advantage of these models, the units boundaries have to be defined according to some criterion [12, 13, 14, 15]. Ideally, the state space would be partitioned into regions that can be considered the same for learning purpose. For example in HRBF model, units are inserted where the residual error is larger, implicitly partitioning the state space adaptively. A measure of the residual can be the variability of the actions associated to a given state [17]: a large variability value meaning that the state is not sufficiently represented there. An alternative approach is based on running "mental trajectories" through the state space and analysing which units are most activated [18]; it can be less reliable as it depends on the validity of these trajectories. These algorithms are expected to reduce the search time for a successful control policy for the inverted pendulum as well as for other dynamical systems. They also lead to new architectures where the Controller and the Critic sub-systems are subdivided into an active sub-system, which actually does the job, and an organising sub-system, which builds the structure of the active sub-system in real-time.

5 Conclusion

RL has been defined as "the problem faced by an agent that must learn behaviour through trial-and-error interactions with dynamic environment" [1]. Its application to control and identification of systems only partially known is becoming more and more interesting in environments where machines will play a greater role. These algorithms have been shown to be extremely powerful also in learning tasks which are usually associated with human intelligence: e.g. the back-gammon program of Tesauro [19] has been proven proficient at a master level. RL if of great interest also for the Human Motor Control Community as paradigm to learn the complex motion coordination. In this respect the introduction of Adaptive Neural Networks as constituents of the Critic and the Controller sub-systems can help to improve the performance allowing to distribute a limited amount of resources to get the best coverage of the state space.

References

1. L.P. Kaelbling, M.L. Littman and A.W. Moore, Reinforcement Learning: A Survey, *J. Artificial Intelligence Research* 4, (1996) 237-285.
2. R.J. Williams, Simple statistical gradient-following algorithms for connectionist reinforcement learning, *Machine Learning*, 8:3, (1992), 29-256.
3. R.S. Sutton, Learning to predict by the method of temporal differences, Machine Learning, 3:1, (1988), 9-44.
4. W.T. Miller III, R.S. Sutton and P.J. Werbos (Eds.), Neural Networks for Control, (1992) MIT Press.
5. A.G. Barto, R.S. Sutton, and C.W. Anderson, Neuronlike adaptive elements that can solve difficult learning problems. *IEEE Trans. Syst. Man and Cybern.* SMC-13, (1983), 834-846.
6. D. Michie and R.A. Chambers, "BOXES: An experiment in adaptive control", *Machine Intelligence,* 2, E. Dale and D. Michies, Eds. Edinburgh: Oliver and Boyd, (1968), 237-152.
7. B. Widrow, N.K. Gupta and S. Maitra, Punish/reward: learning with a critic in adaptive threshold systems, *IEEE Trans. Syst. Man Cybern.* SMC-3, (1973), 455-465.
8. M.I. Jordan and D.E. Rumelhart, Forward models: Supervised learning with a distal teacher. *Cognitive Science,* 16, (1992), 307-354.
9. N.A. Borghese and M. Arbib, Generation of Temporal Sequences using Local Dynamic Programming, *Neural Networks* 8:1, (1995), 39-54.
10. C.W. Anderson, Learning to control an inverted pendulum using neural networks. *IEEE Control Systems Magazine* (1989) 31-37.
11. H.A. Berenji and P. Khedkar, Learning and tuning fuzzy logic controllers through reinforcement. *IEEE Trans. on Neural Networks*, 3:5 (1992), 724-740.
12. N.A. Borghese and S. Ferrari, Hierarchical RBF networks in function approximation. NeuroComputing, (1998).
13. M. Cannon and J.E. Slotine, Space-Frequency localized basis function networks for nonlinear system estimation and control, *Neurocomputing* 9 (1995) 293-342.
14. B. Fritzke, Growing Cell Structures: A Self-organizing Network for Unsupervised and Supervised Learning, *Neural Networks* 7 (1994) 1441-1460.
15. T.M. Martinetz, S.G. Berkovich and K.J. Schulten, Neural-Gas" Network for Vector Quantization and its Application to Time-Series Prediction, *IEEE Trans. Neural Networks* 4:4 (1993) 558-568.
16. A. Baraldi and E. Alpayd, Simplified ART: *A new class of ART algorithms*, ICSI TR-98-004, (1998), Berkeley CA.
17. A.W. Moore, Variable resolution dynamic programming: Efficiently learning action maps in multivariate real-valued space. *In Proc. Eight International Machine Learning Workshop*, (1991).
18. V. Gullapalli, A stochastic reinforcement learning algorithm for learning real-valued functions, *Neural Networks* 3, (1990) 671-692.
19. G. Tesauro, TD-Gammon, a self-teaching backgammon program which achieves master-level play. *Neural Computation,* 6(2), (1994) 215-219.

Mean-Field Neural Networks for Antenna Array Diagnostics

G. Castaldi, V. Pierro, I. M. Pinto
D.I^3.E., University of Salerno,
via Ponte Don Melillo, I-84084 Fisciano(SA), ITALY
voice: +39 89 96 4302, fax: +39 89 96 4218
E-mail: pinto@vaxsa.csied.unisa.it

1 Summary

The identification of faulty elements in *large* (hundreds to thousands of elements) antenna arrays, e.g., radiotelescopes, from measurement of the radiation field is an important nasty problem in Applied Electromagnetics [1], [2]. Given the set of (complex) feeding currents and radiator effective heights, the state of a faulty array can be represented by a point in the lattice $\{0,1\}^N$, where N is the number of radiators, and the states "0" and "1" denote a faulty and a working antenna, respectively. One is led to the minimization of a functional, measuring the distance between two $M-$dimensional vectors, representing the measured data, and the fields (intensities) produced by all possible states of the array, respectively. The measured data can be either fields (complex, amplitude and phase), or intensities (real, squared amplitudes), leading respectively to linear and nonlinear minimization problems.

Hence, depending on whether intensity (nonlinear, easier to perform) or field (linear) measures are chosen, Newton [3] or Gauss-Seidel [4] methods can be used to solve the problem. Both are affected by inherent limitations. Convergency to spurious (local) minima for Newton's method, and ill-conditioning for Gauss-Seidel method show up as the distance between the source and the measurement plane is increased.

During the last few years, a number of hopefully more effective tools for (global) minimization problems where many local minima do exist have been developed, including both stochastic (genetic algorithms [5], simulated annealing [6], Boltzmann machines [6]) and deterministic (mean field neural networks [7]) methods.

In this paper we adopt Vidyasagar's [8] (nonlinear) mean field neural network, which can be regarded as a mean-field version of Boltzmann machine, in view of its comparative speed, to solve the antenna array diagnostic problem. An asynchronous, discrete implementation of the (originally continuous) Vidyasagar's mean field neural network is introduced, which allows to define a Lyapounov function, whereby the convergency to a stationary point can be proven. For a suitable choice of the neural threshold function, the equilibrium state is shown to be unique.

The performance of the algorithm is investigated for different measurement weight factors, and varying values of the ratio d/D, d being the distance between the source and the measurement plane, and D the array diameter. It is shown that a judicious choice of the weights leads in a few steps to the correct solution with probability one as $d/D \to 0$.

2 . Antenna Array Diagnostics with Linear and Non-Linear Measures

The following functionals are to be minimized according to whether field or intensity measurements are available:

$$f(\vec{s}) = \sum_{r=1}^{M} \alpha_r \left| \left[\sum_{i=1}^{N} \vec{E}_i(\vec{x}_r) s_i \right] - \vec{\mathcal{E}}(\vec{x}_r) \right|^2 \tag{1}$$

$$f(\vec{s}) = \sum_{r=1}^{M} \alpha_r \left\{ \sum_{i=1,j=1}^{N,N} \vec{E}_i(\vec{x}_r) \cdot \vec{E}_j^*(\vec{x}_r) s_i s_j - \mathcal{I}(\vec{x}_r) \right\}^2 \tag{2}$$

where $\vec{s} = \{s_1, \ldots, s_N\} \in \{0,1\}^N$ represents the (unknown) state of the array; $[\vec{x}_r,\ r = 1, \ldots, M]$ is the set of measurement positions, $\vec{\mathcal{E}}(\vec{x}_r)$ are the (complex, frequency domain) measured field values, $\mathcal{I}(\vec{x}_r)$ are the (real) measured field-intensity values, and $\vec{E}_i(\vec{x}_r)$ are the (complex, frequency domain) fields produced by the $i - th$ radiator. The α_r denote the relative weights of the $r - th$ measurement, and $*$ denotes complex conjugation.

By noting that if $\vec{s} \in \{0,1\}^N$ then $s_i = s_i^q, \forall q \in \aleph$, the functions $f(\vec{s})$ in (1), (2) can be rewritten as multilinear ($2nd$ and $4th$ order, respectively) polynomials in $s_1, \ldots, s_N$ taking on the same values as $f(\vec{s})$ on the points of the lattice $\{0,1\}^N$. These polynomials will be henceforth denoted as $H(\vec{s})$.

In order to find the mimimum of $f(s)$ on $\{0,1\}^N$, one can use the simulated annealing technique, whereby to each transition from one node of $\{0,1\}^N$ to another one associates a (Boltzmann) probability, depending on the associated variation of H, and a control parameter β^{-1} (the temperature), which is decreased during the process according to a suitable schedule. An alternative implementation of simulated annealing is the stochastic Boltzmann-machine neural-network [7]. Simulated annealing, however, can be pretty slow, the associated computational burdern being exponential with the problem size. In order to speed-up the computation, it has been suggested to evolve the statistical average of the state, instead of the state itself, computed using Boltzmann or some other likely suitable distribution law. The resulting mean-field neural-network can be quite faster. Loosely speaking, the above averaging procedure wipes out the many local (irrelevant) minima, leaving only one fixed point, which hopefully corresponds to the sought *global* minimum of $f(s)$ on the state lattice[1].

3 . Neural Network Properties

The non-linear mean field neural network, introduced by M. Vidyasagar [8] and used in this paper is described by:

$$\begin{cases} \frac{du_i}{dt} + u_i = -\frac{\partial H}{\partial s_i}, \\ s_i = g(\beta u_i), \quad i = 1, \ldots, N \quad , \end{cases} \tag{3}$$

where $H(\cdot)$ is the multilinear polynomial defined above, $\vec{u}$ and $\vec{s}$ denote the potential and the state of the net, respectively[2], $g[\cdot]$ is a strictly increasing C^∞ threshold function such that:

$$\lim_{x \to -\infty} g(x) = 0, \quad \lim_{x \to +\infty} g(x) = 1, \quad g(0) = 0.5, \tag{4}$$

and β a (real, nonnegative) parameter. In view of (4), one has $\vec{s} \in [0,1]^N$, whereas for any *finite* β $\vec{s} \in]0,1[^N$

As a consequence of La Salle and Cohen-Grossberg theorems [9], the system (3) can be shown to be *totally stable*, i.e., every solution of (3) evolves into a fixed point $\in]0,1[^N$ [7].

We shall now prove that (3) has an *unique* fixed point in $]0,1[^N$, under suitable assumptions on $g(\cdot)$.

The fixed points of (3) are defined by:

$$s_i - g\left[-\beta \frac{\partial H}{\partial s_i}\right] = 0. \tag{5}$$

and are readily seen to correspond to the *stationary* points of the Lyapounov function:

$$\Lambda(\vec{s}) = H(\vec{s}) + \sum_{i=1}^{N} \beta^{-1} \int_0^{s_i} g^{-1}(v)\, dv. \tag{6}$$

[1]It should be noted that while the annealing procedure and its stochastic Boltzmann-machine implementation can be proven to converge with probability one to the global minimum in the limit of an infinite number of transitions, provided the temperature is decreased no faster than logarithmically [6], a similar convergency theorem for the mean-field neural network does not exist, to the best of our knowledge. Convergency (in probability) is however observed in a wide variety of applications [8], including the examples discussed below.

[2]Note that the s_i in (3) can take *any* value in $[0,1]$.

where $\nabla\Lambda = 0$, which are further the *only* fixed points of (3).
Using the standard tools from ODE theory, it is possible to prove [10] that:

Lemma I The solutions of the differential system (3) converge to the local *minima* of Λ, but for a set of initial conditions of measure zero. •

We can further readily prove that the minima of Λ are located on $\{0,1\}^N$ for $\beta \to \infty$, where they coincide with the minima of f, using the following lemmas:

Lemma II The minima of H in $[0,1]^N$ are located on $\{0,1\}^N$, where by construction $H(\vec{s}) = f(\vec{s})$.

Proof -This follows trivially from the fact that $\forall i$ the function

$$H\left(s_i, \left\{s_j = s_j^{(eq.)}, j \neq i\right\}\right)$$

is linear in s_i, and thus its second derivative vanishes identically. Hence by Weierstrass theorem it takes its minimum in $[0,1]$ on the boundary $\{0,1\}$.•

Finally, as a trivial consequence of (6),

Lemma III The minima of Λ and H in $[0,1]^N$ coincide for $\beta \to \infty$. •

It can be concluded that for *almost* every initial condition, the system (3) evolves in the limit $\beta \to \infty$ to a minimum of $f(s) \in \{0,1\}^N$.

We can prove that, under suitable assumptions for $g(\cdot)$ only a single (global) minimum exists.
Indeed, for $\beta = 0$ the system (3) has the unique solution $s_i = 0.5$, the Jacobian of the transformation (5) is equal to the matrix identity, and hence **(a)** the associated Leray-Schauder's index i_{LS} is equal to one [11]. Further, the operators:

$$\epsilon_i(\beta, \vec{s}) = s_i - g\left[-\beta \frac{\partial H}{\partial s_i}\right], \tag{7}$$

make up an homotopy on $\{0,1\}^N$. Homotopic invariance implies **(b)**:

$$i_{LS}\left\{\vec{\epsilon}(\beta, \vec{s}),\ \{0,1\}^N\right\} = i_{LS}\left\{\vec{\epsilon}(0, \vec{s}),\ \{0,1\}^N\right\} = 1 \tag{8}$$

On the other hand, a well known theorem due to Cronin (1953) and Schwartz (1963) states that if $f : \bar{A} \subseteq X \to X$ is a compact operator on a (complex)

Banach space A, then **(c)** the number of its fixed point cannot exceed $i_{LS}(f, A)$.

Thus, if (and not only if !) $g(\cdot)$ is an analytic function such that $\vec{\epsilon} \neq 0$ in a neighbourhood of C^N infinity, then the analytic continuation of the mapping (3) fulfills the hypotheses of the above theorem, and **(a)**, **(b)** and **(c)** imply that (3) has a unique fixed point $\vec{s}^{(eq.)}$ in $[0,1]^N$.

4 . Discrete Implementation

We introduce the possibly simplest discrete numerical implementation of (3), viz.:

$$\begin{cases} u_i^{(n+1)} = (1-\Delta)u_i^{(n)} - \Delta \left(\dfrac{\partial H}{\partial s_i}\right)_{\vec{s}=\vec{s}^{(n)}} \\ s_i^{(n)} = g[\beta u_i^{(n)}], \quad i = 1, \ldots, N. \end{cases} \tag{9}$$

The above discretization is *consistent* in the sense that one can immediately prove **(a)** that *any* fixed point of (9) is also a fixed point of (3). We shall now prove **(b)** that the *asynchronous* implementation of (9) is totally stable, and thus necessarily evolves into the unique fixed point of (3).

Lemma - The trajectories of the discrete neural network (9) are bounded.

Proof - Let $r^{(n)} = |u^{(n)}|^2$. Then, the gradient of H being bounded by M,

$$r^{(n+1)} - r^{(n)} \leq \sum_{i=1}^{N} \Delta(2-\Delta)[u_i^{(n)}]^2 + 2M\Delta|1-\Delta||u_i^{(n)}| + M^2\Delta^2 \tag{10}$$

If $0 < \Delta < 2$, all solutions are contained in a ball of radius $2\Delta(2-\Delta)^{-1}M$, the l.h.s. of (10) being negative outside. This is just a simple (discrete) version of Cohen-Grossberg theorem.•

Lemma -. For any $\phi : R \rightarrow R$ continuous and non decreasing one has:

$$\phi(s_i)(s_{i+1} - s_i) \leq \int_{s_i}^{s_{i+1}} \phi(s)ds \leq \phi(s_{i+1})(s_{i+1} - s_i). \tag{11}$$

Proof -Equation (11) follows trivially from the hypothesis using the integral mean value theorem.•

We are now in a position to prove that:

Theorem -. The asynchronous implementation of (9) is totally stable.

Proof -. We define a Lyapunov's function Λ by analogy with the continuous case. The variation of Λ after updating the $i-th$ state only is:

$$\Lambda(\vec{s}^{\,(n+1)})-\Lambda(\vec{s}^{\,(n)}) = \left(\frac{\partial H}{\partial s_i}\right)_{\vec{s}=\vec{s}^{\,(n)}} (s_i^{(n+1)}-s_i^{(n)})+\frac{1}{\beta}\int_{s_i^{(n)}}^{s_i^{(n+1)}} g^{-1}(s)ds. \quad (12)$$

From the previous lemma and Bolzano's theorem, $\exists t \in [0,1]$:

$$\int_{s_i^{(n)}}^{s_i^{(n+1)}} g^{-1}(s)ds = (1-t)(s_i^{(n+1)} - s_i^{(n)})g^{-1}(s_i^{(n)})+$$

$$+t(s_i^{(n+1)} - s_i^{(n)})g^{-1}(s_i^{(n+1)}), \quad (13)$$

whence:

$$\Lambda(\vec{s}^{\,(n+1)}) - \Lambda(\vec{s}^{\,(n)}) = \left(t - \Delta^{-1}\right)\left(u_i^{(n+1)} - u_i^{(n)}\right)\left(s_i^{(n+1)} - s_i^{(n)}\right). \quad (14)$$

In view of the monotonic dependance between u and s, the Lyapunov's function decreases iff $t-\Delta^{-1} \leq 0$, i.e., iff Δ belongs to $(0,1)$. Furthermore, (14) implies:

$$-\Delta^{-1}\left(u_i^{(n+1)} - u_i^{(n)}\right)\left(s_i^{(n+1)} - s_i^{(n)}\right) \leq$$

$$\leq \Lambda(\vec{s}^{\,(n+1)}) - \Lambda(\vec{s}^{\,(n)}) \leq$$

$$\leq \frac{\Delta - 1}{\Delta}\left(u_i^{(n+1)} - u_i^{(n)}\right)\left(s_i^{(n+1)} - s_i^{(n)}\right). \quad (15)$$

Thus, in view of equations (10) and (15), La-Salle theorem ensures that the asynchronous map (9) has at least one fixed point. •

5 . Implementation Hints

In the following we touch briefly the problem of the choice of the parameters β and $\{\alpha_r,\}$. Following [13], where the general problem of determining the transition range $(\beta_{min}, \beta_{max})$ which corresponds to the onset and completion of the migration of the sought fixed point of (3) from $s_i \approx 0.5$ to $\{0,1\}^N$ is considered, we define for the linear (field measurement) case:

$$\beta_{min} = 2max_i\left(\sum_{j=1}^{N}\left|\frac{\partial^2 H}{\partial s_i \partial s_j}\right|\right)^{-1} \quad (16)$$

$$\beta_{max} = 2max_{i \neq j} \left(\left| \frac{\partial^2 H}{\partial s_i \partial s_j} \right| \right)^{-1} . \tag{17}$$

Instead of seeking an optimum schedule for updating β in the above interval, we found that the computationally cheapest strategy consisting in determining first a tentative solution using β_{min}, and then refining the solution using β_{max}, works pretty well [3].

A further degree of freedom in the algorithm is related to the choice of the weight factors α_r in (1), (2). Two main alternative philosophies can be envisaged, namely measurement normalization, viz.:

$$\alpha_r = |\vec{\mathcal{E}}(\vec{x}_r)|^{-2} \tag{18}$$

$$\alpha_r = \mathcal{I}(\vec{x}_r)^{-2} \tag{19}$$

(to be used in (1) and (2) respectively) and operator normalization, viz.:

$$\alpha_r = \left[\sum_{i=1}^{N} |\vec{E}_i(\vec{x}_r)|^2 \right]^{-1} \tag{20}$$

$$\alpha_r = \left[\sum_{i=1,,j=1}^{N} |\vec{E}_i(\vec{x}_r) \cdot \vec{E}_j^*(\vec{x}_r)|^2 \right]^{-1} , \tag{21}$$

(to be used in (1) and (2) respectively). Preliminary numerical experiments showed that measurement normalization performs rather poorly. All results presented in the next section refer accordingly to operator normalization.

6 . Numerical Results

In numerical simulations, we used the discrete implementation of the neural network, proposed above. Sample numerical results are shown in *Fig.s 1* to *4*, for (complex) field measurements. Typical running times on a PentiumPro200 desktop workstation are shown in *Fig. 1.* It is found that the computational burden goes much as the cube of the problem size N.

For an array of N antennas, a minimum number of N independent measurements is obviously required. In practical situations, it is expedient to introduce some redundancy. In our numerical experiments we used a redundancy factor of four, i.e., $2N$ (complex) field measurements in (1). We expect our diagnostic

[3]We can proceed by analogy also for the nonlinear case, where the Hessian components depend on $\vec{s}$, by using $s_i = 0.5$ in (16), and $s_i = 0$ in (17) [13].

algorithm to fail if the number of radiators N in the array exceeds the number N_c of degrees of freedom of the measured field [12]. One has:

$$N_c = \frac{2}{\pi}\,(k_0 R_a)\,\frac{R_m}{d} \tag{22}$$

where R_a and R_m are the (minimum) radii of the antenna and measurement domains, d is the (minimum) distance between the two, and k_0 the free-space wavenumber. In our numerical experiments we took $R_a = R_m = D/2$ and assumed $(\lambda/2)-$equispaced antennas, viz., $D = N\lambda/2$. Hence:

$$\frac{N}{N_c} = \frac{2d}{D} \tag{23}$$

The typical dependence of the success rate from $2d/D$ is displayed in *Fig. 2*, for the operator normalization choice (20). The success rate drops down steeply to zero, as expected, as soon as $2d/D \approx 1$, i.e., $N \approx N_c$, while keeping essentially equal to one for $2d/D < 1$.

The effects of (random, uniform) errors in the radiators' positions, and of some (random, gaussian) noise in the field measurement are shown in *Fig.s 3* and *4*, respectively.

It is concluded that the proposed algorithm is fairly reliable and robust.

7 . Hints for Future Research

We are presently investigating the issue of parallelization of (9) through a study of the properties of the corresponding synchronous implementation. Preliminary results based on the parallel processing facilities at CNR-IRSIP look promising.

This work has been done in partial fulfillment of G. Castaldi Ph.D. programme, and has been sponsored in part by MURST.

References

[1] L. Gattoufi et al., "Matrix Method for Near-Field Diagnostic Technique of Phased Array", IEEE Int.l Symp. on Phased Array Systems and Technology (Boston, 1996), pp. 52-57.

[2] R.G. Yaccarino, Y. Rahmat-Samii, "Microwave Antenna Imaging and Phaseless Reconstruction", Int.l Journal of Imaging Systems and Technol., **199**, pp. 396-406, 1997.

[3] A.B. Demidovich, *Numerical Methods*, MIR, Moscow, 1976.

[4] A. Burlisch, B. Stohr, *Numerical Analysis*, Springer, 1979.

[5] D.B. Fogel, "An Introduction to Simulated Evolutionary Optimization", IEEE Trans. on Neural Networks, **NN-5**, pp. 3-14, 1994.

[6] E.H.L. Aarts and J. Korst, *Simulated Annealing and Boltzmann Machines*, Wiley, 1990.

[7] S. Haykin, *Neural Networks*, Mac Millan, 1994.

[8] M. Vidyasagar, "Minimum-Seeking Property of Analog Neural Network with Multilinear Objective Functions", IEEE Trans. on Automatic Control, **AC-40**, pp. 1359-1375, 1995.

[9] M. Cohen, S. Grossberg, "Absolute Stability of Global Pattern Formation and Parallel Memory Storage by Competitive Neural Networks" vol. **SMC-13**, 1983.

[10] J.J. Hopfield, "Neurons with Graded Response have Collective Computational Properties like Those of Ywo-State Neurons", Proc. Nat.l Acad. Sci. USA, **81**, 3088-3092, 1984.

[11] E. Zeidler, *Nonlinear Functional Analysis and its Applications - I*, Springer-Verlag, 1986.

[12] O.M. Bucci and G. Franceschetti, "On the Degrees of Freedom of Scattered Fields", IEEE Trans. on Antennas Propagat., **AP-37**, 918-926, 1989.

[13] B.W.Lee and B.J.Sheu, "Paralleled Hardware Annealing for Optimal Solutions on Electronic Neural Network" IEEE Trans. on Neural Network, **NN-4**, pp. 588-598, 1993.

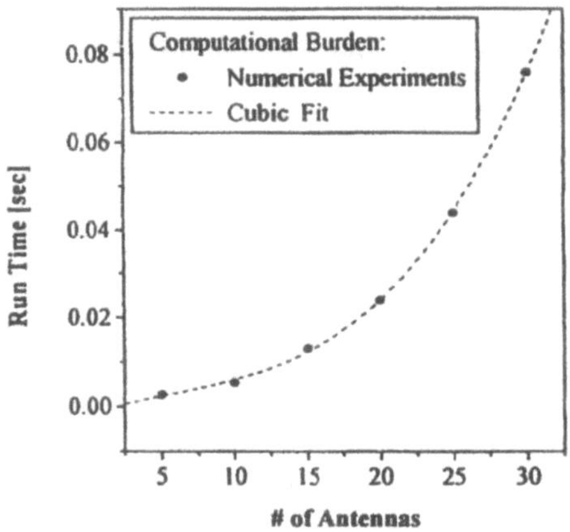

Fig. 1 - Computational burden.

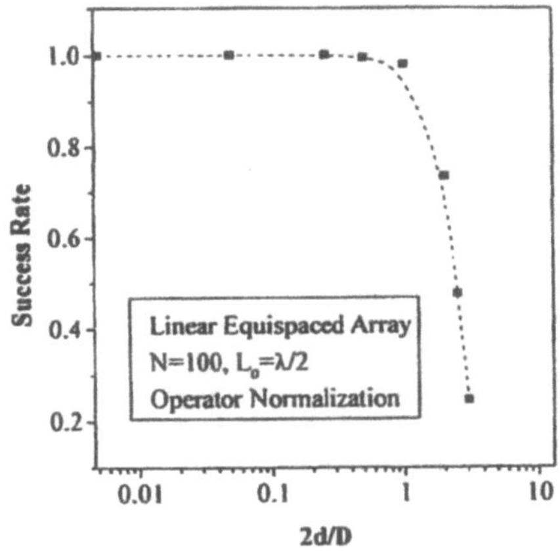

Fig. 2 - Success rate vs 2d/D.

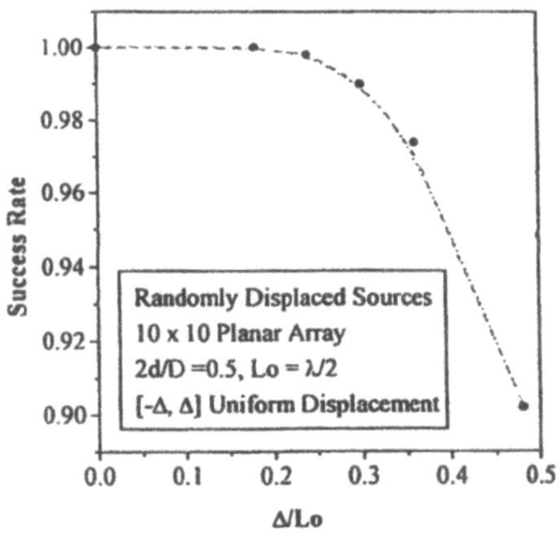

Fig. 3 - Success rate – randomly displaced sources.

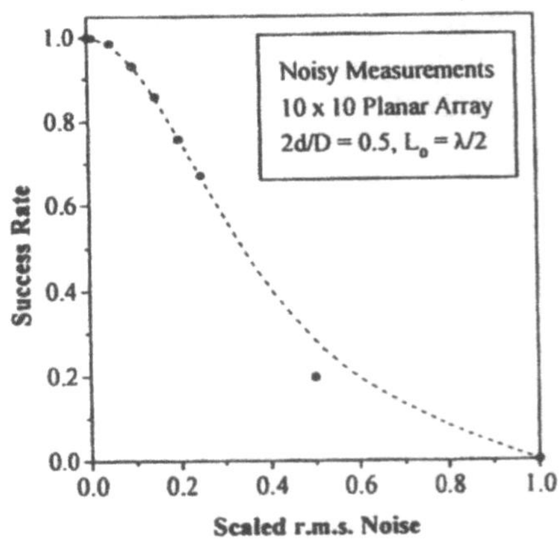

Fig. 4 - Success rate – noisy measurements.

Initial Letter Spotting as a Complementary Feature for Lexical Filtering of Cursive Words

Anna Maria Colla * – Andrea Lorenzon ◇
* Elsag Bailey - Un'Azienda Finmeccanica S.p.A.
Via G. Puccini, 2 - 16154 Genova (ITALY)
◇ Dip. Informatica e Scienze dell'Informazione, Università di Genova
Via Dodecaneso, 33 - 16146 Genova (ITALY)

Abstract
This paper describes a method for lexical filtering to be employed mainly (but not exclusively) in cursive word recognition. This method operates at word level by selecting a subset of the original vocabulary based on a simple and robust characterization of the image of the unknown word. The features adopted for word representation are basically morphological (ascenders, descenders, loops) or structural (hypothesized number of letters). In this work we have added another feature to improve the efficiency of the lexical filter, namely the recognition of the initial letter. In order to classify the initial letter reliably, without segmentation or ad-hoc training, we applied an MLP-based character spotting method.

1 Introduction

Cursive handwriting recognition is generally deemed much more complex than handprinted recognition, as cursive bears more distinctive personal marks and is influenced in greater way by writing speed, mood, type of writing device (pen, pencil), quality of paper, etc. State-of-the-art off-line cursive recognizers are of two main types: "holistic" word recognizers and recognizers based on character classification. The latter have a traditional structure, similar to the one for handprinted or even typed writing: pre-processing includes line segmentation into characters or graphemes, which are separately size normalized, deslanted, variously characterized by structural or morphological features (either global or local), and finally classified, often by applying some "elastic matching".

When the lexicon pertaining to the given application is known, such as for instance in postal applications, the recognizer efficiency can be greatly enhanced by a previous reduction of the search space in the vocabulary, by applying rules about context or a "lexical filter" [1,2]. The latter is a method working at word level which selects a subset of the original vocabulary based on some coarse measurement or characterization of the image of the unknown word. The features adopted to represent words must be simple to extract and robust enough to be meaningful over different calligraphic styles. Building on the methods developed and the results obtained in [7] new features are examined and tested to enrich the handwriting representation in order to implement a more powerful and efficient lexical filter. Among such features, the literature suggests that the initial letter can be a very efficient hint, provided it is reliably recognized (see [3] and references). However, the recognition of the initial letter cannot be reliably accomplished without a good character recognizer. In our proposed test, where the word images pertain to city names, the initial letter is always uppercase and can be either really cursive, or, quite often, handprinted. In order to classify the initials, we tried the MLP-based character spotting method described in [6] which is compatible with objectives and methodology of lexical filters.

In the following we describe the general methodology to implement a lexical filter, the adopted morphological and structural features, and the newly added feature consisting of an attempt to recognize the initial letter. Finally we present the preliminary results obtained on a small subset of CEDAR database.

2 A general methodology for a Lexical Filter

A Lexical Filter is a word pre-classifier working on a set of features which characterize a single word either morphologically (in detail) or structurally (as a whole). A Lexical Filter allows a partial categorization of the input image, by discarding all the words in the full vocabulary which do not exhibit such features.

There are two components in a Lexical Filter: **feature extraction** and **lexical representation** [8]. The former operates on the image of the unknown word and tries to characterize it, the latter is applied to derive a symbolic representation of the original lexicon based on the set of extracted features (described in next section).

The unknown word is represented by an ordered list of the features discovered in the input image (from left to right, for growing abscissae). We built a symbolic representation for each word in the lexicon as the sequence of representations of each letter in the word. Each letter is in turn represented by a grammar in BNF form. In such grammar the features are the terminal symbols, represented as follows:

a: `ascender`
d: `descender`
0: `loop located between upper and lower baseline`
1: `loop located above the upper baseline`
2: `loop located below the baseline`
3: `loop located above the baseline`
4: `loop located below the upper baseline`

Each letter can have several alternative representations according to the peculiar calligraphic style of the writer. For instance, cursive letters **a**, **b** and **f** can be represented as follows:

```
a  ::=  0
b  ::=  a;0 | 0;a | 3
f  ::=  4;3 | 3;4 | 3;d | d;3 | 4;a | a;4 | d;a | a;d | 2;3 | 3;2
```

To obtain the list of candidate words, the ordered sequence of features extracted from the image of the unknown input word is compared by a typical Dynamic Programming scheme with the "symbolic" feature vector representing each word in the lexicon. The adopted method is the so-called **Trace Method** [4]: each element (i.e. feature) can correspond to *at most* one element (no feature in the input word can be cancelled), and no swap is permitted.
A list of word hypotheses is issued. Each hypothesis can be scored with the "confidence" of the extracted features to improve efficiency and reliability.

3 The Features

In our work a word is a binary image represented by a set of structural and morphological features derived from the literature. It has been shown [3] that a combination of seven features, partly structural and partly morphological, can reduce the dimension of the list of candidate words to an average of 4 words for over 50% of a 15,000 word lexicon. Some of these features are not easy to extract in data as noisy as the ones we tested, especially as we wanted to avoid pre-segmentation into "characters". Therefore we initially started with a subset consisting of the simplest structural feature (hypothesized number of letters in the word) and the most straightforward morphological features, namely ascenders and descenders. Another type of morphological features, the loops [1], were added. Finally, we tried to classify the initial letter without segmenting it, by adopting the MLP-based character spotting method described in [6]. More details are supplied in the following.

3.1 Morphological Features

In our approach, where there is no explicit segmentation into characters, the word is the basic unit, and pre-processing comprises noise filtering and baseline straightening. Noise filtering consists of the following: point-like noise filtering, removal of fixed structures, contour smoothing, filling or reconstruction of small areas impaired by the image acquisition process. Baseline straightening is performed by modelling the baseline by one or more contiguous line segments. No further pre-processing is applied to the straightened word, and feature extraction is performed.
Morphological features are derived by analyzing the shape of the writing locally. The morphological features we adopted are ascenders, descenders and internal loops.

Ascender: approximately vertical "thin" and "long"[1] stroke in the upper part of a character lying above the upper ruler line [2].

[1] **"Thin":** with respect to the average thickness of writing. **"Long"** with respect to the average vertical size of words.

[2] **Ruler-Lines**: the **lower** ruler line or **baseline** is the line ideally subtended to the lower profile of the written word or line; the **upper** ruler line symmetrically is the line ideally subtended to the upper profile (lower and upper profile are drawn taking only the characters with no ascenders and no descenders into account). *See an example in Figure 1.*

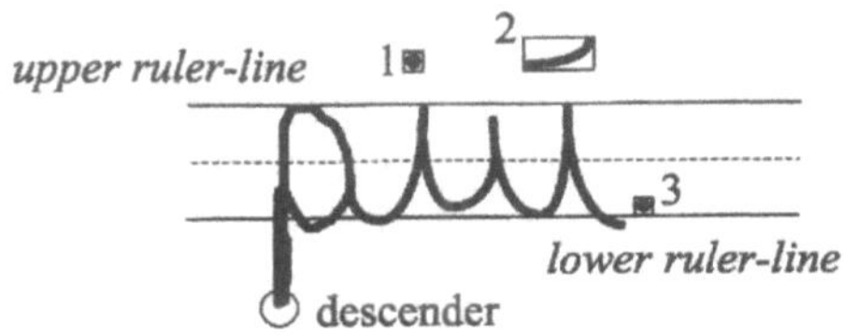

Figure 1: *Word image with estimated baseline and upper ruler line.*

Descender: approximately vertical "thin" and "long" stroke in the lower part of a character lying below the lower ruler line [2] .
Loop: connected component composed of black pixels surrounding a void (white) area. An **Internal Loop** is characterized by its position with respect to the lower and upper rules lines.

As such features may or may not be located within the word image, the dimension of the feature vector is typically variable.

3.2 Structural Features: Character Range

The first step of lexical filtering is performed by rejecting all the words in the lexicon which are too short (too few letters) or too long (too many letters) with respect to the input word. The range of word dimension in letters is computed according to the following rules:

- the **lower limit** is the maximum among:
  - the number of *meaningful* connected components (diacritic signs **(1)** are not taken into account, as well as small components (noise) or components with specific height-to-width ratios, i.e. accents **(2)** and punctuation marks **(3)**: see Figure 1);
  - a fraction of the total number of extracted morphological features;
  - a function of the total number of profile local minima and maxima;
- the **upper limit** is a fraction of the total number of profile local minima and maxima augmented by a constant.

3.3 A Complementary Feature: Initial Letters

An hypothesis about the initial letter is applied after the letter range to further restrict the list of candidate words. The morphological features are applied last. The initial letter of the unknown word is tentatively classified by an adaptation of the character spotting procedure described in [6]. Character spotting is performed by an MLP-based classifier operating on a fixed size sliding window scanning horizontally the normalized word image. The leftmost portion of the image, presumably containing the initial (capital) letter, is normalized to the vertical size of the sliding window in a linear way while keeping its original aspect ratio. Scanning is carried out exhaustively, i.e. the window shifts with

Figure 2: *Example of morphological feature extraction and initial letter spotting.*

a regular step within the current image portion (Figure 2).
At each window position a 26-class MLP classifier [5], trained on *uppercase handprint*, issues an hypothesis. The sequence of MLP outputs is composed according to majority rules. This allows to reject poor classification hypotheses, hence to apply this complementary feature only when it is reliable.

4 Experimental results

Test images were derived from CEDAR US city names database. This database consists of **cursive** or **mixed cursive** single words, with a rather low average image quality, written by several writers, originated from mail processed by Buffalo post office. Image acquisition was performed by an Eikonix EC850 CCD scanner, at 300 dpi resolution in 8-bit greyscale (however, we are currently working on binarized images).
As we had only a net trained for handprinted uppercase letters available, preliminary tests were performed on a small subset of CEDAR database (119 images). The results are summarized in Table 1, where we compare the percentual lexicon reduction obtained by exploiting only the features described in sections 3.1 and 3.2 (second column) and by adding also the initial letter (third column). The figures represent the number of cases for which the lexical filter produced a list of candidate words containing less than a given fraction of the whole lexicon (601 entries). With the initial letter hypothesis, the output was unique for 7 words, and at most 5 words for 37 cases. Morphological features

Candidate list dimension	Without initial letter spotting	With initial letter spotting
< 5%	19	96
< 10%	29	102
< 30%	64	—
< 50%	102	—
OUT	13 *(⋆)*	17 *(7 from ⋆)*

Table 1: *Experimental results on lexicon reduction.*

and initials produced complementary errors, respectively 13 and 10.

5 Conclusions

In this paper we presented a lexical filter for cursive words based on simple features for word characterization. A further feature, the initial letter classified by MLP-based spotting, appears capable to produce a noticeable improvement in the results obtained by just employing the basic morphological and structural features. Although the results presented refer to a very small test, they are quite encouraging. By improving the features used for character representation and by further training the MLP net on a larger set of examples, including cursive letters, we expect to be able to process a larger number of cases. Exploiting complementarity in errors should reduce them. Such improvements are under way and will be tested in full size experiments.

References

[1] Leroy A. Progressive Lexicon Reduction for On-Line Handwriting. Progress in Handwriting Recognition, pp. 399–406, World Scientific, 1997

[2] Madhvanath S. and Govindaraju V. Holistic Lexicon Reduction. IWFHR, pp. 71–81, 1993

[3] Bellaby GJ., Evett LJ. and Powalka RK. Coping with Ambiguity and Error in Cursive Script Recognition. Progress in Handwriting Recognition, pp. 337–344, World Scientific, 1997

[4] Hall PAV. and Dowling GR. Approximate string matching. Computing Surveys 12 (4), pp. 381–402, 1980

[5] Colla AM. and Pedrazzi P. Single and Coupled Neural Handprinted Character Classifiers. In: M. Marinaro and P.G. Morasso (Ed.s), ICANN 94 - Proc. of the Int. Conf. on ARTIFICIAL NEURAL NETWORKS. Sorrento (Italy), vol. II, pp. 969–972, Springer-Verlag, 1994

[6] Camastra F., Cepollina E. and Colla AM. Word Recognition by MLP-based Character Spotting and Dynamic Programming. In M. Marinaro and R. Tagliaferri (Ed.s), Neural Nets. WIRN VIETRI – 97, Proc. 9th Italian Workshop on Neural Nets, Vietri s.M., Italy, 22–24 May 1997, pp. 183–188, Springer-Verlag, 1998

[7] Lorenzon A. Sviluppo di un filtro lessicale per riconoscitori di parole. Tesi di Laurea in Informatica (M.Sci.), Università di Genova, Dec. 1997

[8] Colla AM. and Lorenzon A. A simple lexical filter for two-step recognition of cursive words *(in preparation)*

Visual Person Localization with Dynamic Neural Fields: Towards a Gesture Recognition System

Andrea Corradini*, Ulf-Dietrich Braumann, Anja Brakensiek**, Markus Krabbes**, Hans-Joachim Boehme, Horst-Michael Gross
Department of Neuroinformatics, Technical University of Ilmenau
D-98684 Ilmenau, Fed. Rep. of Germany
{andreac,ulf}@informatik.tu-ilmenau.de

Abstract

For any *visually-based interaction* between persons and acting systems within a real-world environment the *localization* of a user by the system is a necessary condition. The presented work deals with this visual localization problem of a user concretely referred to the *autonomous mobile robot system MILVA* of our department. Since this system is applied under real-world conditions especially for the localization some proper techniques are needed which have an adequate robustness. In our opinion, this requires the combination of several components of saliency towards a multi-cue approach, consisting of structure- and color-based features [2].

This paper introduces one of them: the localization based on a typical *shape of contour*. A simple contour shape prototype model consists of an arrangement of oriented filters doing a piecewise approximation of the upper shape (head, shoulder) of a frontally aligned person. Applying such filter arrangement in a multiresolution manner, this leads to a robust localization of frontally aligned persons even in depth. The central problem of *selecting* the most promising (salient) image region is treated by means of a *three-dimensional dynamic neural field* performing a dynamic winner-take-all process (WTA, [1, 6]).

After a successful localization of a person one can start a more detailed analysis of the *gesture's* meaning: besides the recognition of static gestures we also concentrate on the acquisition and later the recognition of *dynamic gestures*.

1 Introduction

The autonomous mobile robot system MILVA of our department serves not only as an experimental platform for investigations of autonomous robot navigation in real-world environments, but also for work concerning gesture-based interaction between user and robot. It's getting more and more important for both navigation and interaction to tolerate a rising level of complexity of the surroundings the robot system is operating in. The goal should be to operate even without special preparations so that e. g. no remarkable restrictions should appear concerning lighting conditions etc. Our *superior goal* is that MILVA *develops an intelligent behavior* both concerning navigation and interaction using

*supported by: TMR Marie Curie EU Fellowship # ERB FMBI CT 97 2613
**supported by: Thuringian Ministry of Science, Research and Culture: GESTIK-Project

an architecture consisting of elements with neural or neurally inspired mechanisms.

MILVA can develop an active and therewith intelligent detail analysis (attention-based perception, e. g. for the analysis of a gesture) only if there are a priori proper saliency mechanisms available enabling a preselection of a visual image. Consequently, for the interaction with persons we need saliency mechanisms to localize them.

After the introduction of one of the used saliency mechanisms and its selection process with dynamic neural fields we show what follows the localization of a person: a detailed visual analysis of that person. Besides the recognition of static gestures [2] we also concentrate on dynamic gestures. So, the second part of the paper introduces our present results of segmenting regions of motion in visual images by means of the Poggio-Reichardt motion computation model and a following dynamic neural WTA process.

2 Arrangements of Oriented Filters for Feature Extraction

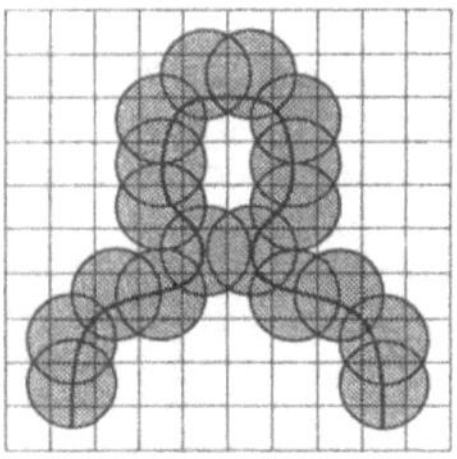

Figure 1: *Schematic sketch of the arrangement of oriented filters.*

This paper introduces a contour shape based approach for saliency based on some physiological considerations as well as on psychophysical effects. Recently, Oren [9] independently developed an interesting similar approach. While our idea refers just to the shape of head and shoulder, his approach considers the complete body of pedestrians.

The visual cortex consists in several parts of cells with oriented receptive fields. Thus, referred to a retinal position broad ranges of the frequency space are covered by a set of oriented filters. A lot of investigations [5, 7] have exposed that there is a high similarity between the profiles of the mentioned receptive fields and two-dimensional Gabor-functions. So, local operations decomposing the visual information with respect to frequency space are made. Psychophysical aspects as good continuation or symmetry (both belonging to the Gestalt laws, [8]) obviously describe effects which necessitate grouping mechanisms. Against this background we conceptualized the approach of *arrangements* of oriented filters. While fig. 1 shows the positions of the filters it does not explicitly show the respective directions. What is important is that we only use four orientations. Since we do not want to discriminate an orientation in the range of $[0, \pi)$ we should cover 45°-sectors of the frequency space. That's why the filters belonging to the arrangement do have exactly that orientation which is the most prominent caused by the contour shape within the filter range[1]. The arrangement is applied by determining

[1]The positioning and orientation selection of the filters is in general an optimization problem. However, since we use a rather coarse discrete grid and very few different orientations it is not as complex so that one would be forced to use a special algorithm. If one defines an energy function there would be on principle no reason against an adaptive neural algorithm.

whether the dominating orientation at each coordinate fits the dominating orientation of a prototype shape. As distance measure we use simply the relative Hamming distance.

Gabor filters are ideal if one considers their sharpness both in position and frequency space, but have high computational costs. There is a way providing a good compromise between computational cost and exactness. This is the approach of inertia tensors [4]. The direction of such tensor corresponds to that of the most dominant oriented visual structure within a certain range. The computational costs are magnitudes less than those of the Gabor filters.

3 Dynamic Neural Fields for Selection

To achieve a good localization a selection mechanism is needed to make a definite choice. This is not limited to a two-dimensional position. Since we use five fine to coarse resolutions we actually can localize persons even in different distances. Therefore, a neural field for selecting the most salient region should be three-dimensional.

What do we understand by dynamic neural fields? Those fields can be described as recurrent nonlinear dynamic systems. Regarding to the selection task we need a dynamic behavior which leads to one local region of active neurons successfully competing against the others, i. e. the formation of one single blob of active neurons as an equilibrium state of the field. The following equation describes the system:

$$\tau \frac{d}{dt} z(\vec{r},t) = -z(\vec{r},t) - c_h h(t) + c_i x(\vec{r},t) + c_l \int_N w(\vec{r}-\vec{r}\,')y(\vec{r}\,',t) d^3\vec{r}\,' \quad (1)$$

Herein $\vec{r}$ denotes the three-dimensional coordinate of a neuron position in the field, $z(\vec{r},t)$ is the activation of a neuron $\vec{r}$ at time t, $y(\vec{r},t)$ is the output activity of this neuron computed as a sigmoidal function, $x(\vec{r},t)$ denotes the external input, $h(t)$ is the global inhibition at time t gathering the activity from each neuron over the entire field and $w(\vec{r}-\vec{r}\,')$ denotes the Mexican-hat function of lateral activation of neuron $\vec{r}$ from the surrounding neighborhood $N \subseteq \mathbb{R}^3$. The constants c_h, c_l and c_i represent parameters of the system.

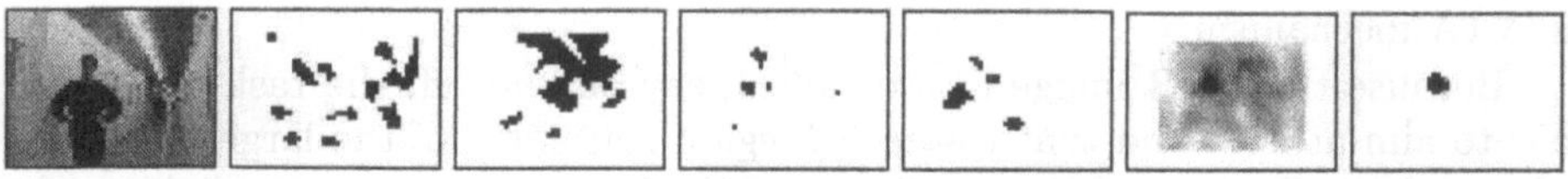

Figure 2: *These pictures illustrate the localization results in an indoor environment of one pyramidal plane (48 × 36 pixels). From left to right: input image, orientation filtering (0°, 45°, 90° and 135°), filter arrangement result, selection within a 3D field of dynamic neurons.*

The results of the system shall be qualitatively illustrated in fig. 2. All the images show the state of the system in a snapshot at that moment when the activity change Δy of the most active neuron became less than 1%. On average, the system takes 11 steps. The expansion of the blob is not restricted to one plane. To give a more precise specification of the distance of a person one should determine the center of the blob and interpolate the distance.

Our presented results are exemplary, the usage of the shape of contour provides one solution for the person localization problem, even under quite different conditions. The novel approach with a three-dimensional dynamic neural field can be assessed as robust method for the selection process.

4 Motion-based segmentation

Our goal is to build a motion-based recognition framework which can detect motion from a sequence of monocular images of a scene in which objects are in motion.

The knowledge of research in neurophysiology and psychophysics has influenced the design of vision systems particularly in recognition, interpretation and description of motion from a time-varying image sequence. In our paper we present an approach for the recognition of motion from image sequences modeling biological vision. Many psychophysical experiments have shown that the human visual system is able to detect, localize and isolate a moving object against a surround on the basis of motion information alone. The relative movement between an object and the ground permits both to identify its boundaries and to detect its presence [10].

Based on behavioral experiments and neurophysiological knowledge of the visual system of the fly *Musca domestica*, Poggio and Reichardt [10] proposed a neural circuitry for the figure-ground discrimination by relative motion. In our work we used a neural circuitry with forward shunting inhibition like the Poggio-Reichardt's one adapted for monocular images. We computed the motion region $R(t)$ between each sequential pair of images at time t and $t+1$ of a certain interval and then we summed over t in order to obtain an cumulative image named binary motion region BMR [3]. It describes those spatial regions within the sequence where motion occurred starting from the beginning of the recognition task.

In order to remove the noisy regions from the detected ones we then applied to the BMR image a process of extraction of relevant regions by a neurally-plausible, iterative competition/cooperation system of artificial neurons; a WTA mechanism.

Because the BMR image is two-valued, say on and off, the task the WTA has to aim at is to find that *connected* region consisting of the largest number of on-values. Because of the conditions of the images to process and the goals to aim at, selecting the actually largest region within the image cannot be accomplished by the neural scheme employed for the localization task[2] but only by a different approach.

This another WTA circuit, referred to as *largest region WTA* (LRWTA),

[2]The dynamic neural field introduced in the previous section cannot treat this problem because the blobs (fig. 2) cannot exceed some certain size (equilibrium) but the motion regions we want to segment now might be quite different depending on the kind of motion. Moreover we need a different approach dealing with binary inputs.

is composed of a single layer of N excitatory neurons of identical type each corresponding with one single pixel of the image. The neurons are mutually connected within their neighborhood defined by excitatory synaptic strength depending on the definition of connected region[3]. We consider pixels as neighbored if they have a common edge, i. e. we used a 4-neighborhood.

In order to find the actually largest region within the image each neuron of the correspondent neural field has to collect the excitatory activity of its neighboring neurons. For the synaptic strength of each neuron we chose a $n \times n$ separable kernel satisfying the property of symmetry, separability and unimodality. To avoid attenuation or amplification the sum of the elements of the kernel is required to be 1. Because a 4-neighborhood cannot be defined through a separable kernel we circumvented this problem with the introduction of a proper function $f(\cdot)$ of the activity. For a 3×3 separable kernel originated by $\vec{w} = [a, b, a]$ and under the condition $4a < b$ we defined it as follows:

$$f(z_i(t)) = \begin{cases} 0: & z_i(t) \leq 4a^2 + b^2 \\ z_i(t): & z_i(t) > 4a^2 + b^2 \end{cases} \tag{2}$$

The dynamic of the field can be described by the following equation:

$$z_j(t+1) = \frac{\sum\limits_i w_{ji} f(z_i(t))}{\max\limits_{j \in \text{Field}} \{z_j(t)\}} \cdot I_j \tag{3}$$

where $z_j(t)$ is the inner state of the neuron j at time t, I_j is the correspondent initial input value of the BMR image, w_{ij} are the synaptic strengths and $f(\cdot)$ is a linear function which depends on the choice of the neighborhood. In addition, each element receives the original binary input I_j. It avoids that original cells without activity can contribute to the activity of the bounded region. The iterative computation of eq. 3 is stopped as soon as the change of two following states is below a certain threshold (stopping criterion).

Figure 3: *Demonstration of the functioning of the LRWTA. From left to right: initial and final image of a sequence, BMR image from the sequence, result of the first step of the competition/cooperation task (largest region was selected, others are temporary suppressed), time courses of the activities of all neurons until the selection of the largest region was finished.*

Beginning from the winning neuron the underlying region is segmented and its area is computed. When the measure of this area is less than a fixed percentage of the largest hitherto segmented region, it is considered a noisy region and the selection task is stopped. Otherwise the region is inhibited and a new

[3]In our paper a region is considered connected when we can reach any pixel within it only by moving to neighboring pixels.

competitive process begins. The selected motion region has to be permanently inhibited in order to avoid that it is again selected in the next competition task. Without inhibition the currently selected region would be always selected again (it is the largest!) and would not permit smaller regions to win the competition.

5 Conclusions and Future Work

Our contour-shape-based approach is one of the saliency components besides others like facial structure and skin color. In a series of tests under several conditions (indoor, outdoor) we obtained a quite robust person localization. For computational reasons, the used image resolution was highly reduced so that the distances of the persons had to remain rather close.

What does the results of the localization task mean for the development of a gesture recognition system? In our opinion, the analysis of the motion region could provide a promising way for the analysis of dynamic gestures. From the motion template we can first extract a feature vector and then recognize it by means of a statistical classifier (e. g. NN or HMM). A good head localization task is therefore essential for a further feature extraction step. It allows to calculate a feature vector relating to the relative head position and regardless to the user's position within the image (translation invariance).

References

[1] Amari S.: Dynamics of Pattern Formation in Lateral-Inhibition Type Neural Fields. *Biol. Cyb.*, 27:77–87, 1977.

[2] Boehme H.-J., Braumann U.-D., Brakensiek A., Corradini A., Krabbes M. and Gross H.-M.: User Localization for Visually-based Human-Machine-Interaction, In *Proc. of the Third IEEE Int. Conf. on Automatic Face and Gesture Recognition*, pp. 486–491, 1998

[3] Davis, J. W. and Bobick, A. F.: The Representation and Recognition of Action Using Temporal Templates. In *Proc. of the IEEE Computer Society Conference on Comp. Vis. and Patt. Rec.*, pp. 928–934, 1997.

[4] Jähne, B.: *Digital Image Processing* Springer-Verlag, 1995.

[5] Jones, J. P. and Palmer, L. A.: An Evaluation of the Two-Dimensional Gabor Filter Model of Simple Receptive Fields in Cat Striate Cortex. *J. Neurophysiol.*, 58(6):1233–1258, 1987.

[6] Kaski, S. and Kohonen, T.: WTA Networks for Physiological Models of Competitive Learning. *Neural Networks*, 7(6/7):973–984, 1994.

[7] Koenderink, J. J. and v. Doorn, A. J.: Receptive Field Families. *Biol. Cyb.*, 63:291–297, 1990.

[8] Koffka, K.: *Principles of the Gestalt Psychology.* Brace & World, 1935.

[9] Oren, M., Papageorgiou, C., Sinha, P., Osuna, E., and Poggio, T.: Pedestrian Detection Using Wavelet Templates. In *Proc. of the IEEE Computer Society Conference on Comp. Vis. and Patt. Rec.*, pp. 193–199, 1997.

[10] Reichardt, W., Poggio, T., and Hausen, K.: Figure-Ground Discrimination by Relative Movement in the Visual System of the Fly – Part II: Towards a Neural Circuitry. *Biol. Cyb.*, 46(suppl.):1–30, 1983.

Phoneme Classification using a Rasta-PLP preprocessing algorithm and a Time Delay Neural Network: Performance Studies *

A. Esposito[a,c],
E.C. Ezin [b],
[a] International Institute for Advanced Scientific Studies (I.I.A.S.S.) Vietri sul Mare Salerno Italy

[b] Institut de Mathématiques et de Sciences Physiques (IMSP), B.P. 613 Porto-Novo, Bénin, West Africa

[c] INFM unitá di Salerno

Abstract

An automatic system for classifying the English stops [b, d, g, p, t, k] which uses a preprocessing technique based on a modified Rasta-PLP algorithm and a classification algorithm based on a simplified Time Delay Neural Network (TDNN) architecture is proposed. Phonemes, extracted from the TIMIT-NIST database, and produced by 73 speakers were used to train and test the system. The work is intended to study three different aspects of the problem: First, what role play the the preprocessing phase in the performances of the net? Second, what is the optimal number of neurons which balance the trade-off between net performance and computational time? Third, the optimal learning rate must be found through trial and error processes or can be found as a function of the input data? To this aim experiments to tune the preprocessing parameters, the optimal number of hidden neurons in the TDNN, and the learning rate have been performed. Classification percentages on the test data equal to 92.9 for [b], 91.8 for [d], 92.4 for [g], 80.3 for [p], 90.8 for [t], and 94.2 for [k] have been achieved.

1 Introduction

The aim of the research on automatic speech recognition is to use speech as input to a computer. As early as the 1950s, simple recognizers had been built, yielding credible performance. However, it was soon found that the techniques used in these systems were not easily adaptable to sophisticated systems, because several problems emerged that significantly degraded recognition performance. One of these problems was how to recognize phonemes. There were

***Supported by World Lab., and IMSP.**

some good reasons to approach the phoneme recognition problem. Phoneme recognition of speech utterances offers a promising avenue towards large vocabulary recognition of isolated, connected and continous speech. The attraction is that if accurate identification of vowels and consonants could be made, the recognition of words could be resolved automatically. There have been several attempts to implement phoneme recognition using standard techniques based on Hidden Markov Models (HMM) and new techniques based on Artificial Neural Networks (ANN) (see [1, 2] for a review on HMM and ANN respectively).

The work in the area of phoneme recognition using neural networks started with Waibel *et al.* [3] and was justified by the their potential of providing massive parallelism, adaptation and new algorithmic approaches to the problem. The initial studies demonstrated that multilayer neural nets, with time delay units, provide excellent discriminations on small samples of pre-segmented words (consonants and vowels) which are otherwise hard to discriminate. However, to obtain such results Waibel *et al.* they used an elaborated network architecture and a specific set of data based on utterances produced by a small number of speakers. Other attempts to realize similar performance using a different database (like TIMIT) and even phonetic units (like vowels) which are much more easy to discriminate than consonants did not give the same results (see [4] for a survey on phoneme recognition using several preprocessing methods and neural network algorithms). Due to these difficulties researchers tried to recognize speech signals which are larger and less variable than phonemes (like words). For these speech signals it is easier to find some stable acoustic attributes which can be used to allow a good learning and classification process (see [5]). Phoneme recognition still remains a difficult problem to deal with. The interest in phoneme recognition comes up again in these last years, due to the intrinsic problems of automatic speech recognition systems based on Hidden Markov Models. These models, infact, even though perform very well, are limited by the fact that they need a lot of training data and their performances degrades if new data are considered. The general idea behind this renovate interest was that if systems are able to recognize phonemes, the amount of training data will be reduced. Some recent works on phoneme recognition are intended just to prove this hypotesis. It is possible to obtain some good performances, combining the acoustic knowledge of the speech signal and powerful computational models such as Time Delay Neural Networks. To this aim the work of Gee-Swee Poo [19] reports, in classifying Mandarin phonemes, percentage of classification above the 95%.

Because of these promising results, we decided to approach again the phoneme recognition problem using data extracted from the multispeaker continuous TIMIT database. We exploit a new preprocessing algorithm based on the Rasta-PLP algorithm (see [6, 7]) and a neural classification approach based on TDNN. Moreover, this work also exploits the knowledge of the acoustic and perceptual features which significantly characterize the stop consonants (see section 3). Combining the three aspects mentioned above, we obtained classification percentages on the test data which are better than those obtained in the previous works of Zue *et al.*, Bengio *et al.*, and Flammia [8, 9, 10], which also tried to recognize stop consonants extracted from TIMIT. We shall show how the modification of some parameters in the preprocessing Rasta-PLP algorithm will allows to capture acoustic and perceptual stop's features, and obtain significant improvements in their recognition. Moreover, we shall report experimental results which allow to define an upper bound on the number of neurons in the first hidden layer of a TDNN. Above this bound the net performance decreases and the computational time for its training increases. Furthermore, we propose a learning rate function which avoids the trial and error processes needed to establish a good learning rate value to train the net on the current data.

2 The speech data structure

The consonants [p, t, k, b, d, g] used to train and test the net were extracted from TIMIT-NIST database. All data files in the two directories (TIMIT/TRAIN and TIMIT/TEST) are considered. The TIMIT-NIST database is composed of English sentences produced by speakers from different USA regions. Each speaker read ten English sentences. Each sentence was labeled phoneme by phoneme. A total of 730 different sentences were used to collect the consonants for the training and the testing set.

Our data (the stop consonants [b, d, g, p, t, k]) are extracted from such sentences. The sentences are different from one speaker to another. The training data used were produced by 38 speakers (24 males and 14 females). The testing data were produced by 35 speakers (22 males and 13 females). The training data were produced by speakers from the same USA region (drl in TIMIT). The testing data were produced by speakers from different USA regions (dr1-dr2-dr3 in TIMIT) and then, with a different accent. The sentences from which the consonants were extracted were different, both in the training and the testing data. Table 1 summarizes the size of the training and testing data for each stop consonant.

3 Extraction of acoustic and perceptual stop's features

The speech signal was processed using the Rasta-PLP algorithm (see [7]), due to its better performance over Linear Prediction and Perceptual Linear Prediction (PLP) (see [6]) algorithms. PLP analysis uses concepts from psychophysics of hearing in order to derive an estimate of the auditory spectrum. Rasta-PLP add a spectral band-pass operation to PLP analysis in order to obtain acoustic features which are robust to distortions. Each speech segment was processed through several steps which include critical-band spectral resolution, equal-loudness curve, intensity-loudness conversion, inverse Discrete Fourier transform, autoregressive coefficients, all-pole model.

We modified some parameter values of this algorithm in order to capture some of the invariant and transitional cues (see section 3 below for definition) of the stop phonemes. Indeed, the speech signal was sampled at 16 kHz instead of 20 kHz as in the original algorithm. The speech signal was weighted by a Hamming window 10 msec long (20 msec in the original analysis) and each frame of a phonetic segment overlapped with the preceding ones by 5 msec (10 msec originally). The other Rasta-PLP parameters were kept to their default values. This analysis provided nine acoustic features (8 coefficients plus gain) at the rate of 100 frames/sec. All stop consonants [b, d, g, p, t, k] were processed in the same way.

The modification of the preprocessing parameter values can be justified taking into account the acoustic features of stop consonants. Blumstein and Stevens [11], through a set of perceptual experiments, suggested that useful information about the stop consonant identity can be found mostly in the short time interval that follows the release. For this reason we used an analysis window and an overlapping step both shorter than their original values. Moreover, we used a lower sampling rate because, in running speech, frequencies above 8kHz did not give significant acoustic information (Stevens, personal communication).

We used Rasta-PLP as preprocessing algorithm because, as said above, it uses concepts from psychophysics of hearing in order to derive an estimate of the auditory spectrum. Which are the cues that our auditory apparatus uses to discriminate between speech sounds? Perceptual experiments, performed by Cole and Scott [12], suggested that our auditory apparatus uses *invariant* and *transitional* cues. An *invariant* cue, as it was defined by Cole and Scott, is an acoustic cue which accompanies a particular phoneme in any vowel environment. It is an acoustic feature that is present whenever a particular phoneme is produced during speech, and provides information about its phonetic identity. *Transitional* cues were defined as acoustic cues which accompany a particular phoneme in a specific phonetic environment. Both are used by listeners to discriminate between phonemes. The identification of a particular stop involves identification of either invariant or transitional cues, depending upon the position of the consonant in the syllable and the position of the syllable in the utterance.

The major cue which distinguishes stop consonants from the others in ongoing speech is silence. Stop consonants are always preceded by a closure interval prior to the release, and while voicing may continue throughout this interval in some environments there is a period of relative silence along the remaining of the energy spectrum which lasts at least 10-20 msec. The silent interval is an essential cue for the identification of a stop consonant (see [12]). Stop consonants are released with an explosive burst when produced before a vowel in a monosyllable. For voiced [b, d, g] the burst may be followed by a short open interval (5-10 msec in running speech), while for voiceless stops [p, t, k] the burst is followed (in initial syllable position) by a period of aspiration. The explosive burst is a transitional cue for stop consonants and it might help to discriminate them in some vowel environments (see [12]). Defined as the time interval between the burst onset and the voicing, Voice Onset Time (VOT) reliably discriminates between voiced and voiceless stops (see [13]) in initial position of isolated words and in short sentences. However, in other phonetic environments, VOT values were found to be compressed for both voiced and voiceless stops, and the separation between the two categories was generally less sharp (see [14]). The Rasta-PLP algorithm is capable of extracting such cues from the raw speech signal because it derive an estimate of the auditory spectrum, where such cues are enhanced. Cole and Scott's experiments also showed that some stop consonants occurring in the word initial position cannot be substituted by other consonants if the perception has to be left unchanged, except for [g] which could be replaced by [b] whenever it occurs. When a stop consonant occurs in any other position, all the voiced stops could be replaced by [d] whereas all the voiceless stops could be replaced by [t] without modifying the perception of the word or the sentence in which they are embedded. The results of these experiments suggested that, in running speech, some speech segments could be changed by production errors or by phonological rules without changing the listener's perception. Moreover, the perception of [b, d] is much more stable than the perception of [g], and the perception of [t, k] is much more stable than the perception of [p]. Since neural networks can only extract features from their inputs in order to realize the recognition process, the phonemes that undergo changes like those mentioned above are not classified so well as the others. This means that extracting the acoustic features from the input patterns, the net will better generalize for [b, d] and [t, k] than for [g] and [p]. Our experimental results, reported through the Tables 2, 3, and 5 in section 5, confirm this hypothesis. Net recognition percentages for [g] and [p] are always lower than those for [b, d, t, k].

4 The time dependent neural classifier

To perform the classification task, we used Time-Delay Neural Networks, which have turned out to be very suitable for phoneme recognition [15]. TDNNs try to capture the relationships among sequences of acoustical events under translation in the time of the signal window that is being examinated.

Our net is constituted of an input layer of nine components, a first hidden layer (in which the number of neurons was varied through the set of the experiments performed), a second hidden layer of six neurons, and an output layer of six neurons. The basic unit of TDNNs is a neuron that has been modified by introducing delays for each connection. The input layer is activated by a vector of nine components. The input action on the first hidden layer is seen through two component delay units, which were passed over the input components. Likewise, in the second-hidden layer, each neuron sees the first hidden layer through two component delay units, stepped by one frame at a time. Each of the output units, corresponding to [b], [d], [g], [p], [t], [k], is connected to the second hidden layer. The output layer performs, over the time, an integration, of the nodes' outputs in the second hidden layer. The delays are chosen short in order to take into account the way we processed the input data (short window analysis). This net operates similarly to that described by Waibel *et al.* and by Waibel [3, 16, 15] . However, the net is simpler in its structure compared to the Waibel's net. Indeed, the net input is a vector of nine components while the Waibel net input was a matrix of 16×15 components. Moreover, to discriminate [b, d, g, p, t, k] Waibel [15] used two TDNNs, one for [b, d, g] and another for [p, t, k]. Classification performance was very good in such case, but the resulting net architecture was greatly complicated.

Our network was trained using an on-line back-propagation algorithm (see [17]. Bengio [4] showed that on-line learning is most appropriate for automatic speech recognition applications. The weights are updated after each pattern presentation without using the momentum term. The momentum does not seem to be useful in on-line learning algorithms (see [4, 18]). The learning rate, after a series of trial and error processes was set to $\eta = 0.03$. This value gave the better net performances in terms of computational time and classification percentage.

The backpropagation is a gradient descent algorithm of the Mean Square Error (MSE) as a function of the weights. It works through two major phases: the forward and backward phase. The activation function chosen is the sigmoid one, due to its mathematical properties (see [20, 21]). Its use allows the output information to be available to the units in earlier layers so that they can have their weights adjusted. This adjustment is performed during the backward phase and its goal is to decrease the net output error in time.

5 Recognition experiments

A set of experiments was carried out in order to test different hypothesis. In all the experiments performed the input data were normalized so as to span over [-1 1].

5.1 Changes in the preprocessing parameter values

The first experiment was carried out in order to test if our modifications of some Rasta-PLP parameters allowed to capture significant information about the phonemes we wanted to recognize. We run the net with an architecture of 9-24-6-6 (i.e. 9 units in the input layer, 24 units in the first hidden layer, 6 units in the second hidden layer, and 6 units in the output layer) for several epochs. Tables 2 and 3 report respectively the net performance with the original Rasta-PLP (i.e. with a sampling rate of 20 kHz, an analysis window duration of 20 msec, and a window step of 10 msec) and the modified Rasta-PLP algorithm (i.e. with a sampling rate of 16 kHz, an analysis window duration of 10 msec, and a window step of 5 msec). It is interesting to note that after a certain number of epochs, the overall net performance did not improve significantly even though the number of epochs was increased. For example, in table 2 for [b, d, g] and in table 3 for [p, t, k] the correct classification percentage after 1800 epochs is better than the correct classification percentage after 2600 epochs. However, it is evident from tables 2, and 3 that whathever was the number of epochs, the correct classification percentage always improved when the modified Rasta-PLP algorithm was used. Moreover, these results confirm that the way the data are processed plays an important role on the subsequent net performance (see also [4]). As a consequence of these results, all the training and testing data for the subsequent experiments were processed with the modified Rasta-PLP algorithm.

5.2 Optimal number of neurons for improved performance

It is largely accepted that the net performance improves when the number of hidden neurons increases (see [20, 23, 24]). There is, however, an upper bound on the hidden neuron number which depend on the current task to which the net is devoted to (see [25]). The net gives, for such a bound, the minimum generalization error (The generalization error is the net error on an independent test set). Our goal in carrying out this set of experiments is to propose a net architecture that allows us to define (for our current task) an upper bound on the number of neurons in the first hidden layer. This bound should give the best compromise between the net's classification performance and the computational training time. We expect that the net performance with a number of hidden neurons above this bound should not pay for the increased computational training time. Therefore, in our experiments we only varied the number of neurons in the first hidden layer. Table 4 reports the overall net performance for this set of experiments. All the nets were trained on 1600 epochs. As it is possible to see, the net performance improved as thenumber of hidden neurons in the first hidden layer increased, together with the computational time required for the training. Our experimental data (see Table 4) also showed that the classification percentages reached a maximum value (for a certain number of hidden neurons) and then started to decrease. In order to determine the performance function which fits our experimental data (that is the performance as a function of the number of hidden in the first hidden layer), we used the interpolating theory and Lagrange's formula (see [26]) on our experimental data. The result plot shows that a 9-256-6-6 net architecture is optimal for our current task.

Finally, we used such a net architecture in the training and testing phase. The net was trained for 1600 epochs. The results are reported in Table 5 both for the training and testing data, and for each consonant.

Our generalization error on the independent test data set was 9.6%, which gave a significant improvement compared to that reported by Zue *et al.*, Bengio *et al.*, and Flammia [8, 9, 10]. These authors obtained a generalization error of 35%, 30.7 %, and 24.9% respectively using more sophisticated preprocessing features and more complicated network architectures. It is important to note that even though the number of epochs was increased the net performance did not improve (see Tables 2 and 3) due to the overtraining phenomenon. Indeed, it has been show (see [22]) that such phenomenon appears after a certain number of epochs depending on the training data and the net architecture. The data reported in Tables 2 and 3 confirm this hypothesis; in Table 2 the net performance decreased after 2600 epochs, in Tables 3 the training data have been changed and net performance still increased after 2600 epochs.

5.3 Optimal learning rate

Once we found the right way to preprocess the data and the right number of hidden neurons to achieve a good phoneme recognition task, we performed another set of experiments in order to test a general way to set up the learning rate value. The subsequent experiments were performed using the optimal net architecture found in the above section and data processed with the modified Rasta-PLP algorithm.

Generally, in the backpropagation training algorithm, the learning rate value is randomly changed until, through trial and error processes, an appropriate value for the current task was found. This is a hard process which relies only on the researchers' experience. We tried a general way to set up the learning rate value defining it as a function of the net output error on the current input. For a better understanding, let be

$$\beta = \max_k \delta_k^{(pat)}$$

be valued on all output neurons, where

$$\delta_k^{(pat)} = y_k(1 - y_k)(pat_k - y_k), \quad k = 1, \ldots, 6$$

is the error on the current pattern *pat* computed over the output node k; pat_k is the k-th component of the input pattern; y_k is the output of the node k. Then the learning rate $\eta^{(pat)}$ for the current input *pat* was set equal to

$$\eta^{(pat)} = \exp^{-\beta}$$

and it was changed after each epoch. The experimental data showed that with a such learning rate, the convergence was never reached and local minimum problems appeared. Mathematically, this effect could be explained by the fact that large and small errors appeared randomly during the training and cannot be predictable. Consequently, the MSE undergoes to great oscillations from one training cycle to another because of the negative β sign.

In the second experiment we define the learning rate as a decreasing function of the number of current epochs; i.e.

$$\eta = \frac{1}{M}(1 - \sum_{j=1}^{Ne} \frac{1}{2^j})e^{-|\beta|}$$

where M is a free parameter. It could be used to decrease the initial learning rate value when it is too high for the classification task to which the net is devoted. In our case M is equal to 1. Ne is the current number of epochs (Ne defines the learning rate as a function of the current number of epochs), and

$$\beta = \max_k \delta_k^{(pat)}$$

was valued on all the output neurons. With such a learning rate the net MSE value after 1600 epochs was lower than that obtained with $\eta = 0.03$ (MSE=424.041992 against MSE=463.101379). However, the net classification performance was roughly the same (the classification percentage over all the consonants and after 1600 epochs was 88.06%) and the computational time increased. Nevertheless, such a choice avoided trial and error processes in defining the learning rate value and speeds up the net convergence.

6 Concluding Remarks

In the present paper we report a set of experiments which allowed us to define some parameter values for the Rasta-PLP preprocessing algorithm in order to capture significant acoustic information for English stop consonants. Then we found an optimal number of hidden neurons in a TDNN architecture devoted to classify these phonemes. Finally, we defined the learning rate as a function of the net output error over the time. Such a learning rate, even though it did not improve the net performance with respect to the previous one, avoided random choices and speeded up the net convergence. Further work is in progress for classifying other phonemes than stops.

ACKNOWLEDGMENTS: We are grateful to M. Marinaro, M. Frixione, A.Petrosino for suggestions and comments.

References

[1] L. R. Rabiner, A Tutorial on Hidden Markov Models and selected applications in speech recognition, in *Reading in Speech Recognition*, A. Waibel, K. F. Lee Ed's, Morgan Kaufmann Publishers, 267-295, (1992).

[2] R. P. Lippmann, Review of neural networks for speech recognition, in *Reading in Speech Recognition*, A. Waibel, K. F. Lee Ed's, Morgan Kaufmann Publishers, 374-392, (1992).

[3] A. Waibel, T. Hanazawa, G. Hinton, K. Shikano, K. J. Lang, phoneme recognition using time-delay neural networks, *IEEE Trans. on Acoustics, Speech and Signal Processing*, **ASSP-37(3)**, 328-339, (1989).

[4] Y. Bengio, Artificial neural networks and their application to sequence recognition, *Ph.D. Thesis*, McGill University, Montreal, Canada, (1991).

[5] H. Sakoe, R. Isotani, K. Yoshida, K. Iso, T. Watanabe, Speaker-independent word recognition using dynamic programming neural networks, in *Proc. of ICASSP-89*, Glasgow, UK, 29-32 (1989).

[6] H. Hermansky, Perceptual linear predictive (PLP) analysis of speech, *JASA*, **87(4)**, 1738-1752, (1990).

[7] H. Hermansky, N. Morgan, RASTA processing of speech, *IEEE Trans. on Speech and Audio Processing*, **2(4)**, 578-589, (1994).

[8] V. Zue, S. Seneff, J. Glass, Speech database development: TIMIT and beyond, *Speech Communication*, **9(4)**, 351-356, (1990).

[9] Y. Bengio, R. De Mori, G. Flammia, H. Kompe, Phonetically motivated acoustic parameters for continuous speech recognition using artificial neural networks, in *Proc. Eurospeech91*, Genova, Italy, 551-554, (1991).

[10] G. Flammia, Speaker independent consonant recognition in continuous speech with distinctive phonetic features, *M.Sc. Thesis*, McGill University, Montreal, Canada, (1991).

[11] S. E. Blumstein, K. N. Stevens, Perceptual invariance and onset spectra for the stop consonants in different vowel environments, *JASA*,**67(2)**, 648-662, (1980).

[12] R. A. Cole and B. Scott, Toward a theory on speech perception, *Psychological Review*, **81(4)**, 348-374, (1974).

[13] L. Lisker, A.S. Abramson, A cross language study of voicing in initial stops: acoustical measurements, *Word*, **20**, 384-422, (1964).

[14] L. Lisker, A.S. Abramson, Some effects of context on voice onset time in English stops, *Language and Speech*, **10(3)**, 1-28, (1964).

[15] A. Waibel, Connectionist glue: modular design of neural speech systems, *Proc. of the 1988 Connectionist Models Summer School*, Morgan Kaufmann, (1988).

[16] A. Waibel, H. Sawai, K. Shikano, 1989, Consonant recognition by modular construction of large phonemic time-delay neural networks, *Proc. ICASSP-89*, Glasgow, UK, 112-115, (1989).

[17] D. E. Rumelhart, J. L. McClelland, Parallel distributed processing: explorations in the microstructure of cognition, MIT Press, (1986).

[18] L. Bottou, F. Fogelman Soulie, P. Blanchet, J. S. Lienard, Speaker-independent isolated digit recognition: multilayer perceptrons vs. dynamic time warping, *Neural Networks*, **3(4)**, 453-456, (1990).

[19] Gee-Swee Poo, Large vocabulary Mandarin Final recognition based on Two-Level Time Delay Neural Networks (TLTDNN), *Speech Communication*, **22**, n.1, 17-24, (1997).

[20] G. Cybenko, Approximation by superpositions of a sigmoidal function, *Tech. Report No. 856*, Urbana, IL: University of Illinois Urbana-Champaign, Dept. of Electrical and Computer Engineering, (1988).

[21] S. W. Ellacott, Aspects of the numerical analysis of neural networks, *Acta Numerica*, **3**, 145-202, (1994).

[22] R. S. Sutton, Two problems with backpropagation and other steepest descent learning procedures for networks, in *Proc. of 8th Conf. of Cog. Scie. Soc.*, 823-831 (1986).

[23] A. R. Gallant, H. White, There exists a neural network that does not make avoidable mistakes, in *Proc. of ICNN-88*, San Diego, CA, **1**, 657-664, (1988).

[24] K. Hornik, M. Stinchcombe, H. White, Multilayer feedforward networks are universal approximators, *Neural Networks*, **2**, 359-366, (1989).

[25] J. E. Moody, Note on generalization, regularization, and architecture selection in nonlinear learning systems, in *Proc. of IEEE/NNSP-91*, Los Alamitos, CA, 29-32 (1991).

[26] W. H. Press, B. P. Flannery, S. A. Teukolsky, W. T. Vetterling, Numerical recipes: ..., Cambridge University Press, (1988).

Table 1: Size of the training and testing data

Phoneme	Training	Testing
[*b*]	183	176
[*d*]	300	265
[*g*]	166	157
[*p*]	211	188
[*t*]	329	326
[*k*]	389	352
Total	1578	1464

Table 2: Net classification percentages (on the training data) for each consonant using the original Rasta-PLP algorithm and a 9-24-6-6 TDNN net architecture. On the row are reported the consonants.On the columns is first reported the number of epochs on which the net was trained and then the percentage of correct classification. As shown, in some cases, above a certain number of epochs the net performance decreased (see text for further details). The last row reports the correct classification percentage over all the consonants.

Epochs	200	1000	1800	2600
[*b*]	67.1	73.2	69.1	65.0
[*d*]	59.0	56.3	63.1	70.0
[*g*]	32.1	31.3	33.0	35.0
[*p*]	34.1	45.1	47.4	53.1
[*t*]	80.0	84.2	84.2	81.5
[*k*]	62.2	67.1	67.0	66.3
Overall	46.0	59.5	60.6	61.6

Table 3: Net classification percentages on the training data (reported on the columns as a function of the number of epochs) for each consonant using the modified Rasta-PLP algorithm and a 9-24-6-6 TDNN net architecture. On the row are reported the consonants. The first number on ecah column indicated the number of epochs. As shown, there is a significant improvement only changing some Rasta-PLP parameters.

Epochs	200	1000	1800	2600
[*b*]	74.3	73.2	69.4	73.4
[*d*]	76.0	81.3	79.1	76.0
[*g*]	54.0	45.2	49.1	58.2
[*p*]	51.0	63.0	65.0	64.2
[*t*]	86.3	89.4	88.2	88.5
[*k*]	79.4	77.4	84.1	82.0
Overall	70.2	72.2	72.5	73.7

Table 4: Net classification percentage (on the training data) averaged over all the consonants for TDNN net architectures with a different number of hidden neurons in the first layer. On the first row is reported the number of the nodes in the first hidden layer; on the second row the percentage of correct classification.

24	34	64	128	256	350
72	73	82	85	88	86

Table 5: Classification percentages on the training and testing data respectively with a 9-256-6-6 TDNN net architecture.

Cons.	*%Correct*	
	Training	*Testing*
[*b*]	93.7	92.9
[*d*]	92.3	91.8
[*g*]	77.5	92.4
[*p*]	78.4	80.3
[*t*]	92.4	90.8
[*k*]	94.4	94.2
Overall	88.1	90.4

A Neural System for Remote Sensing Multispectral Image Classification

Márcio Leandro Gonçalves
PUC • MINAS – Av. Padre Francis Cletus Cox, 1661 Cep 37701-355
Poços de Caldas - MG – Brasil

Abstract

This work presents a system for Remote Sensing (RS) multispectral image classification based on Artificial Neural Networks (ANN), aiming at two objectives, namely: searching of techniques for improving the performance in the classification task and to exploit the advantages of unsupervised learning for feature extraction. The system is divided in two phases: *feature extraction* by the Kohonen Self -Organizing Map (SOM) and *classification* by a Multilayer Perceptron (MLP) network, trained by a learning algorithm which uses 2nd-order information exactly calculated. To evaluate the efficiency of this classification scheme, a comparative analysis with the maximum likelihood algorithm, conventionally used for RS multispectral images classification, is realized.

1 Introduction

Since the resurgence of interest in the middle eighties, ANN have shown its efficacy and versatility in a wide range of applications. In particular, successful applications in RS image classification have already been reported [1][4], showing superior results to those of conventional statistical classification approaches.

Among the advantages of ANN over conventional statistical methods one can point out the non-necessity of *a priori* knowledge of the probabilistic data model, since ANN have the ability to learn the data distribution properties during the training phase, as well as the ability to generalize and to incorporate non-statistical information and knowledge that may be potentially valuable.

The most relevant restrictions on a broader utilization of ANN refer basically to its present performance limitations, due to the slow convergence of standard *backpropagation* training algorithm, that is normally used, and to the amount of adjustments to the training parameters [1][4].

To improve the performance of ANN techniques this work investigates the utilization of an advanced training algorithm for the MLP networks.

Besides improving the performance in the classification task by this approach, another objective to be searched in this work is to exploit the ANN potential for the task of unsupervised feature extraction.

In this way a neural system was proposed, dividing the classification problem in two phases: a phase for feature extraction from the RS image by Kohonen's Self-Organizing Map (SOM) and a phase for classification, using a Multi-Layer Perceptron (MLP) network.

2 Feature Extraction by Kohonen Map

The objective of the feature extraction phase is to identify the spectral classes present in the image and to define the set of correspondent samples to be used in the classification phase afterwards.

There is no well-developed theory for feature extraction, mostly features are application-oriented and often found by heuristic methods and interactive data analysis.

One of the approaches is the use of competitive learning resulting in data clustering. An example is Kohonen's Self-Organizing Map (SOM) [2].

In this approach was generated an auxiliary visual tool from the SOM, denominated *Kohonen Clusters Map* (KCM), which enables to identify the spectral classes present in the image through the visualization of the clusters generated by SOM.

2.1 Kohonen Clusters Map (KCM)

In this approach, 3x3 pixel windows taken from the original image were used as training patterns for the SOM. These patterns were randomly and uniformly obtained from all over the image and presented as input vectors to the SOM. Since the SOM has the property of arranging its weight vectors in rectangular or hexagonal *grids* [2], and considering that both input data vectors and weight vectors have the same dimension, this enables to generate an *image of the weight grid of the SOM.*

The resulting grid image, after the unsupervised learning by the SOM, was denominated *Kohonen Clusters Map* (KCM). Figure 2 (a) shows an example of a rectangular KCM generated from the test image (Figure 1).

The KCM produces a visual auxiliary tool for the task of identifying and selecting the spectral classes present in the image and their correspondent training samples, which will be used afterwards in the phase of neural classification. The KCM has some important characteristics that enable these tasks:

- As the SOM performs a clustering on the training patterns, it's possible to visualize the spectral classes present in the original image through the clusters obtained in the KCM;
- The SOM property of preserving the topological relationships among the input data vectors is reflected in the KCM property of preserving these relationships among the clusters so obtained, in terms of distances among them. Clusters that are close to each other in KCM represent land cover classes which posses similar spectral features;

• The SOM property of preserving the probability distributions found in the input data can be verified in the KCM, where higher frequency spectral classes in the input data will be mapped onto bigger regions in the KCM.

3 MLP for Classification

Having selected the desired classes and their correspondent samples from KCM, the objective of the second phase in our proposed system is to perform the final classification of the image using a Multilayer Perceptron (MLP) network.

The MLP network has one hidden layer. The number of neurons per layer varies according to the number of classes and to the size of selected samples to perform the training.

In this search for alternatives to the poor performance of standard backpropagation, normally pointed out as the main drawback to a broader utilization in RS image classification, this work used an advanced training algorithm for the MLP network.

The MLP learning algorithm used here is an improved version of the Scaled Conjugate Gradient (SCG) algorithm presented in [5].

The SCG is a learning algorithm for ANN developed from the general conjugate gradient methods for optimization problems, which are well suited to handle large-scale problems in an effective way. The SCG method makes use of both first- and second-order information to determine the parameters in the learning process, but Moller uses an approximation for the 2nd-order information.

Pearlmutter [6] solved the problem concerning the computation and storage of the Hessian matrix, which is responsible for the 2nd-order information, in a surprisingly and efficient way. The use of the exact computation of 2nd order information in the SCG learning algorithm results in a learning algorithm that presents a final convergence much quicker than the standard backpropagation. This algorithm was used for the MLP network training.

4 Experimental Results

A TM/LANDSAT image composed by bands 7, 4 and 5, respectively associated to the colors red, green and blue, was used to test the efficiency of the proposed neural system (figure 1). The image has 512x512 pixels, and it corresponds to the crossing region of rivers Pardo and Mogi (São Paulo state, Brazil, longitude 48° west and 21° south) and was taken on August 8th of 1990.

Figure 1: Landsat TM image.

The SOM used in the first phase was composed of 600 neurons, geometrically distributed over a rectangular grid of 20 rows by 30 columns. The KCM obtained after the SOM training is shown in figure 2.

After the SOM training, 5 land cover classes were easily identified from the generated KCM with the assistance of an expert: nude soil, water, humid soil, growing crop and vegetation. The samples correspondent to these classes were selected near the corners and in the center of the KCM. Each sample for the neural classification was composed by a 3x3 pixel window.

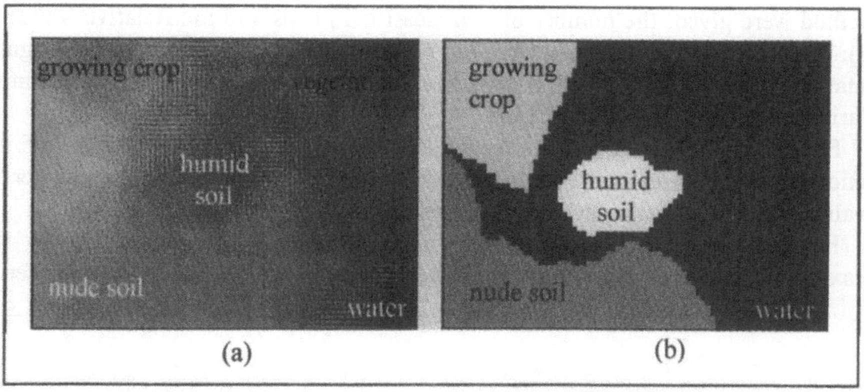

Figure 2: (a) KCM from test image; (b) Classified KCM.

These samples were presented in the training of the MLP network. The network has 12 neurons in the hidden layer and 5 neurons in the output layer, correspondent to the land cover classes chosen for this test classification.

error	backpropagation	SCG
1e-3	1606	239
1e-4	8029	343
1e-5	60625	364
1e-6	542720	439

Table 1: MLP training performance.

The table 1 shows the performance in terms of number of epochs in the training process between the standard backpropagation learning algorithm and the SCG algorithm with the exact computation of second order information.

To evaluate now the classification performance of the proposed system its results were compared to those obtained by the Maximum Likelihood algorithm, a statistical classification method widely used. The samples used for classification by the Maximum Likelihood algorithm were 5x5 pixel windows taken directly from the original image as is conventionally done. The performance classification results between both methods are shown on table 2.

	neural network	maximum likelihood
pixels of training samples	81	425
nonclassified pixels	37641	48862
relative variance	0.5214	0.3747
kappa coefficient	59	

Table 2: Classification performance.

First it's shown the total of pixels from the samples that were needed for both methods. Then two measures for evaluating the classification performance for each method were given: the number of non-classified pixels and the Relative Variance [3]. The Relative Variance describes the percentage of variability from the original data which are explained by the classification. In this way the bigger the Relative Variance the better the classification.

Finally the table shows the kappa coefficient, which is a measure of the concordance between two data sets, it's another commonly used parameter to evaluate the classification accuracy of satellite images.

Figure 3 show the classified test image by the neural system and by the Maximum Likelihood algorithm respectively with the 5 land cover classes chosen.

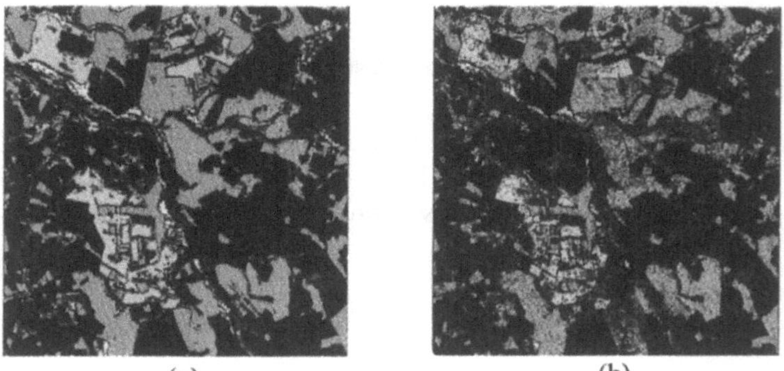

(a) (b)

Figure 3: (a) ANN classification; (b) Maximum Likelihood classification.

5 Conclusions

The advantages of the proposed neural system may be better understood taking into account that ANN classification is commonly done by a single MLP network, where the feature extraction task, i.e., the selection of classes and samples is done directly from the original image by an expert. With the modularization of the system, the feature extraction task was performed by SOM, generating an auxiliary visual tool, the KCM, which provides useful information regarding the representativity, the distribution and the similarity of spectral classes. It makes possible the identification and selection by visual inspection of the spectral classes and its respective training samples, which will be used in the classification phase.

Figure 2(b) shows the classified KCM itself, where one can verify the separation of classes, building *decision regions* for the classification. The non-classified regions, in black, may indicate the presence of distinct classes, which may be interactively incorporated in finer and more comprehensive selections of classes.

In this work was also proposed an alternative for improving the performance of ANN in terms of training time, with the use of an advanced learning algorithm for the MLP network, a version of the SCG algorithm instead of the standard backpropagation.

The positive results achieved show the feasibility to search of new techniques for reaching increasingly better performances as shown in table 1. As for the classification performance itself, the proposed neural system showed superior performance over a well known statistical classification method (table 2). The results obtained for neural classification associated with its performance improvement in terms of time motivate therefore to continue and to expand research efforts in this area, since still only a limited part of the power of the neural nets is actually being utilized.

References

[1] Foody, G. M.; McCulloch, M. B.; Yates, W. B.. *Classification of Remotely Sensed Data by an Artificial Neural Network.* Photogrametric Engineering and Remote Sensing, v. 61, n. 4, pp. 391- 401, 1995.

[2] T. Kohonen. Self-Organization and Associative Memory. Springer-Verlag, New York, 1988.

[3] R. A. Johnson, D. W. Wichern. Applied Multivariate Statistical Analysis, Prentice-Hall, New Jersey, 1982.

[4] I. Kanellopoulos, A .Varfis., G. G. Wilkinson, J. Mégier. Land-cover Discrimination in SPOT HRV Imagery using an Artificial Neural Network. International Journal of Remote Sensing, 13(5), pp. 917-924, 1992.

[5] M. F. Moller. A Scaled Conjugate Gradient Algorithm for Fast Supervised Learning. Neural Networks, 6(4), pp. 525-533, 1993.

[6] B. A. Pearlmutter. Fast Exact Multiplication by the Hessian. Neural Computation, 6(1), pp. 147-160, 1994.

Invariance of WNNs Performance for Pattern Recognition Testing Translated Image Input Spaces

Stanislao Lauria
Department of Cybernetics, The University of Reading
Reading U.K.

Dr. Richard Mitchell
Department of Cybernetics, The University of Reading
Reading U.K.

Abstract

Weightless Neural Networks (WNNs) have been used in pattern recognition vision systems for many years. In operation, a series of images of an object are presented to the network, each being processed suitably and effectively stored in a memory called discriminator. Then, when another image is shown to the system, it is processed in a similar manner and the system reports how similar it is to the taught images The operation of these networks requires that binary values be produced from the input data (processing data). In this paper, the network generalisation properties,that is recognising unseen images, are studied for translated images.

1 Introduction

Weightless neural networks, implemented in hardware by Aleksander and Albrow [1], have significant advantages over traditional methods of simple pattern recognition. A complete introduction to the n-tuple technique of pattern classification is given in [1]. In general, WNNs transform the input space into a binary space, and subsequently sample this transformed space by a set of K logical nodes each having N address lines (n-tuple). The set of K neurons, implemented using standard RAM (Random Access Memories), forms a *discriminator*. The basic operation is to sample n Boolean values from the binary input vector; these form a tuple which is used to address a location in one RAM. The network transformations can then be described as follows.

Firstly, the integer values of the input vector are transformed to a binary format. The individual co-ordinates are transformed indipendently. It is desiderable that the Hamming distance between the resulting binary patterns are proportional to the original distance between co-ordinates. Several codes, illustrated later, can provide this. Secondly, the next stage mapping concerns the combination of the individual co-ordinate patterns into a single binary vector. Detailed nature of the binary space sampling determinates the exact dependence of tuple distances on the Hamming distance betweens patterns. The mapping should aim to make the binary vector distances directly dependent on the input space distances. The generalisation properties of the network require that the overlap between responses produced for differing points in the input space decrease as these points become more distant as showed in [5].

When the network is being taught an image, a '1' is written into each address locates in each RAM neuron. Several versions of a given image are taught in this manner, so the system can recognise an image similar to, but not identical to, an image it has been taught. During the analysis phase, when an input image is presented, tuples are formed as in the learning phase and a count is made of the number of 1's in the address locations. Different classes of input can be taught into the same discriminator, and the system can report whether the discriminator 'recognises' any input. However, it will not be able to discriminate between the classes. If recognition and discrimination is required, then a network with many discriminators is used and each class of input is taught into its own discriminator. Then, when an input is being analysed, tuples are formed and passed to the respective RAMs of all the discriminators, and the input is 'recognised' as being from the class whose discriminator was found to have the most number of '1's.

Although for change of light conditions it has been demonstrated by the authors the existence of a succesfull way of solving both the problem of input encoding and the relationship between generalisation and n-tuple dimension [2, 3], little research has yet been done when the input space also includes 'translated' images (translated input space). Therefore, a software simulator has been used [2] in order to study the best configuration for the network matching both the sharpness caracterisics and generalisation properties when a translated input space is considered. Two different approach have been used. In the first one (Sec. 2.2), these comparisons are achieved by testing different sets of translated images and looking at the success in terms of pattern recognition, for different input data processors. In the second one (Sec. 2.3), a correlation between the type of processing data used and the relative Hamming distance obtained, has been studied. Indeed, according to [4], the behaviour of these networks is heavily dependent on the Hamming distance of a test pattern to those in the training set. Therefore, once the data are transformed in binary values, a comparison between different data processers can be achieved from the analysis of the Hamming distances obtained.

2 Experiments

2.1 The input space

The aim of this research is to investigate the response of a weightless network when comparing translated images. The set used is a given 'central' image and a set of images 'shifted' around it. In the tests 25 images of 64x64 pixels were sampled from the overall image of 128x128 pixels (one being the 'central' position and the other 24 as the 'shifted' ones around it). Moreover, it was decided to choose 4 different objects and record their appearance under 7 different lighting conditions using a standard video camera, so that the total number of images for each disk box is 175. Then the input space is defined by 700 of images (175*4). Then, when each image was tested, the data were processed by the software simulator described in [3], in which various configurations of weightless network can be built.

2.2 Data processors

The images being analogue contain grey level, not binary data. Therefore some processing is needed to convert the grey level values into the Boolean states required to define the address input to the RAMs. In this section the three different methods tested are presented.

The simplest process is to threshold the input, sampling a true or false condition depending on the magnitude of the analogue input relative to a reference level. However, a large amount of information is lost using this method, while the logical address patterns input to the RAM will vary greatly according to the changes of overall amplitude of the source. Indeed, for the same image, but with different conditions of light, the tuple so formed can be very different from the value previously stored. As a consequence, the change of light causes the network not to recognise the image previously learnt.

The second method of processing level data is to use 'thermometer coding' [1]. This involves converting each sampled point of the input into an array of Boolean values, where the greater the amplitude of the source, the more Boolean values are true. This is equivalent to maintaining a number of threshold systems, each with a different threshold. However, it requires more memory and it is slower than the conventional method.

A different technique has been developed at the University of Reading, called Minchinton cells [3]. These are processing blocks connected between the input and the RAMs. Each cell samples data from the input data, and computes a local function on its sample. The samples are randomly distributed over the input data. The cell function used in this experiment is defined as: $V[x1] > V[x2]$, where V is the input data and V[x] is the value at position x, with x chosen randomly. This cell returns a Boolean true if the value at position x1 in the input data exceeds that at position x2. Thus the cell gives an output dependent upon the relative amplitude of a point V[x] to the other points in the input, and not the absolute amplitude. This gives better tolerance to changes in overall lighting, as shown by the following argument. Suppose the lighting of the input data increases. It is likely that the values at the two random positions will increase, and this could cause problems if simple thresholding were used. However, the two values are likely to increase by similar amount, hence the difference between the values will hardly change and so the output of the cell will not change. Moreover, generalisation properties of a standard weightless system [1] can compensate eventual changes. This type of Minchinton cells can, like weightless networks, be easily implemented in hardware, for example by using the modular hardware system developed at Reading University [3], or using a software simulator such as an original one developed by the authors [2]. The performances of the three processing methods were obtained in two ways. First using a WNN on data, and second using Hamming distance analysis. These are described in the next two session.

2.3 Test 1: WNNs recognition

In this test for each processing method, the simulator was configured as a WNN with four discriminators (one for each disk box) and the 7 'central' images from each class (i.e. each disk box) were trained in the associated discriminator. The WNN has been then tested for the all input space (that is the WNN assigned to each image a given class). Finally, for each of the 24 'shifted' position, the average value, of the network responses for the 28 images associated to that 'shifted' position (that is 7 shifted images for each of the 4 disk box), has been then obtained.

2.4 Test 2: Hamming distances

In this test each of the microfeatures is processed and the tuples are formed. The tuples for each of the 'central' microfeatures were then compared with the corresponding tuples of its relative 'shifted' microfeatures; the difference bqetween these is the Hamming distance. This test was performed comparing the 7 'central' ones with the relative sets of 24 'shifted' images for the 4 disk boxes. Once again, for each of the 24 'shifted' position, the average value has been calculated from the 28 images obtained from the 7 images belonging to each of the 4 different disk boxes.

3 Results and Discussions

3.1 WNNs recognition

As all the images tested have been correctly classified in the appropriate class from all the three configurations, attention has been focused on the response values. Fig 1(a) shows the response of the network, for each processing method, when the image presented to the network is shifted. The y-axis is the percentage recognition, the x-axis is the amount by which the image is shifted. An attempt was made to fit a smooth curve, for each method, through these points. Such curves are shown in fig 1(b). An approximated curve is fitted. Each point is the average obtained from 28 values (7 values for each of the 4 disk boxes), and the errors for the averaged values are smaller then 15% (not showed in the graph)

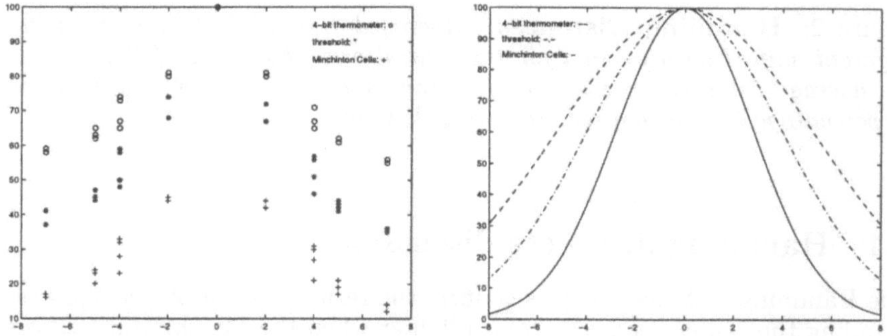

(a) (b)

Figure 1: **WNNs Recognition.**

3.2 WNNs recognition discussions

The three different configuration of WNNs were able to discriminate correctly. In the tests the responses for the shifted images, from a given class, are always an order of magnitude bigger than that ones given from images belonging to other classes. Of interest is to consider the gaussian curves fitted to the response data (Fig. 1(b)). The standard deviation of the Minchinton Cells response is smallest, that for the 4-bit thermometer code. This means that the Minchinton Cell method is less tolerant of positional shifts in the data.

3.3 Hamming distances

Hamming distance values for the 24 different shifted positions and for the central one are shown in Fig 2 (a) (b) and (c) for Threshold and 4-bit, Minchinton cells and thermometer respectively. For each of the 25 positions (1 'central' and 24 'shifted') the average value is represented by a gray square.

The gray level corresponds to a comparison between two images is the Hamming distance value in percentage: i.e. the amount of pixels having different Boolean values between the two images. As comparisons are between the 'central' image and the 'shifted' ones, the central square is the value of the 'central' comparison with itself, whereas the other squares represent the values in function of the shifting from the central position.

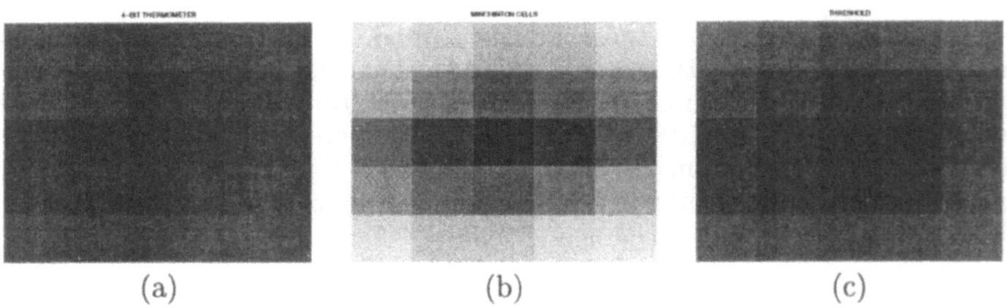

Figure 2: **Hamming Distances.** *Averaged Hamming distances for the 24 different shifted position and for the central one. For each of the 25 position the average value is represented by a gray square. The gray intesity indicates in percentage the normalized Hamming distance.*

3.4 Hamming distances discussions

The Hamming distance analysis confirm the results. In Fig 2, the squares surrounding the center square are much lighter when the Minchinton Cell is used then when the other methods are used. When the graph exhibits a dark central square with the other ones lighter, and the lighter the squares are and the greater is the shifting from the central position, then an increase of sharpness characteristic is exhibited by the network, as cofirmed with the previous test. Conversely, when the graph exhibits dark squares with little difference between the central ones and the others, then the network shows higher generalization properties. Hence, Minchinton cells provide the best sharpness properties showing a better contrast between the central images and the shifted ones, whereas 4 bit thermometer and threshold have a better generalisation characteristic.

4 Conclusion

The use of Weightless Neural Network for pattern recognition of translated images has been investigated. In particular, attention has been focused on comparisons between the performances of different ways to pre-process the data input. The results shown that the Minchinton Cell processer is less tolerant than the other methods when the images are shifted. This result is obtained using WNNs and confirmed by Hamming distances. If the network is being used to recognise images, this may be a problem. However, if the system is being used to found the position of an object within an image this property is useful.

In conclusion, when the networks have been trained with the smallest training set that does not affect the discrimination performances between different classes (i.e. using only the central images) the sharpest characteristic has been exhibited by Minchinton Cells. But, if few shifted images are added in the training set, in theory, it is possible to obtain an increase in the generalization properties (i.e. a decrease in the sharpness carachteristic). In this sense then, Minchinton Cells show a better performance having a wider range of variability in the sharpness/generalization characteristic. Hence, further research is necessary to confirm Minchinton cells performance when shifted images are included in the training set.

References

[1] Aleksander I., Stonham T.J. A guide to pattern recognition using RAM. **IEE Proc. E, Comput Dig. Tech** 2(1).pp. 29-40. 1979

[2] Lauria S.,Advances in WNN, First Year Report, SL-RJM-96-2,1996

[3] Mitchell R.J., Bishop J. M., Box S.K., Hawker J. F. Comparison for grey level data in WWN. **Proc WNNW'95**, pp123-128, Univ. of York. UK 1995.

[4] Aleksander I., Microcircuit learning nets.**Electronic letters.**(1970). Vol. 6 n.5 p134-136.

[5] Stonham T. J., Improved Hamming distance analysis for digital neural net. **Electronis Letters**, 1970 , vol 6, n. 5, pp. 155-156,

The CALMA project: Computer Assisted Library for MAmmography

V.M.Marzulli
INFN and Physics Department
Pisa, Italy

Abstract

CALMA is a collaboration among the INFN sections of Bologna, Pisa, Torino, Udine and Trento and the Hospitals of Bari, Bologna, Livorno, Udine and Trento. We aim to obtain a CAD (Computer Assisted Diagnosis) system in screening mammography based on neural networks. After a description of the project, some preliminary results are shown for the classification of a peculiar kind of mammal disease (microcalcification clusters).

1 Introduction: a small overview of the problem

Traditional mammography, using screen and film, is widely recognized as the only imaging modality for the early detection of breast cancer in asymptomatic women [1]. In this way, it has been shown possible to reduce significantly breast cancer mortality. Currently this accounts for 32% of cancer incidence and 18% of cancer mortality for women in the United States. 5 years survival rates are 93% for localized cancers, falling down for regional (72%) and distant diseases (18%). This shows how the early detection of breast cancer is the main element in the reduction of breast cancer mortality.

Typically, sensitivity ranges from 73% to 88%, while specificity are of the order of 83% to 96% [1] [2].

These relevant figures are shown to increase of 7% - 10% and 2% - 5% respectively if the image is analyzed independently by two radiologists: in particular sensitivity increases up to 14% if the two radiologists come from the same "training". The aim of CALMA is to improve these merit figures of diagnosis behaving as "the second radiologist" in the screening.

In addition to this, using a computer based criterion of decision would guarantee a better control on mistakes due to the fact that the radiologist might be (for any reason) not totally "efficient" for the diagnosis. Other important advantages are the speed of the method and the reproducibility of the diagnosis.

In many cases an early breast cancer diagnosis follows from the detection of a cluster of microcalcifications, i.e. groups of small but very brilliant objects. As microcalcifications may show up late in the developement of the disease,

[1]Sensitivity is defined as the fraction of recognized cases of illness over the total number of ill patients. Specificity is the ratio of recognized healthy patients over the total of healthy

a way to attack earlier the problem would be to increase the capability of detecting tumorous masses caracterized by a very low contrast but peculiar shapes, such as the so called "spiculated lesions" ([2]).

Our group is mainly involved in the detection and diagnosis of the spiculated lesions, but we are directing our investigations as well in the field of microcalcifications.

2 The DATABASE

The first step to be accomplished is to collect a set of useful images to train the system, i.e. a DATABASE is needed of mammographic images of 50-100 microns per pixel and with 4096 grey levels, each one described by a homogeneous and coherent diagnosis.

We think that 20000 to 50000 images with these requirements can be collected in a reasonable amount of time.

At this moment, in two months from the beginning of the project, we own 600 mammographic images. Out of these, 200 show a pathological behaviour. These numbers are increasing, and we aim to complete the database in a 3 years time.

Images come from the collaborating hospitals where radiographies are scanned (up to now with a 85μ granularity) and transferred to CD-ROM. Given the average dimensions of the images, without any compression, 70 to 80 images can be written on a single CD-ROM. CDs are hosted in a juke-box controlled by a "Origin 200" Silicon Graphics server. The juke-box can contain up to 240 CD-ROMS, with an average access time of 3 seconds.

At the moment, other similar databases, public and free, are the MIAS database (London, with 320 images, 8-bits resolution and 200μ granularity) and the Nijmegen database (the Netherlands, wth 40 images, 12-bits resolution, 100μ granularity). In addition to these free images other can be bought. Among the private databases we have to remember the largest existing at the moment, the Chicago one, with more than 10000 images.

3 Possible approaches

The large dimensions of digitized images forbid a direct computational approach to the mammal image as a whole: a typical digital mammographic image is made of 1800 × 3000 2-bytes words (assuming a 12-bits digitization) which is by far a large amount of information. Also thinking of a neural network approach, the number of inputs cannot be as large as 6 million units.

Some software needs then to be produced in such a way to pre-select some suspicious regions. This pre-selection needs of course to keep always all the pathological regions, reduce of some orders of magnitude the problem to be solved by the next step producing a limited amount of false positives.

At this level our strategy is to follow three different approaches in order to evaluate which would be the most suited to attack the problem:

Figure 1: Detail of microcalcification cluster

1) The pre-selection provides a series of tools to choose one or more regions of the mammography (each one of 40×40 to 160×160 pixels, i.e. 1500 to 25000 numbers) and transforms the information into a much smaller set of figures, from 100 to 400 numbers. Some of the algorithms used to perform this reduction are described later. The disadvange (and the risk) of this method is in the lost of information at the very first step. On the other hand, dealing with a few hundred numbers would make affordable (in terms of time and dimension of the learning sample) the use of a supervised learning neural network.

2) The pre-selection just indicates the relevant regions (as large as indicated above) and later a supervised learning is tried on these reduced, but still rather large images. Here the advantage is that no information is lost at the pre-selection level, but a lot of effort is needed in the learning step. A large amount of consistent images is needed.

3) The pre-selection indicates the regions, and images are passed, large as they are, to an unsupervised network (Kohonen-like, for instance).

The pre-selection algorithms perform standard operations on the images, such as enhancement, filters of various kinds, gradient patterns etc., as well as more dedicated operations which have been thought to cope the peculiar shapes of malignant objects.

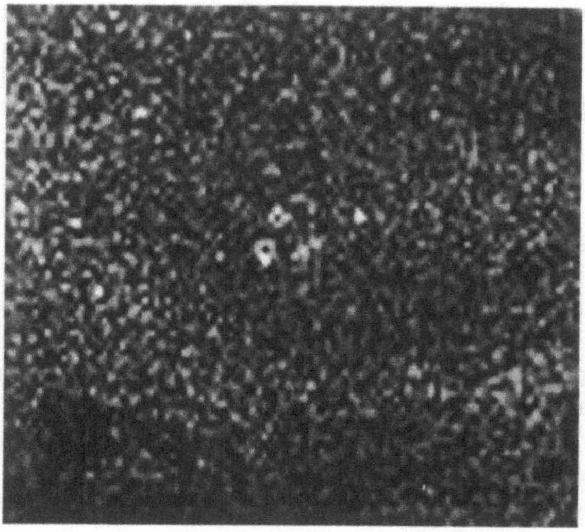

Figure 2: Same detail after enhancement of microcalcifications

Concerning clusters, a tool to recognize them and count the number of microcalcifications inside them is being prepared. Figure 1 shows a microcalcification cluster in a mammography, while figure 2 shows the same detail after a dedicated enhancement.

For spiculated lesions, first of all a local maximum is found, then the image is unrolled into a spiral starting with the pixel of the maximum, and later some fourier analysis is performed trying to figure out some patterns which might be peculiar of the spiculated lesions. Another approach here is to plot for each region the ratio perimeter over area of the pixels above a fixed threshold, as a function of the threshold. We think that the behaviour of this curve might be a significant feature of the problem.

The software is written in C, and is resident on the Origin 200 machine, which has 2 R-10000 180 MHz processors (used serially at the moment) with a 256 Mbyte RAM.

The second part of the job, that is the neural network training will be done using a commercial machine, CNAPPS, as well as exploiting APE [3], a parallel machine developed by INFN.

4 First attempts to solve the problem

Due to the still limited amount of available images of our own (at the moment we are trying to define a "protocol" with a radiologist in order to have a set of coherent diagnosis on images coming from different Hospitals), we started to test our pre-processing software and neural networks for micro-calcifications clusters on a public database (the Nijmegen one, 40 images with desease, with pixel size of 100μm and 12 bits resolution).

At a very first step a set of filter algorithms was tried, and a high-pass filter was chosen. A number of windows of 100×100 pixels were selected with a set of simple cuts on the pixel intensities distributions on the windows. The windows chosen were, in many cases, partially overlapping. In this way we obtained 326 ill and 536 sane regions.

A very simple feed-forward neural network was tryed for this first exercise, trained using the SNNS net simulator package ([4]). Our sample was split in a learning sample (276 ill and 448 sane) to be provided as an input to the net, and a test sample (50 ill and 88 sane), totally independent, to validate the results. The last sample was forced to be so poor by the limited amount of images available.

The net had 18 input, 4 hidden and 1 output neurons. The input variables were 18 features extracted from the regions (raw and filtered), such as maximum intensity, average intensity, variance, number of local maxima between the absolute maximum and the average, average distance among maxima, average area of local maxima, minimum distance of local maxima, average intensity of maxima (absolute maximum excluded).

As a learning alghorithm, the back-propagation learning with momentum was tried, providing best results with $\eta = 0.3$ (learning parameter), $\mu = 0.1$ (momentum), c=0.05 (flat point elimination value), $\delta = 0.1$ (max true/output difference).

5 Preliminary Results and Perspectives

In this way we obtain 80% sensitivity and 90% specificity (with a correct classification of 86%). Of course these figures are affected by a rather large uncertainty because of the poor statistics of our sets.

In the next future we hope to be able to increase the number of images in our database, so to be able to validate our procedures on larger and more significant ensembles.

References

[1] A.G.Haus and M.J.Yaffe, Editors, *A categorical course in physics. Technical aspects of breast imaging*. Radiological Society of North America, 1993, presented at the 79th scientific assembly and annual meeting of RSNA.

[2] R.A.Smith, "Epidemiology of breast cancer" in *A categorical course in physics. Imaging considerations and medical physics responsibilities*, Madison, Wisconsin, 1991. Medical Physics Publishing.

[3] A.Bartoloni et al., Int. J. Mod. Phys. **C4**, 993 (1993).

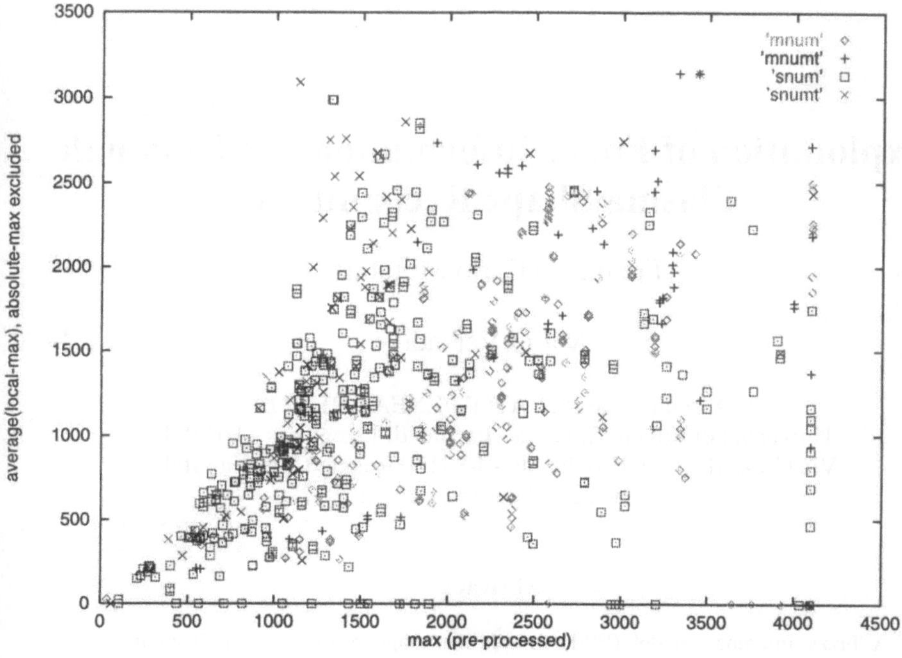

Figure 3: Correlation plot of two input variables to the neural network: squares and crosses represent the sane images (learning and test samples, respectively), diamonds and "+" are the ill images

[4] A.Zell et al., "SNNS - Stuttgart Neural Network Simulator - V4.1", University of Stuttgart Report No. 6/95 (1995).

Exploitation of Fuzzy Information for Tokamak Plasma Shape Recognition

Francesco Carlo Morabito

Mario Versaci

Associazione EURATOM/ENEA/CREATE
Università di Reggio Calabria - Facoltà di Ingegneria - DIMET
Via Graziella, Loc. Feo di Vito - I-89100 Reggio Calabria , Italy
morabito@unirc.it

Abstract

A fuzzy inference model (FIM) for plasma shape recognition applications is presented. The model is directly extracted from a data set of examples of the problem without using any learning procedure. The most relevant advantages of the FIM are: 1) the solution of the problem can be expressed in terms of very simple as well as explainable rules, and 2) a very limited number of inputs is required to obtain a sufficient estimation accuracy. The first objective overcomes one of the most limitations of neural network models. The second one has a strong impact on the throughput time in real time applications. The resulting model can be tuned by varying the parameters of the membership functions (centres and variances of the gaussian functions) in order to best fit the data set distribution. The qualitative analysis of the data set may also capture relevant insight on some difficult aspect of the problem, like its basic ill-posedness and the detection of category transition. The results presented in this paper regards a benchmark database of simulated plasma equilibria in the ASDEX-Upgrade machine. The main conclusion is that a FIM is an efficient tool for real time analysis of magnetic data in tokamak reactors.

1 Introduction

In recent years neural computing has emerged as a successful framework for fast analysis of multi-channel data in plasma shape recognition [3]. We wish to use here the non-conventional term "shape recognition" rather than "model fitting" to point out that, starting from the reading of a set of measurements (e.g. magnetic fields), we does not try to determine the value of the field in a different point. This interpolation (curve fitting) problem is easily solved by a simple linear regression in the remaining measurements [2]. Our goal is here to associate to a particular pattern of fields, the geometric configuration of the plasma column in the vacuum chamber of a tokamak reactor which presumably generates that pattern. It is clear that if we suppose to know just one measurement, the problem is strongly ill-posed because there are a lot of configurations producing the same value of the measurement in any possible sensor location. The simultaneous use of a pattern of measurements may partially solve the problem provided we are able to correlate the information among measurements. This is basically the main reason for using a neural network approach. The basic idea is to replace the complicated procedure in the analytical physical model by a relatively simple non linear mapping, which is also suitable for fast data processing. This mapping gives approximated plasma parameters as a function of the measurements.

The basic assumption is that the plasma configuration is reproducible starting from a limited number of parameters, each expressed in the form of a non linear mapping. The accuracy of the representation obtained by neural models is generally sufficient for preliminary analysis and feedback control [3, 6]. However, it is known that the neural models will act as black boxes, whose performance may hardly be explained. This limitation sounds often unreasonable. Indeed, the typical inputs of the model are samples of an underlying magnetic flux or field function and this imply that the values of the measurements are properly correlated. An expert may easily carry out considerations about the pattern of measurements from which to infer a reasonable plasma configuration. The problem would be to translate experience into models. On the other hand, if this is the case, it would be inappropriate to attempt of extracting any information from the data by using a blindly operating learning algorithm. This can be easily understood if one think to the well-known concepts of random initialisation of the model and arbitrary selection of the network architecture (i.e., number of layers, nodes, activation functions, ...). The results of this unprincipled perspective of proposing neural models are the lack of general acceptation of the conclusions [4]. For example, polynomial fitting is commonly preferred in plasma physics (in terms of functional parameterisation [2, 9].

2 The ASDEX-Upgrade database [1]

The identification of the boundary of a plasma current column evolving in the vacuum chamber of a tokamak starts from the knowledge of a set of measurements recorded at appropriate locations by means of selected sensors [2, 9]. The sensors (flux loops and magnetic probes) are located along some contour of the vacuum chamber [1, 9]. The database for training the ANNs has been generated by means of the Garching Equilibrium code [2] which represents a numerical model of the experiment. Each record of this database includes a large number of equilibrium parameters for a given converged equilibrium and a set of corresponding magnetic measurements. The set of 34 measurements for each equilibrium, which form the input vector for both FP and ANN training procedures, consists of 19 dB_θ/dt probes, 12 flux differences coming from 13 poloidal flux loops, 2 saddle loop fluxes, and the current flowing in the in-vessel passive stabilizer. The criterion used to generate the sample cases is described in detail in [2, 9], the principal aim being to cover the entire range of possible experimental configurations.

The generated set of plasma equilibria, totalling about 4000 cases, belong to four possible categories of plasma configurations, namely upper and lower X-point, inner and outer limiter. Figure 1 shows the typical inner limiter configuration, which will be analysed in the following.

3 Extraction of qualitative information from the database

The main goal of this paper is to show how information qualitatively extracted from the database can drive the construction of simple yet efficient models for solving plasma identification problem [8]. The basic approach is to analyse the database with visualisation tools for highlighting some expected properties. One of the most relevant problem in real time tracking of the discharge is the possibility for the plasma column of passing from a configuration to another one. This implies a first derivative discontinuity of the mapping plasma parameter- magnetic measurement [9]. In order to show how the pattern of measurement vary during the transition among categories, we plot a set of 25 patterns per category. From bottom to top, Fig. 2 reports a grey-scale representation of the typical patterns (1-25 inner limiter, 26-50 outer limiter, 51-75 upper divertor, 76-100 upper divertor). We may say, by a qualitative inspection of the plot, that the categories may be characterised as follows:

	1st block of Meas. (1-7)	2nd block of Meas. (8-14)	3rd block of Meas. (20-27)	4th block of Meas. (28-33)
Inner Limiter	LOW	HIGH	VERY LOW	HIGH
Outer Limiter	HIGH	LOW	LOW	MEDIUM
Upper Divertor	HIGH	MEDIUM	MEDIUM	MEDIUM
Lower Divertor	MEDIUM	LOW	MEDIUM	LOW

This means that we can give a visual interpretation of the classification problem in terms of fuzzy variables and their respective fuzzy values. From the reading of the pattern, we may simply express a guess about the membership to a kind of category. To avoid the use of a large number of input variables, we can use some features extracted from the block of measurements, e.g., the mean value and the standard deviation of each block. In this paper, we focuses mainly on the regression of a plasma parameters, thus we does not investigate more deeply this classification aspect.

Fig. 3 shows the distributions of a selected set of measurements within the "inner limiter" category (750 equilibria) and the horizontal minor radius of the plasma column (Ahor) . The Figures clearly show that the identification problem is ill-posed. However, it is possible to draw some general rules of behaviour of the variables. These "rules" reflect physical considerations about the location and size of the plasma column. For example, since the 6th measurement represent the difference between two fluxes measured in two symmetrical locations with respect to the Z=0 plane of the machine, if the corresponding measurement is ZERO, the plasma column is basically centered in the R-Z plane. On the other hand, for physical consideration, the plasma has a large probability of showing a large horizontal minor radius when symmetrically centered. In other words, plasmas located either on the top or on the bottom of the vacuum chamber are typically very small. This kind of comments may be extended with the aim of generating a data base of rules (an inference engine) which may be used for estimating the parameter of interest.

4 The fuzzy inference model

The core of the FIM is the database of rules which qualitatively describe the physical model. The construction of the model follows the following steps:

1) determination and fuzzification of the input measurements of interest and of the plasma parameter to be regressed;
2) determination of the rules with single-antecedent, which basically derives by inspection from the curves of the type of Fig. 3c and 3d;
3) derivation of rules with "and" connectives between two input variables (trough the use of patterns like Fig. 3a);
4) determination of multiple input rules (three variables).

Prior knowledge of the problem can be useful in determining proper combination of inputs which help in reducing ill-posedness of the task (e.g. Fig. 3b, in which a difference between fluxes #5 and #6 is considered). Once defined the type of membership functions (in this problem we use Gaussian functions), the FIM gives the estimation of the plasma parameters within the database. By analysing the curves of error (i.e. estimation error vs. input variable), it is possible to roughly tune the model and also to discover regions in which there is a need for some additional rule. These rules are typically of the type with connectives between antecedents.

A different approach to the FIM construction can be based on the fuzzy curve concept [5-7]. However, the fuzzy curve relates just one input to one output and thus, it does not allow the determination of combined antecedents. In addition, the fuzzy curves show typically a very limited variance, which limits the capability of extracting rules.

For the estimation of the Ahor variable in the inner limiter database, we have used 5 single antecedent rules per input variables (totalling 15 rules), 10 rules with "and" connective between two antecedents, and 5 rules with three combined input variables. The full fuzzy inference engine has thus 30 rules. The number of rules is slightly reduced by considering rules with "or" connective and also by introducing a criterion for deciding the "credibility" of the rules. The algorithm for measuring the credibility was proposed in [10], and basically assigns a value in the range 0 to 1.0 to each rule which measure how correctly the rule describes the behavior of the system around a particular operating point.

The physical estimation error (root mean squared error) of the FIM model based on 4 inputs is in the order of 1.5 cm which has to be compared to 0.8 cm of a neural network model based on 34 input measurements. It is now clear the strong impact which can have the fuzzy approach. It is to be noted that a polynomial (quadratic) model on the first 14 principal components obtained by eigen-analysis of the database of training is able to reach an error of about 0.4 cm. However, the number of degrees of freedom of the NN model (sigmoidal hidden layer and linear output) is 433, while the polynomial model requires 582 free parameters. As a consequence, the FIM appears to be very parsimonious with respect the other models.

5 Adaptive correction of the model

The performance of the FIM may be considered quite satisfactory if one think that no learning procedures have been used. It is of course possible to introduce some adaptation of the FIM in various respects [6, 7]. For example, the centres and standard deviations of the Gaussian functions can be adaptively changed in order to fit the training data. It is yet possible to train a NN to learn the estimation error based on the limited number of inputs used for the FIM. Finally, the model can be improved by presenting the rules to an expert and, thus, by introducing novel rules as well as corrections to the previous rules. Indeed, we have applied the fuzzy identification techniques to the whole ASDEX database by achieving very interesting results which confirm the convenience of using this approach. With reference to the estimation of the Ahor parameter, we achieve an RMS error of 1.5 cm by using only 7 rules (5 inputs), or, alternatively, an error of 0.65 cm with 15 rules (6 inputs).

6 Concluding Remarks

The main limitation of the present analysis comes from the fact that the measurements are considered not affected by noise. The impact of noise is to be analysed in terms of distorsion on the fuzzy rules activations and credibilities. The problem will be assessed in future works.

References

[1] The ASDEX-Upgrade Team, in Fusion Technology, Proc.13th Symp. Varese, Pergamon,Press, 1, 203, 1984.

[2] P.J.McCarthy, "*An Integrated Data Interpretation System for Tokamak Discharges*", Ph.D. dissertation, Cork University, 1992.

[3] E.Coccorese, R.Martone, and F.C.Morabito, "*Identification of Non-Circular Plasma Equilibria using a Neural Network Approach*", Nuclear Fusion, Vol. 34, n.10, Oct 1994, pp.1349-63.

[4] C. Bishop, *Neural Networks for Pattern Recognition*, Clarendon Press, Oxford, 1995.

[5] F.C.Morabito, M.Versaci, "*The use of fuzzy curves for the reconstruction of the plasma shape and selection of magnetic sensors*", Fusion Technology, Proceedings of the 19th Symp. on Fusion Technology, SOFT'96, C. Varandas and F. Serra, Eds., Lisbon, Portugal, 16-20 September 1996, pp.937-940;

[6] F.C.Morabito, M.Campolo, "*Are Hybrid Fuzzy-Neural Systems Actually Useful in Plasma Engineering?*", Proc. of the 9th Italian Workshop on Neural Networks, M.Marinaro and R.Tagliaferri Eds., WIRN'97, Vietri S/M, 22-24 May 1997;

[7] F.C.Morabito, M.Versaci, "*A Fuzzy-Neural Approach to Real Time Plasma Boundary Reconstruction in Tokamak Reactors*", IEEE ICNN Intl. Confer. on Neural Networks, Houston, Texas, 9-12 June 1997, Vol.1, pp.43-47;

[8] F.C.Morabito, "*A Soft Computing Approach to Plasma Evolution Tracking in Tokamaks*", SPIE Conference on Applications of Soft Computing, San Diego, CA, 28-29 July 1997, Vol. 3165, pp. 231- 242

[9] P.J. McCarthy, F.C. Morabito, "*Function Parameterization and Artificial Neural Networks for equilibrium parameters recovery in Asdex-Upgrade: a*

comparison", *Journal of Applied Electromagnetics and Mechanics*, Vol.8, 1997, pp.343-373.

[10] C. X. Xu and Y.Z. Lu, *"Fuzzy Model Identification and Self-Learning for Dynamic Systems"*, IEEE Trans. Syst., Man, Cyber., Vol. SMC-17, July 1987, pp. 683-689.

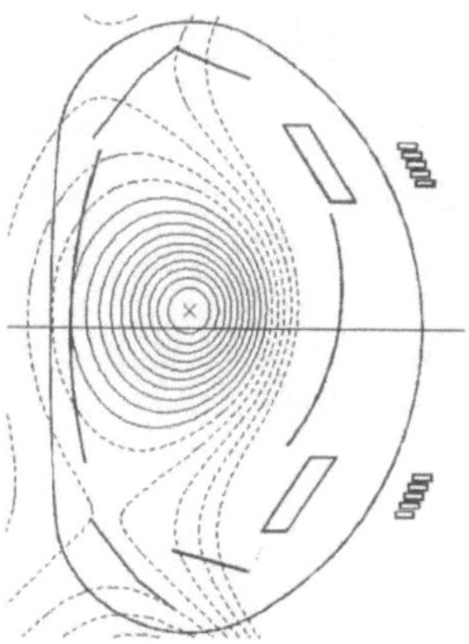

Fig.1:The "Inner Limiter" configuration.

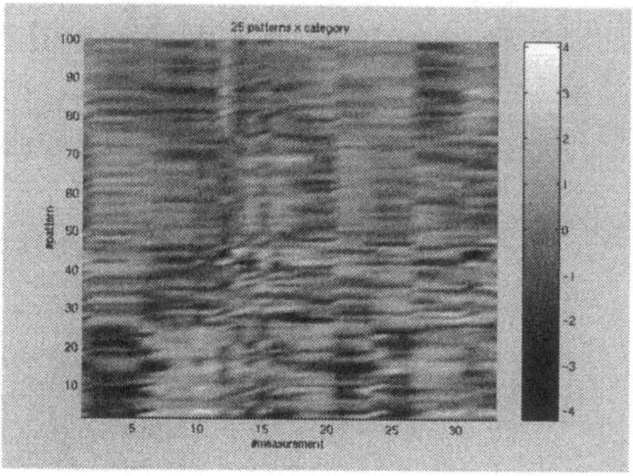

Fig.2: Distribution of the patterns for different categories.

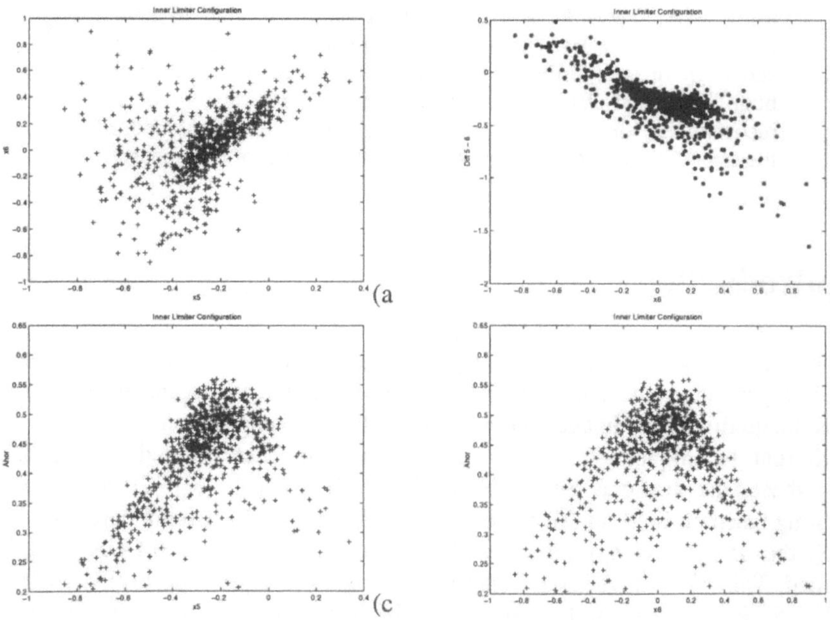

Fig.3:

a) Distribution of the measurements within the database;

b) distribution of (X5-X6) vs. X6;

distribution horizontal minor plasma radius vs. measurements #5 c) and #6 d).

New Results in Image Compression through M.A.I.A. Neural Networks

Giuseppe Pappalardo
Dipartimento di Matematica - Università di Catania
Catania – Italia

Domenico Rosaci, Giuseppe M.L. Sarnè
DIMET - Facoltà di Ingegneria - Università di Reggio Calabria
Reggio Calabria – Italia

Abstract

In this paper some results achieved in the image compression field by using neural networks are presented. A particular kind of neural network has been used that allows compressing both sets of images and single images; furthermore, the compression can be effected with both Black and White and coloured images. The quality of the reconstructed image and the compression ratio are functions both of the error defined in the learning stage and the number of used images.

1 Introduction

The need of quickly compressing single images or films with high compression ratios is a pressing requirement in fields as telecommunications and informatics as well as the quality of the image reconstruction after the compression.

Different techniques of image compression have been studied and realized; among these, the artificial Neural Networks (NN) [3], [4], [5], [7] constitute an interesting research field; in this paper the results obtained by using a particular kind of NN, the *Modulated Asynchronous Information Arrangement* (MAIA) NN, are presented. This NN allows using high ratios of signal-to-noise ratio (SNR) and peak value of SNR that result independent from both the subject and the nature of the image; this latter can be a Black/White (B/W) [11] image or a coloured one (at this time a set of 25 images with $16.7 \cdot 10^6$ colours can be used).

The compression ratio depends only on the number of stored images on the same NN. The quality of the reconstructed image depends on the error planned in the learning stage and is inversely correlated to the number of stored images. The speed of the process is practically constant when the error varies.

The paper is organized as follows. In Section 2 the image compression generality is formulated; in Section 3 the MAIA NN model is proposed. In section 4 the experimental results are discussed in terms of quality of the Compression/Decompression (C/D) process, error term and number of compressed images. A comparison is also made with two well known methods: JPEG and MPEG 4.

2 The Image Compression. Generality

As known, an image can be represented through a matrix of points (pixels); since a numerical value expressed in bytes can be associated to each pixel, an image can be represented by a *mxn* matrix of bytes ($\mathbf{I}_{m,n}$). The compression can be thought as a transformation F such that: $\mathbf{I}_{m,n} \rightarrow \mathbf{I}_{d^*}$ where $\mathbf{I}_{d^*}$ is a vector $dx1$, with $d<mxn$ [14]. Naturally, *d* has to be smaller than *mxn* as much as it is possible compatibly with the loss of information. The ratio $(mxn)/d$ denotes the compression degree that can be obtained.

When an image has been codified (compression) in the vector $\mathbf{I}_{d^*}$, it can be reconstructed (decompression) through a suitable transformation of $\mathbf{I}_{d^*}$. After the decompression, a check needs in order to establish the goodness of the reconstructed image, i.e. if there was any loss of information. The goodness of the reconstruction can be evaluated through some indicators: the signal-to-noise ratio (SNR) and the peak value of SNR.

In this paper *k* images-matrix *nxn* compressed into an image-vector $kn+\lambda n^2$ are proposed (λ is a coefficient <1). The square dimension of the pixel matrix of the image does not affect the generality of the method proposed in the following.

3 The MAIA Model

The transformation F defined in section 2 can be obtained by using the MAIA NN [10], based on the asynchronism of the input data processing . The problem to be resolved has a non linear nature, then the non linear version of the MAIA NN has been used for the C/D experiment.

As a general remark, in the MAIA NN the formal neuron is identified with a tension generator and the synapses with an RLC electrical circuit [10]. The "transfer function" T associated to the neuron has the function of converting an input i(t) time-variable in an output u(t)=T(i(t)) still time-variable [9], [13].

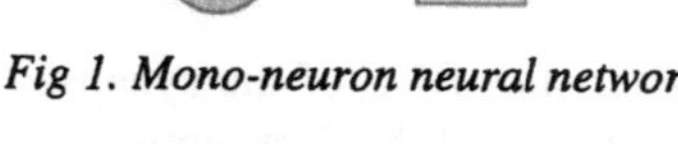

Fig 1. Mono-neuron neural network

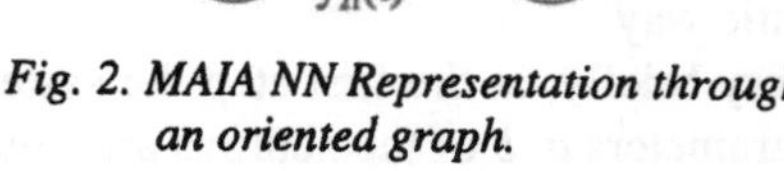

Fig. 2. MAIA NN Representation through an oriented graph.

The behaviour of the MAIA NN depends on the nature of the input i(t). If the input is transformed in a sinusoidal function of the form:

$$i_n(t)=i_n \mathrm{sen}(\omega_0 t+\alpha_n) \qquad (1)$$

where i_n is the amplitude (i.e. the original input), ω_0 the initial frequency and α_n the initial characteristic impedance angle, the steady state solution of the equations underlying the system is [10]:

$$y_m(t)=y_m \text{sen}(\omega_0 t+\alpha_n+\phi_m) \tag{2}$$

with:

$$y_m = Z_m i_n \; ; \; Z_m = 1 \Bigg/ \sqrt{R^2+\left(\omega_0 L-\frac{1}{\omega_0 C}\right)^2} \; ; \; \phi_m = -\text{tg}^{-1}\frac{\omega_0 L-\frac{1}{\omega_0 C}}{R} \tag{3}$$

y_m is the amplitude of the sinusoidal output, Z_m the circuit impedance and ϕ_m the characteristic impedance angle produced by the circuit. As said before, the synapses are identified with electrical circuits, then the previous parameters are referred to synapses. Particularly, the synapsis *m* coming out from a neuron has got as weight the impedance Z_m of the electrical circuit.

In the very simple case of a mono-neuron NN (fig.1), the transfer function is the identity one (i.e. u(t)=T(i(t))=i(t)), since this form guarantees the identification of the mono-neuron NN with an electrical circuit [10]. In the general case, more linear mono-neuron systems, giving rise to different NNs, can be considered. In fig.2, for example, the input to neurons is equal to the sum of the external inputs i(t) and the information y(t) coming from the synapses ending in it [1]. The Kirchoff rule applied to the system of fig.2 allows resolving the system.

In the case of non linear problems, a suitable transfer function has been introduced [10]:

$$T(i)=\text{tgh}(ai+b) \tag{4}$$

where *a* and *b* are parameters to be calibrated together with the synapsis characteristics. The solution of the non linear MAIA NN can be obtained through the following relationship [10]:

$$\overline{\mathbf{Y}} = \mathbf{W}(\omega_0)\overline{\mathbf{I}} \tag{5}$$

The vectors of estimated data $\overline{\mathbf{Y}}$ and $\overline{\mathbf{I}}$ in the previous relationship are linked to the sinusoidal transformation y(t) and i(t) as follows:

$$y_i(t) = \text{Im}\left(\overline{y}_i e^{j\omega_0 t}\right), \quad i_i(t) = \text{Im}\left(\overline{i}_i e^{j\alpha_0}\right) \tag{6}$$

The learning strategy is realized by updating *a*, *b* in (4) and RLC in (3), and this is obtained through a function S(t), called *neuron state*, given by the sigmoid function. This function is associated to each neuron and if a different initial state is set for each neuron, an effective learning can be guaranteed.

However, for non linear problems the MAIA NN can be resolved only in a heuristic way.

Very briefly, at the first step the system characteristics RLC of the synapses and the parameters *a, b* of the neurons are suitably initialized and the first input pair (i, y) is processed by choosing an initial values ω_0^1. At the presentation of the pattern p on a synapsis m, at a time t_m^p, the estimated value $\overline{y}_m^p$ is obtained; this value is compared with the expected value (y_m^p), belonging to the training set. The difference $\varepsilon_m^p = \overline{y}_m^p - y_m^p$ between the two values gives the *m*-synapsis error relative to the pattern p that is used to modify the system characteristics RLC, the elements of $W_m(\omega_0^p)$ and the parameters a and b. Two methods can be chosen in order to obtain the above modification, the Hebb [6] or the Grossberg [9] rule. Usually, the first one is used; for example the updating of R is $\Delta R_m^p = \Pi\ S_n S_k$, where Π is the learning rate (belonging to the range [0, 0.1], n and k are the starting and ending neurons of the synapsis m. In a similar way all the other parameters can be computed. The updating of the synapses and neuron parameters is obtained at a different time t_m^p and for this reason this kind of NN is called *Asynchronous.*

The computing complexity of the MAIA algorithm is linear as regards the synapses and patterns, and it is inversely related to the desired error: it belongs to the pseudo-polynomial in $1/\varepsilon$ set of algorithms.

4 Compression/Decompression through a MAIA NN

The set of experiments proposed in this paper was addressed to verify the performance of the MAIA NN in the C/D of set of images in an exhaustive way, and compatibly with the learning modalities.

Some variations with respect to previous experiments have been introduced. These variations, however, does not concern the topology of the MAIA NN for the C/D of images [10], which still consists in a completely connected graph, i.e. each neuron is linked to each other, included itself. Then, if the image to be compressed has a nxn dimension, the NN must have n neurons and n^2 synapses.

In the compression stage, given a set of images the process will start from an arbitrary configuration ω_0, for each image the bytes associated to each pixel of the examined image will be stored in the synapses and it will be used as weight value for searching the synapses frequencies that allow the NN to converge to the desired value of error; when this test will be satisfied for each image, the NN no longer will modify its parameters, and the process will converge to a solution. For each image in the set of the k images presented, the n outputs to the neurons and the λn^2 values of the synapses frequencies represent the compression I_{d*} of the set of images $I_{n,n}$.

In the decompression stage, the same topology of NN has been used; in this stage the synapses weights and all the variable terms are randomly initialised and updated in subsequent MAIA iterations. The procedure is an inverse process as regards to the compression stage, in which given the synapsis weights, the synapses frequencies and the outputs to the neurons are estimated. In fact, in the case of decompression, the synapses frequencies and the outputs to the neurons are used in order to obtain the NN convergence weights. The weights obtained over the synapses, in this case, represent the numerical value of the pixels of the initial image. Then, the decompression decodes I_{d*}, which represents the image set, in the initial k images $I_{n,n}$.

The number of compressible images only depends on the desired quality and the planned ratio of compression. In these experiments, a maximum number of 25 images to be compressed is considered, since the performance of the MAIA NN decreases very quickly when a larger number is considered.

In terms of learning error, values smaller than 10^{-3} do not improve the performance of the MAIA NN, while the computing times increase appreciably.

The experiment is formed by 2500 tests with sets from 1 to 25 images, and learning error from 0.001% to 1% with step of 0.01%. Obviously, 2500 values of SNR and peak of SNR have been obtained varying from 38.69 dB to 10.57 dB (for the SNR values, fig.3) and from 40.29 dB to 13.94 dB (for the of SNR values, fig.4).

Figs 3 and 4 show that the NN performances decrease rather quickly when the set of images and the learning error value increase. As regards the compression degree, it is a function of only the number of processed images and varies from 1:1.3 for 1 image to 1:27.2 for sets of 25 images. If a post processor (or a generic compression program without loss of information) is used the values are respectively 1:2.1 and 1:24.

The learning procedure for the compression stage on a computer with a Pentium 100 processor varies from 4 seconds for one image to 100 seconds for 25 images, while the decompression varies form 0.5 to 13 seconds.

Tab 1 shows the compression ratios, the SNR values and the peak of SNR for 1, 12 and 25 images obtained by using the MAIA NN, JPEG (without loss of information and minimum compression ratio and with loss of information and maximum compression ratio), and MPEG 4.

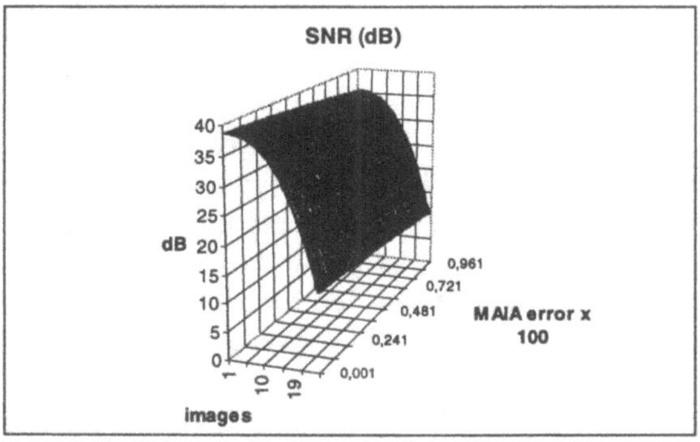

fig. 3

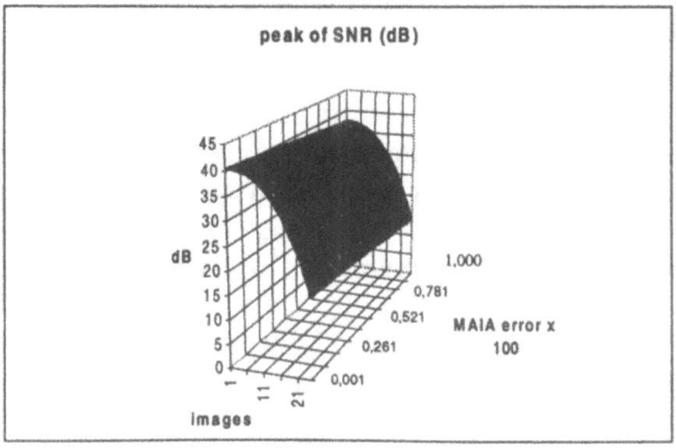

fig. 4

Therefore, the C/D of images obtained through a MAIA NN is competitive as regards JPEG with medium size set of images (12 images) while its performances are not very interesting in all other cases.

Model	1image		12 images		25 images	
	Comp. rat.	SNR / pSNR (dB)	Comp. rat.	SNR / pSNR (dB)	Comp. rat.	SNR / pSNR (dB)
MAIA	1:0,99	38,69/40,29	1:10,97	34,81/36,57	1:20,92	13,75/16,74
JPEG	1:2,79÷12	0 loss ÷28,15/30,0	1:2,79÷12	0 loss ÷28,15/30,0	1:2,79÷12	0 loss ÷28,15/30,0
MPEG4			1:64	26,66/28,42	1:60	26,12/28,33

Tab.1. Comparison of C/D models.

References

[1] Arbib M. The metaphorical brain: an introduction to cybernetic as artificial intelligence and brain theory. Wiley-Interscience, New York, 1972

[2] Ballestrino A, Celentano G. Teoria dei sistemi, vol III. (*in italian*) Liguori Editore, Napoli, 1982

[3] Carrato S, Marsi S, Sicuranza G, Ramponi G. Two and three dimensional data coding. Proc. 1 Italian Workshop Digital Image Processing by NN. Fond. Bordoni, Roma, 1993.

[4] Cottrell G, Munro P, Zipser D. Image Compression by BP: an example of extended programming. Advances in cognitive science 1988.

[5] Gelenbe E, Sunger M. Random nets and image compression. Proc. ICANN 94, Springer-Verlag, London, 1994, pp. 1117-1120.

[6] Hebb D. The organization of behaviour: a neuropsycological theory, New York, Wiley 1949

[7] Kangas J. Increasing the error tolerance in transmission of vector quantized images by self-organising map. Proc. ICANN '95. EC2&Cie, 1995, 1:pp.287-291

[8] Mc Culloch W, Pitts W. A logical calculus of the ideas immanent in nervous activity, Neurocomputing: foundation of research, MIT Press, Cambridge, 1988

[9] Nilsson NJ. Learning machines: foundation of trainable pattern classifying system. Mc Graw-Hill, London, 1965

[10] Pappalardo G, Rosaci D, Sarnè GML. MAIA unsupervisioned neural network in the compression and decompression of images. Proc. WIRN 96, 1996, pp 217-222.

[11] Piazza F, Smerilli S, Uncini A, Griffo M, Zunino R. – Fast spline neural networks for image compression. In: Proc. WIRN 96, 1996, pp 223-230

[12] Ruberti A., Isidori A. Teoria dei sistemi. (*in italian*) Boringhieri, Torino, 1979

[13] Rumelhart D, McClealland JL and the PDP Research Group. Parallel distributed processing, vol 1, MIT Press, Cambridge, 1986

[14] Shannon CE, Weaver W. – La teoria matematica delle comunicazioni. (*in italian*) ETAS Libri, Milano, 1971

[15] Zadeh LA, Desoer CA. Linear System Theory, McGraw-Hill, New York, 1953

The Effect of Public Sector on the Financial Sector: An NN Approach in a View of Complexity.

Maria Rosaria Alfano
Massimo Salzano
Dipartimento di Scienze Economiche, Università di Salerno†
Salerno, Italy
salzamas@netlab.it

Abstract

In the existing literature there are many works which use NN for forecasting the financial market. Generally, these are centred on technical analysis and/or endogenous variations.

A different approach is based on the use of "fundamentals". In this latter approach, the effect of fiscal variables between the fundamentals is disregarded.

We considered both technical-endogenous and fundamental variables at the same time for the forecasting. Particularly, we used the payment due by taxpayers (estimating the daily's burden) as explicative for differences in the level of Comit30 index. The sensibility of financial behaviour to the tax burden seemed to be characterised by a "threshold" effect; that is to say a complex - non-linear behaviour.

Final forecasting reached is quite good. It is possible to demonstrate that the correct consideration of exogenous shocks, together with endogenous behaviour of the financial market, could be the key passage for a correct use of neural networks in complex systems. More, the importance of a proper non-linear consideration of the interrelation between fiscal and financial sectors was highlighted.
Keywords: Economics, Public Finance, Neural Net, Complexity, Chaos, and Financial sector

1 Introduction

The analysis of the financial data is not an exact science. Fundamentally because: a) not all factors affecting the financial sector are faultless known; b) their interrelation is too complex to be described by a mathematical model. However, it now seems that Neural Networks (NN) can decipher patterns in the data and help with such analysis[1].

† 1st and 2nd paragraphs have been prepared by Alfano, the remaining by Salzano, of course, conclusions are cooperative results.

1 Mei Lin e Frank C. Lin (1993).

Although, many studies have been conducted using this new tool, results are not completely satisfactory. This is perhaps due to the fact NNs are only recently becoming employed in the financial and economic community and up until now, have not been very diffuse in this context. As consequence, many of the works are oriented towards looking for the effect of endogenous behaviour; even when "fundamentals" are considered, fiscal variables are generally disregarded.

In 1987, Lapedes and Farber showed that backpropagation NNs could be used for forecasting a strongly chaotic time series and more accurate results could be obtained compared with conventional methods such as those of linear forecast. Their result was more accurate than that obtainable with any other conventional prediction method.

In the case of forecasting the Dow Jones Industrial Average, since Dow Jones varies considerably, it is better to use variations rather than amounts as inputs. With smaller values the net is more sensitive to small changes

In the First International Conference on Artificial Intelligence Applications on Wall Street, in the October 1991, Bowe presented his work on financial forecasts for a day in advance. A NN was trained and the prices of the stocks were compared with the anticipated prices using the standard deviation as a measure of accuracy of forecast. Utans and Moody applied risk forecast to anticipate likely corporate bond ratings. Jang and Others described a method for short period forecasting of stock negotiated on the Taiwan stock market by using two sets of NN one "feedforward", and one "multilayered". Margarita obtained good results about the shares of Fiat in the Milan's Stock Exchange.

It should be noted that in economic and financial literature the use of NN is generally more concentrated for the application of "technical" analysis and not for analysis based on "fundamentals". This is due to the fact that the fundamentals are generally the base of a linear approach which is commonly rejected by researchers interested in NN. Such an approach generally considered correct is not appropriate, in the case of economic system. In fact, it seems to over-simplify economic reality. As many researchers have pointed out, the understanding and forecasting of the behaviour of financial and economic sectors must be based on a complex approach in which both endogenous and exogenous variables need to be considered[2].

The amount of money, drained from the market by the public sector, in an exactly predictable timetable, is a strong exogenous shock variable which must be taken into account for the correct forecasting of financial behaviour. The amount of disposable money within the financial sector decreases according to this timetable, thereby reducing the market energy. So in that period the possibility of a financial decrease is greater. Thus, generally speaking it is possible to know in advance when those external shocks will arrive. This information will increase the accuracy of forecasts.

[2] Even the more recent works on application of NN to financial sector avoid to consider the fiscal variables. See for example: Haefke, C.; Helmenstein, C. (1996); Shaaf, M. (1996); Hutchinson, J.M.; Poggio, T.; Lo, A.W.(1994), Refenes, A. (ed. by) (1995).

2 The Data and the NN used

The considered period was from 1/1/1991 to 31/12/1996. For the Italian financial market the official data about Comit30 was used. For the fiscal sector data from the "Ministero delle Finanze" was considered.

In the paper only tax receipts from IRPEF, IRPEG and IVA were considered because they represented the most relevant variation in total yields, month by month.

As this is month data from the "Tesoreria", it had to be considered relative to the correct month in which a substantial fiscal event was present in the last day of the period. The fiscal timetable was used. On this base, estimated the amount of fiscal burden that the economic system had supported on any given day of the period was calculated.

Figures 1 and 2 show the existence of a strong link between the index of the financial market and the tax receipt. The analysis of the data indicates that this relation is non-linear. In fact, there is a threshold value, dependent on economic, social and political variables, under which the tax burden does not influence financial behaviour. Due to the existence of this threshold, the fiscal phenomena had to be considered complex (in the graphs only the threshold values of tax burden and corresponding financial values are represented).

It became evident that a strong sensibility to fiscal events became apparent only after a certain total amount of fiscal burden. An amount which was flexible during the period. Overall, it amounted to 7800 mld for the first period and 11400 mld for the second.

This threshold suggests that the original historical series should be split into two: the first starting on January 1st 91 and finished on May 19th 93, the second beginning 20th May 93 and ending 31st December 96.

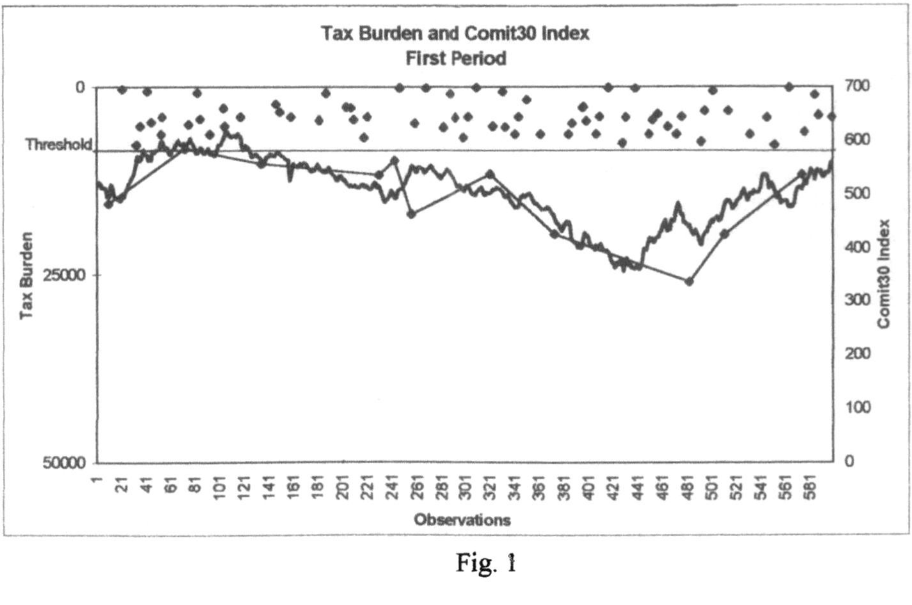

Fig. 1

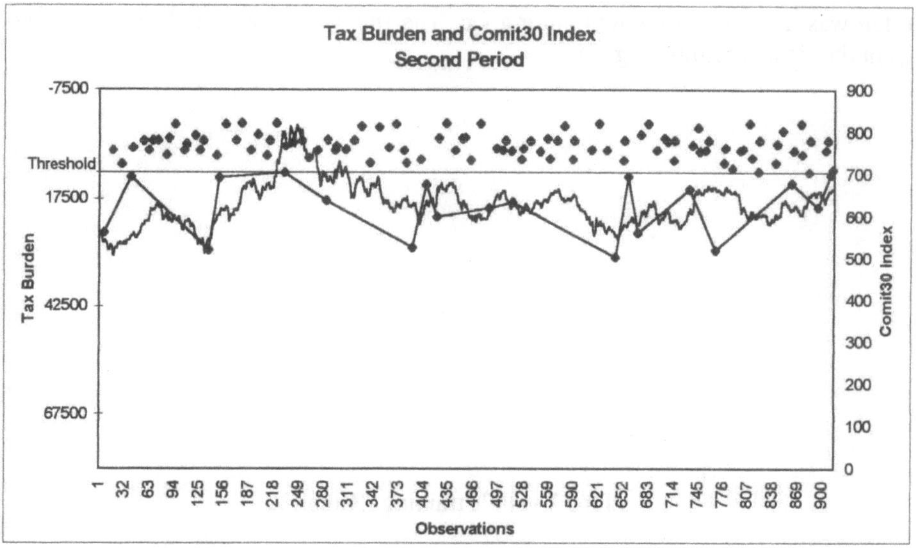

Fig. 2

From the figures it was possible to ascertain that, during the same periods, the exogenous shocks (free trade of capital, crisis of political and social values, change of interest rates, war in the Persian Gulf etc.) could modify the system's endogenous behaviour for a while; after which the system would return to its original trend. The final period, after March 1996 is characterised by a less precise behaviour pattern perhaps due to the starting point of 15/3/1996 contracts, when all contracts had to be immediately paid (and not at the end of financial month).

For NN a very simple shell (NNDT114 - Levernberg-Marquardt) was used as the final end was not that of showing any particular characteristic of NN, but, rather, to use it as a tool for highlighting the fiscal - financial relationship. For technical information an NN with 4 hidden neurons, one feedback and a penalty of 0.01 was used.

3 The financial application

Italian data by Comit30 Index was used. The considered period was from 1993 to 31/08/1994; with only working days considered. Initially, the possibility of forecasting differences relative to the previous day's trading was evaluated. The results were not very interesting (they are not shown).

Subsequently, to analyse the capacity of NN to explain our financial data, 3 main NNs were tried on monthly data: i) only with the level of Comit30 index (one observation lag); ii) with both the level of Comit30 (same lag) and tax payments; iii) only with the latter values. As can be seen from Fig. 3, when using only the level of Comit30 index, the result was not very interesting. In fact the rms was 0.1139 and the training showed a 1-observation lag. Both the levels of Comit30 (same lag) and tax payments gave good results was quite good (Fig. 4). The rms was 0.0712. This was the direct consequence of using tax burden. In fact, even in the case where only the tax burden was used the result was quite good. The rms was 0.0871 and there were no lags in the trained results (Fig. 5).

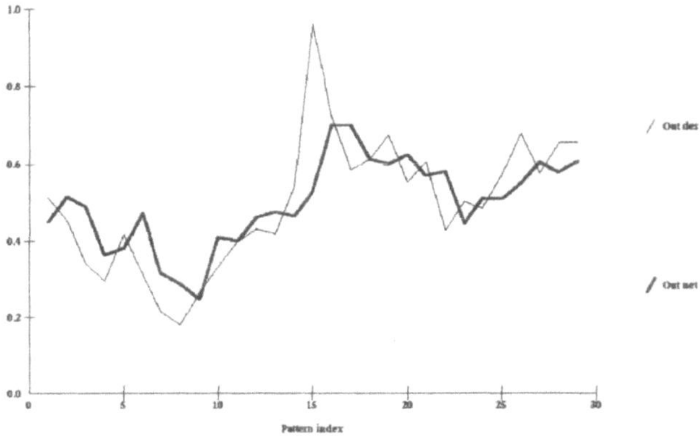

Fig. 3 - NN's training - Comit30

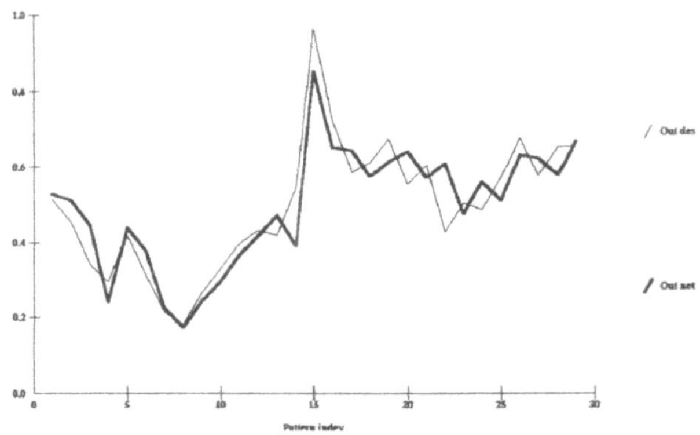

Fig. 4 - NN's training - Comit30 and tax burden

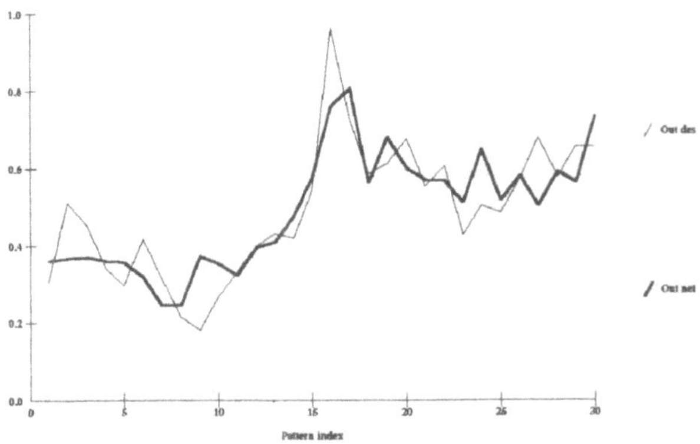

Fig. 5 - NN's training - tax burden

The results seem very interesting, but more work must be done for a better integration of fiscal external information. In fact, the training and forecasting for the complete set of observation (daily data) are shown in Fig. 6.a and Fig. 6.b; obviously, this is obtained after a new training in which the NN has not seen the period that have to be forecasted. And the forecast still presents lag.

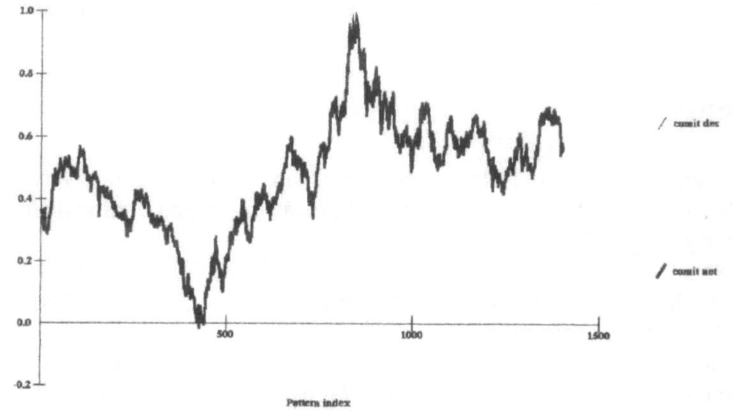

Fig. 6.a - Training with the period 1991-96

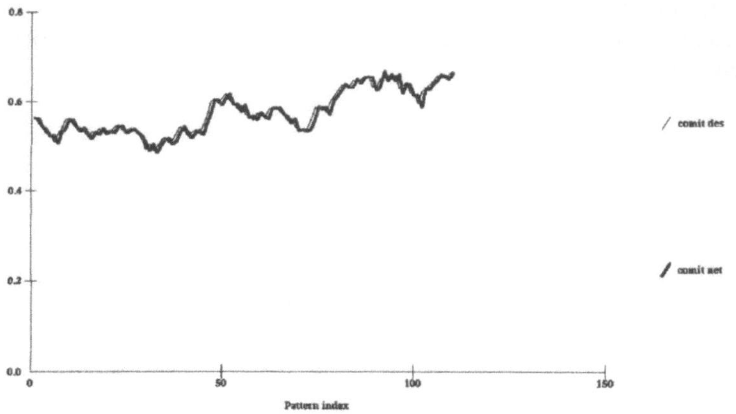

Fig. 6.b Forecast for 114 final observations

4 Main results

The increase in the accuracy of forecasting of financial market obtained from the use of fiscal data was very important. This case seems to be under - estimated in literature when compared against the obvious consideration that in periods preceding data of strong tax burden, economic agents are not prone to invest in the financial market but vice versa, to disinvest, thereby causing a possible reduction in share prices.

As this phenomena occurs within a complex scenario the use of the usual linear methodology would be incapable of capturing it. Using a simple NN, it is demonstrable that the shocks caused by the fiscal burden have a strong possibility of clarifying and explaining financial variations.

Moreover, it seems that a mixed approach of both "Technical" and "Fundamental" variables is an interesting way of applying NN to the financial sector.

It is not possible to arrive at any complete theoretical conclusions. In fact, as the *fiscal burden for each day had to be estimated too few observations were evaluated.*

References

[1] Haefke,-Christian; Helmenstein,-Christian (1996): Neural Networks in the Capital Markets: An Application to Index Forecasting, Computational-Economics; 9 (1), February, pages 37-50.

[2] Shaaf,-Mohamad (1996): A Neural Network and Econometric Comparison of the Relative Importance of Fiscal and Monetary Actions, Studies-in-Economics-and-Finance; 17(1), Fall, pages 69-87.

[3] Hutchinson,-James-M.; Poggio,-Tomaso; Lo,-Andrew-W. (1994): A Nonparametric Approach to Pricing and Hedging Derivative Securities Via Learning Networks; National Bureau of Economic Research Working Paper: 4718, April.

[4] Mei Lin e Frank C. Lin (1993): Analysis of Financial Data Using Neural Nets, AI, pp. 33-37

[5] Refenes, A. (1995): Neural Networks in the Capital Markets, John Wiley & Sons, New York.

A Hybrid Framework for Indoor Robot Navigation

Edmondo Trentin and Roldano Cattoni
ITC-Irst (Istituto per la Ricerca Scientifica e Tecnologica)
I-38050 Povo (Trento), Italy
{trentin,cattoni}@irst.itc.it

Abstract

This paper introduces a hybrid system for modeling, learning and recognition of sequences of "states" in indoor robot navigation. States are broadly defined as local relevant situations (in the real world) in which the robot happens to be during the navigation. The hybrid is based on parallel Recurrent Neural Networks trained to perform *a-posteriori* state probability estimates of an underlying Hidden Markov Model given a sequence of sensory (e.g. sonar) observations. The approach is suitable for navigation and for map learning. Encouraging experiments of recognition of noisy sequences acquired by a mobile robot equipped with 16 sonars are presented.

Keywords: mobile robot, indoor navigation, learning, sequence recognition, recurrent neural network, Dynamic Programming, Hidden Markov Model.

1 Introduction

Mobile robots are required to accomplish complex tasks like transport missions and map building. Many approaches to these tasks have been proposed in literature: from traditional artificial intelligence techniques [7], through the more unconventional behavior-based methods [3], to the so called "mixed approaches" [9, 4]. All these techniques heavily differ on the way the sensory data are organized and possibly interpreted with respect to a world model. A common approach is that of using a global map encoding at least 2D metric information about the environment; the navigation is then based on the estimation of the robot *position* with respect to the map.

This paper explores a framework in which the building blocks are represented by *sequences* of local relevant situations, which does not contain global metric information. Such relevant situations correspond to portions of the environment easily recognizable by humans: in the case of indoor navigation in structured buildings, example of local relevant situations are corridor intersections, approaching openings or doors. The focus is therefore moved from the metric structures towards higher-level topological information. Founding navigation on the perception of local relevant situations allows a rather easy definition of the main navigation tasks: (1) *navigating* to a prescribed point is described as a sequence of relevant situations, each associated to a specific *behavior*, that is a local motor action accomplished by activating an appropriate sensors-to-actuators command loop; (2) *learning maps* is the process of building a graph-like structure where nodes encode the relevant situations and links represent behaviors needed to navigate from one situation to another; (3) *issuing commands* by a remote operator (teleoperation) benefits from the information of current estimated relevant situation, which helps the remote operator in selecting and monitor behaviors.

This paper introduces a novel approach to *learning* a characterization (model) of such relevant situations from examples. During a supervised training session the proposed system learns the dynamics and the statistics of the stochastic process which characterize the sensory observations through time, when the robot moves and traverses the different situations. The approach is based on the integration of Recurrent Neural Networks (RNNs) [6] within an Hidden Markov Model (HMM) [8], in order to take advantage of the learning capabilities, flexibility and temporal modeling of these paradigms.

The underlying idea is to use the states of the HMM to represent the above-mentioned relevant situations, and to describe the statistics of sensory signals within each situation by way of the emission probabilities attached to the corresponding state (observable part of the model).

Unfortunately, standard HMMs do suffer from some major limitations that can dramatically decrease their applicability to the task at hand (see Section 2.3). One way to overcome these problems is represented by the adoption of Artificial Neural Networks (ANNs). The resulting hybrid architecture can be thought of as an *adaptive* (i.e. learning) model of the environment, where the relevant semantics (conceptual information that, in a human perspective, describes a particular situation) is given by the states of the hidden Markov chain and by the topology defined by the allowed transitions between pairs of states; the time-dependent statistics of sensory data are captured via the connectionist estimates of the emission probabilities associated with the states.

The advantages expected from the present approach can be pointed out as follows:

1. Capability of *learning* from examples the parameters that characterize a model of the environment, taking into account the time-varying nature of sensory data and the sequential relations among states.

2. Adaptivity of the model to environmental changes in time: the proposed learning paradigm can be used in conjunction with an on-line training scheme, e.g. the *on-line gradient descent* [1], to refine parameters after the original training phase without requiring a complete retraining of the system.

3. Capability of dealing with long-term time dependencies among situations (states along paths in the HMM) as well as with short-term sensory sig-

nal evolution. Recognition of relevant situations is not based upon *local* considerations only, but relies on previous history in a significant manner.

4. Robustness against noise, sensory fluctuations and instantaneous transitory variations. The adoption of RNNs is crucial from this point of view, since their internal memory of past observations allows for a robust smoothing (or filtering) mechanism.

5. No need for metric information. Recognition of situations is based only on sonar data, and on short-term and long-term memories of previous observations and previous situations.

2 Framework for the hybrid HMM/RNN system

2.1 HMM for Robot Navigation

We can summarize the elements of a HMM in the following way:

1. a set S of Q states, $S = \{S_1, \ldots, S_Q\}$, which are the distinct values that the discrete, hidden stochastic process can take.

2. An ***initial state*** probability distribution, i.e. $\pi = \{Pr(S_i \mid t = 0), S_i in S\}$, where t is a discrete time index.

3. A probability distribution that characterizes the allowed transitions between states, that is $\mathbf{a}_{ij} = \{Pr(S_j \text{ at time } t \mid S_i \text{ at time } t-1), S_i \in S, S_j \in S\}$ where the ***transition probabilities*** $\mathbf{a}_{ij}$ are assumed independent of time t.

4. An ***observation*** or ***feature*** space $\mathcal{F}$, which is a discrete or continuous universe of all possible observable events (usually a subset of $\mathbf{R}^d$, where d is the dimensionality of the observations).

5. A probability distribution (referred to as ***emission*** or ***output*** probability) that describes the statistical properties of the observations for each state of the model: $\mathbf{b}_\mathbf{x} = \{Pr(\mathbf{x} \mid S_i), S_i \in S, \mathbf{x} \in \mathcal{F}\}$.

HMMs represent a learning paradigm, in the sense that examples of the event that is to be modeled can be collected and used in conjunction with a training algorithm in order to *learn* proper estimates of the probability distributions π, $\mathbf{a}$ and $\mathbf{b}_\mathbf{x}$. The most popular examples of such algorithms are the *forward-backward* (or *Baum-Welch*) [8] and the Viterbi [8] algorithms. Both of them are based on the general ML criterion, i.e. they aim at maximizing the probability of the samples given the model at hand (the Viterbi algorithm concentrates only on the most alike path throughout all the possible sequences of states in the model).

These algorithms belong to the class of ***unsupervised learning*** techniques, since they perform unsupervised parameter estimation of the probability distributions without requiring any prior labeling of individual observations (within the sequences used for training) as belonging to specific states. In the case of robot navigation we can take advantage of the knowledge of the exact physical state that the robot is traversing during the acquisition of the training sequences, by adopting a ***supervised learning*** algorithm (see Section 2.3).

Once training has been accomplished, the HMM can be used for ***decoding*** or ***recognition***. In this study a Dynamic Programming (DP) [1] algorithm, equivalent to a Viterbi decoding, is performed on the outputs of the ANNs in order to recognize the sequence of states that the robot has traversed during the navigation. This is useful for learning maps of the environment, where each sequence of traversed states forms the building block for a global representation of the explored environment. In addition, real-time estimation of probabilities of states given the sequence of observations – performed by the ANNs – can be used to control the robot relying on a probabilistic estimation of the current state.

2.2 The RNN model

In this study samples are temporal sequences of patterns, instead of individual independent samples. To take this temporal dependency into account, RNNs are used. The class of recurrent networks that is considered presents a set of ***status*** or ***context units***, providing a kind of duplicate, or ***backup***, of the input layer. Context units are charged of all the recurrent connections. This permits to keep well separated the different contributions from the current input vector and from the evolving history. The context layer is linked with feedforward connections to the hidden layer, resulting in an extension of the input layer. Self-recurrent connections from each context unit to itself are added at the context layer, sharing a common weight w_s. Direct lateral connections from each input unit to the corresponding context unit also hold, with shared weight w_l. The role of the new units is to keep track of an internal, evolving ***status*** of the network. The latter thus acts in a way which somehow resembles a finite state automaton [5].

The dynamic of context unit k is described in terms of its output o_k at time t, whose equations are

$$\begin{aligned} &o_k(0) = 0 \\ &o_k(t+1) = w_l x_k(t) + w_s o_k(t) \end{aligned} \tag{1}$$

where $o_k(t)$ is the output from unit k at time t (assuming that the sequence of patterns to be processed starts at $t = 0$), w_l is the common weight of lateral connections between input unit k and the context unit itself, $x_k(t)$ is k-th component of t-th pattern in the input sequence.

2.3 Integrating HMMs with RNNs

Attempts to use standard HMMs for robot navigation suffer from some significant limitations, due to theoretical as well as to problem-specific matters. The first, main drawback is represented by the implicit assumption of stationarity of events described by individual emission probabilities attached to the states of the Markov chain. In the present task this requirement is indeed not satisfied, since the platform keeps on moving while traversing a given state, and the sensory signals change in time accordingly. Moreover, fluctuations of the sensory apparatus, environmental noise, people crossing, etc., induce random variations on the signal itself, so that a kind of smoothing over time, or filtering, based on previous inputs is sought. Instead of modeling the distribution of sensor readings in a given state with a (static) probability density function (pdf), the heavily time-varying nature of the sensory observations is exploited by means of dynamic neural networks.

In addition, the combination of ANNs with HMMs is intended here as an attempt to overcome some major intrinsic limitations of standard HMMs (see for instance [2]). Standard CDHMMs, trained with Forward-Backward or Viterbi algorithms, present poor discriminative power among different models, since they are based on the ML criterion, which is itself non-discriminative. Furthermore, classical HMMs rely on strong assumptions on the statistical properties of the phenomenon at hand, for instance modeling the stochastic processes involved as first order Markov chains. Also, the parametric form of the probability density functions that represent the emission probabilities associated with all states is constraining. Given these limitations, the use of RNNs with their discriminative training and their capability to perform non-parametric estimation over whole sequences of patterns can be attempted.

In the present approach we stress the fact that the objective of discriminative training relies on the maximization of the posterior probability of a given sequence of local relevant situations $\mathcal{P}$, i.e. a sequence $q_1, ..., q_L$ of states of the Markov model (the *correct* sequence), given a sequence $X = \mathbf{x}_1, ..., \mathbf{x}_L$ of sensory observations, rather than the usual maximization of the likelihood of the observations given the model.

Repeatedly applying the properties of joint probabilities we can write:

$$Pr(\mathcal{P} \mid X) \quad = Pr(q_1^L \mid X) = \prod_{\ell=1}^{L} Pr(q_\ell \mid X, q_1^{\ell-1}) \tag{2}$$

where, by convention, $Pr(q_1 \mid X, q_0) = Pr(q_1 \mid X)$. Attempts to determine analytical developments for the present formulation does not lead to computationally feasible expressions. [2] proposed a hybrid speech recognition system where a static feedforward neural network was used as estimator of the posterior probabilities of states given a fixed-size window over the observations ([2], page 164, Case 4). In the present research, an approximate version of equation (2) is used, by taking:

$$P'(\mathcal{P} \mid X) = \prod_{\ell=1}^{L} Pr(q_\ell \mid \mathbf{x}_1^\ell) \tag{3}$$

that is to say, for each state of the HMM a network is trained to estimate the state posterior $Pr(q_\ell \mid \mathbf{x}_1^\ell)$ given the partial sensory vector sequence $\mathbf{x}_1^\ell$. This is accomplished using a supervised gradient-descent training algorithm on a recurrent neural network for each one of the states (after training, the output of the nets must be actually divided by the state-prior if an unbalanced training set is used). The training set T_j for the j-th network is obtained by labeling the ℓ-th pattern of each training sequence $\mathbf{x}_1^L$ with the target probability values 1.0 or 0.0 according to the fact that the robot was, or was not, in state q_j at time ℓ during acquisition of sensory signal $\mathbf{x}_\ell$. All the available training sequences $\mathbf{X}$ are used to train all the Q RNNs, with different distributions of the corresponding labels. This instantiates the concept of discriminative training, since models are trained on "positive" as well as on "negative" examples. As pointed out in [2], *ibidem*, this is no longer strictly a HMM, but the general framework is still valid and the standard decoding algorithms (e.g. Viterbi) can still be used.

If the hybrid system is used to learn maps of the environment, e.g. assuming that navigation locally relies on a different paradigm, recognition of whole sequences of states can be accomplished in a batch manner when the robot reaches certain specific checkpoints (such as a landmark or the end of a corridor). The recognition of the sequence can be carried out by a DP algorithm, in order to find the path $\mathcal{P}$ through the states of the HMM which maximizes the quantity $P'(\mathcal{P} \mid X)$ as defined by equation (3), using the outputs of the RNNs as estimates of the posteriors $Pr(q_\ell \mid \mathbf{x}_1^\ell)$ for $\ell = 1, \ldots, L$. This is equivalent to a Viterbi decoding algorithm. Transition probabilities $\mathbf{a}$ can still be introduced in the criterion to be maximized with DP, like in standard HMMs, for instance in order to constrain the search to admissible sequences of states only. Transition probabilities are not learned in the present scheme, but can be computed in advance, according to the topology of the environment and possibly to relative frequencies of physical states within it, and kept fixed.

If the hybrid model is adopted to control the navigation itself, the conditional probabilities $Pr(q_\ell \mid \mathbf{x}_1^\ell)$ of states given the observation sequence $\mathbf{x}_1^\ell$, estimated in real-time by the corresponding RNNs, are used at each time step $t = 1, \ldots, \ell$ to estimate the most alike current state within the on-going path (possibly constrained by using the transition probabilities of the underlying HMM to compute joint probabilities).

3 The mobile robot and the experimental environment

In the present research the robot moves in corridors and open spaces of our Institute; the presence of fixed obstacles and people makes the environment not completely predictable. The vehicle was designed and assembled at IRST; its external dimensions are 0.5 x 0.5 x 0.9 m. Sixteen sonars, frontally and laterally distributed, can detect reflecting surfaces in the range [0.2 m, 4.0 m].

Although the sonar-based sensory apparatus can provide rough range in-

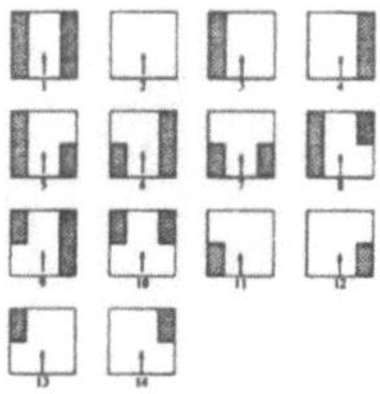

Figure 1: *The chosen relevant situations or* states*: free space is painted in white color, while persistent structured obstacles (like walls) are in gray; the arrow represents the direction of the moving robot.*

formation, relevant situations (or states) at a high level of description (like corridor intersections or approaching openings) can be defined. 14 such states were defined (Figure 1). Not all the transitions between states are allowed; furthermore, some admitted transitions can occur more frequently than others – according to the structure of the environment – so that it is possible to associate with them a probability of occurrence.

Real-world training and test data sets were collected using the mobile robot during indoor navigation. The task was the recognition of sequences of states having arbitrary length. The 14 states introduced above were adopted. 16 sonar readings were acquired at each time instant.

RNNs were first trained to estimate state posterior probabilities in a supervised manner, i.e. relying on a prior labeling of data, accomplished by using the *a priori* knowledge of the correct state at each time during the navigation. Figures 2-4 show the typical curves exhibited by the networks corresponding to an individual state of the HMM, compared with the associated target outputs, as a function of time (discrete time instants are represented along the abscissa). Figure 2 represents the desired output, i.e. a null probability wherever the robot is not in the corresponding state, and a probability of 1.0 when the robot actually is in that state. The target function was smoothed in correspondence with transitions to and from the state. This qualitatively means " the robot is approaching (or leaving) the state". Training of the networks took benefit from this smoothing. Figure 3 shows the output obtained from a static feed-forward net, namely a MLP, trained with the Back-Propagation algorithm. The same architecture chosen for the RNNs was used (except for the context units and the corresponding recurrent connections, of course), along with the same free parameters (e.g. learning rate, number of epochs, momentum rate, etc.). The MLPs were trained to attempt estimation of state posterior probabilities given the *current* observation. The graphics indicates that a static model is not feasible for such an estimation, due to the heavy instantaneous fluctuations of sonar readings through time. The situation is different when RNNs are used. Figure 4 shows the output obtained with the RNN discussed in Section 2.2. The capability to take into account the history of previous observations ("memorized" in the context units), i.e. the estimation of state posteriors at a certain time given the whole sequence of observations, results effective. The modeling accuracy

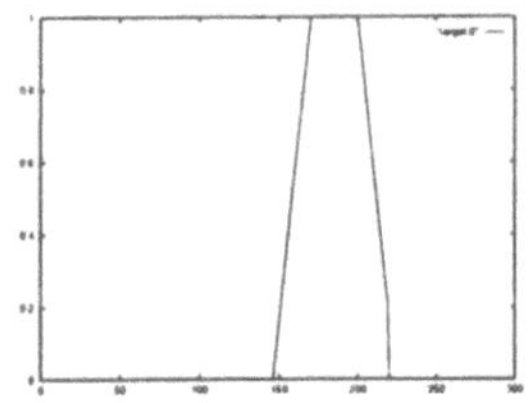

Figure 2: *Target output*

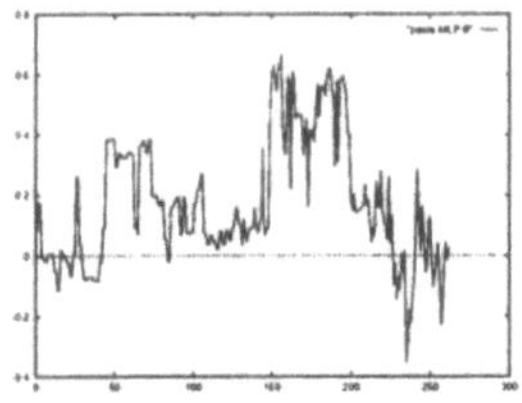

Figure 3: *MLP output*

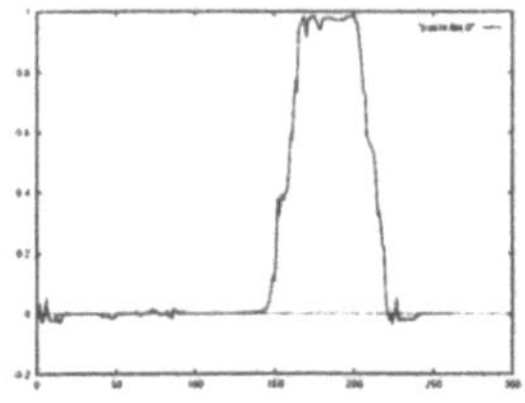

Figure 4: *RNN output*

that can be qualitatively detected from the graphics is indeed encouraging.

Once training was accomplished, the hybrid system was tested on a workstation using the test data acquired on the field. The on-line version discussed above was used to perform recognition of current state in real-time, according to the topology of the underlying HMM and to the estimates of individual state posterior probabilities yielded by the RNNs (the latter ones operated in parallel, each one processing the whole sequence of input patterns), allowing for perfect (100%) recognition.

4 Conclusions

This paper introduced a trainable hybrid HMM/RNN approach suitable for description of sequences of "local relevant situations" (states) traversed by a mobile robot. The approach is flexible, can deal with arbitrary features (16-dim sonar vectors were used in the present work), and can be used as a basis for complex navigation tasks: navigating to a prescribed point, learning maps, perform teleoperation sessions, etc.. The technique is expected to be feasible for learning and recognition of whole sequences of such relevant situations, where recognition relies on a Dynamic Programming algorithm computed over the outputs of the RNNs, as well as for local estimation (in real-time) of state probabilities given the temporal sequence of all sensory observations (and possibly given global statistics on allowed transitions between states). The theory underlying the model, as well as the architecture and dynamics of the RNNs used as nonparametric probability estimators, have been discussed.

Simulation experiments on a workstation with real sonar data resulted encouraging, showing the ability of the RNNs to learn the time-dependent state probability curve much better than static MLPs. The hybrid has also correctly recognized the sequence of states encountered during simulations of navigation on real-world data.

References

[1] Y. Bengio. *Neural Networks for Speech and Sequence Recognition*. International Thompson Computer Press, London, UK, 1996.

[2] H. Bourlard and N. Morgan. *Connectionist Speech Recognition. A Hybrid Approach*, volume 247 of *The Kluwer international series in engineering and computer science*. Kluwer Academic Publishers, Boston, 1994.

[3] R. Brooks. Intelligence without Reason. AI Memo n. 1293, Massachusetts Institute of Technology, Artificial Intelligence Laboratory, April 1991.

[4] R. Cattoni, G. Di Caro, M. Aste, and B. Caprile. Bridging the gap between Planning and Reactivity: a Layered Architecture for Autonomous Indoor Navigation. In *Proceedings of the IEEE/RSJ/GI International Conference on Intelligent Robots and Systems IROS'94*, pages 878–885, Munich, Germany, September 1994.

[5] A. Cleeremans, D. Servan-Schreiber, and J.L. McClelland. Finite State Automata and Simple Recurrent Networks. *Neural Computation*, 1:372–381, 1989.

[6] J. Hertz, A. Krogh, and R. G. Palmer. *Introduction to the Theory of Neural Computation*. Addison Wesley, 1991.

[7] N. J. Nilsson. Shakey the Robot. Technical Report 323, SRI International, April 1984.

[8] La. R. Rabiner. A Tutorial on Hidden Markov Models and Selected Applications in Speech Recognition. *Proceedings of the IEEE*, 77(2):257–286, 1989.

[9] C. E. Thorpe, M. Hebert, T. Kanade, and S. A. Shafer. Vision and Navigation for the Carnegie-Mellon Navlab. *IEEE Transactions on Pattern Analysis and Machine Intelligence*, 10(3):362–373, March 1988.

Section 7
Architectures and Algorithms

Training Feedforward Neural Networks: Convergence and Robustness Analysis

A. Alessandri
CNR–IAN National Research Council, Via De Marini 6
16149 Genova, Italy – e-mail:angelo@ian.ge.cnr.it

M. Maggiore
The Ohio State University, 330 Caldwell Laboratory, 2015 Neil Avenue
OH 43210-1272, USA – e-mail:morgan@avalon.eng.ohio-state.edu

M. Sanguineti
DIST–University of Genoa, Via Opera Pia 13
16145 Genova, Italy – e-mail:marce@dist.unige.it

Abstract

We develop a new algorithm for the learning of feedforward neural networks, by stating the learning process as a parameter estimation problem. We provide an analysis of its convegence and robustness properties. Two different versions of the algorithm are discussed, depending on the way in which the training set is explored during learning. The simulation results, for both classification and function approximation problems, confirm the effectiveness of the proposed algorithm and its advantages with respect to error back–propagation and extended Kalman filter–based learning.

1 Introduction

In this paper, a new algorithm for the learning of feedforward neural networks based on a training set is developed. Standard error back–propagation, despite of its wide diffusion as a training algorithm for such networks, presents some drawbacks: tipically, slow convergence rate, particularly in high-dimensional settings, and arbitrarily chosen learning rate. In the last decade, to overcome these problems various algorithms based on different approaches have been proposed (modifications to the back–propagation, second–order methods, learning based on the exteded Kalman filter, etc.), sometimes introducing other flaws, such as heuristic choice of parameters and computational burden.

By stating the learning process for feedforward neural networks as a parameter estimation problem for a nonlinear system having a layered structure, training algorithms based on the extended Kalman filter (EKF for short) have been developed. They show faster convergence than BP and avoid tuning parameters, at the expense of increased computational burden and memory requirements.

In this paper, after stating the learning of feedforward neural networks as a parameter estimation problem, we develop a new batch learning algorithm, which works according to a sliding–window scheme (only a portion of the data is processed at every time; the past is summarized in one prediction). We discuss two variants of the algorithm: in the first the temporal window is supposed to move one stage at a time, whereas the second considers the case in which it moves d stages at a time. We prove the convergence and robustness properties of the proposed algorithm. Unlike BP, it shows fast convergence and, unlike the EKF–based training, it does not require successive linearizations. Tested on both classification and function approximation problems, it shows a better performance than error back-propagation and training based on EKF.

2 The learning algorithm

2.1 Estimation of the weights

We consider multilayer feedforward neural networks containing only one hidden layer, composed of ν neural units, with the hyperbolic tangent as activation function; the output layer is linear. The function implemented by such a network is denoted by $\underline{\gamma}(\underline{w}, \underline{u})$, where $\underline{w}$ and $\underline{u}$ are, respectively, the input and the weights vector. The data set consists of P input/output pairs $(\underline{u}_t, \underline{y}_t)$, $t = 0, 1, \ldots, P-1$, where $\underline{u}_t \in \Re^m$ and $\underline{y}_t \in \Re^p$ represent, respectively, the input and the desired output to the network. We let the weights of the network constitute the state of the following nonlinear system ($t = 0, 1, \ldots P-1$):

$$\begin{cases} \underline{w}_{t+1} = \underline{w}_t \\ \underline{y}_t = \underline{\gamma}\,(\underline{w}_t, \underline{u}_t) + \underline{\eta}_t \end{cases} \tag{1}$$

where $\underline{w}_0 = \underline{w}_1 = \ldots = \underline{w}_{P-1} \triangleq \underline{w}$ and $\underline{\eta}_t \in \Re^p$ is a random noise whose statistics are unknown, eith $\underline{\eta}_t \in K \subset \Re^p$, where K is a known compact set. We assume that $\underline{w} \in W$ and $\underline{u}_t \in U$, where W and $U$are compact sets. If we consider the data set $(\underline{u}_t, \underline{y}_t)$, $t = 0, 1, \ldots, P-1$, as generated by a process governed by an unknown function $\underline{f} : \Re^m \longrightarrow \Re^p$, i.e., $\underline{y}_t = \underline{f}(\underline{u}_t)$, then $\underline{\eta}_t$ is the error achieved by the approximator $\underline{\gamma}$ in the approximation of $\underline{f}$, in correspondence of the value $\underline{u}_t$. The possibility of writing (1) is guaranteed by the universal approximation properties of feedforward neural networks, provided that the unknown function $\underline{f}$ be sufficiently smooth (for instance, [4]).

The algorithm we propose for the network training is based on a sliding-window state estimator for system (1) [1]. According to this, we estimate the constant parameters vector $\underline{w}$ in such a way as to minimize the cost

$$J_t \triangleq \mu \|\underline{\hat{w}}_t - \underline{\bar{w}}_t\|^2 + \sum_{i=t-N}^{t} \left\|\underline{y}_i - \underline{\gamma}\,(\underline{\hat{w}}_t, \underline{u}_i)\right\|^2 \quad t = N, N+1, \ldots, P-1 \tag{2}$$

where $\underline{\hat{w}}_t$ is the estimate of the weights vector $\underline{w}$ at the time t and $\underline{\bar{w}}_t$ is the "a priori" information on the value of $\underline{\hat{w}}_t$. Minimization of J_t at each temporal stage leads to a sequential state estimator; the optimal estimate $\underline{\hat{w}}_t^\circ$ of $\underline{w}$ at time t, given the "information set" $I_t^N \triangleq \{\underline{y}_{t-N}, \ldots, \underline{y}_t, \underline{u}_{t-N}, \ldots \underline{u}_t, \underline{\bar{w}}_t\}$, is $\underline{\hat{w}}_t^\circ \triangleq \operatorname{argmin}_{\underline{\hat{w}}_t} J_t\left(\underline{\hat{w}}_t, I_t^N\right)$, where $\underline{\bar{w}}_t \triangleq \underline{\hat{w}}_{t-1}^\circ$, and $\underline{\bar{w}}_N$ is the initial prediction. Let W_t° be the sets of the optimal estimates that minimize the cost (2), $Y \triangleq$

$\underline{\gamma}(W,U)$, $Y \subset \Re^p$ and $\underline{H}\left(\underline{w}, \underline{u}_{t-N}^t\right) \triangleq \mathrm{col}\left(\underline{\gamma}\left(\underline{w}, \underline{u}_\tau\right),\ \tau = t-N, \ldots, t\right)$, where $\underline{u}_{t-N}^t \triangleq \mathrm{col}\left(\underline{u}_{t-N}, \underline{u}_{t-N+1}, \ldots, \underline{u}_t\right)$.

2.2 Convergence and robustness analysis

Assumption A. There exists a compact set $\tilde{W}$ such that $\tilde{W} \supseteq W \bigcup \left(\bigcup_{t=0}^{+\infty} W_t^\circ\right)$.

Assumption B. There exists an integer N such that, for any $\underline{u}_{t-N}^t \in U^{N+1}$, the mapping $\underline{H}\left(\underline{w}, \underline{u}_{t-N}^t\right) : \mathcal{W} \longrightarrow \Re^{p(N+1)}$ is an injective immersion, where $\mathcal{W}$ is the closed convex hull of $\tilde{W}$.

Given a symmetric positive definite matrix A, we denote by $\lambda_{\min}(A)$ and $\lambda_{\max}(A)$ its minimum and maximum eigenvalue, respectively; for a generic matrix B, $\|B\|_{\max} \triangleq \|B\| = \sqrt{\lambda_{\max}(B^T B)}$ and $\|B\|_{\min} \triangleq \sqrt{\lambda_{\min}(B^T B)}$. Moreover, let $D(\underline{w}, \underline{u}_{t-N}^t) \triangleq \frac{\partial \underline{H}}{\partial \underline{w}} \in \Re^{p(N+1)\times n}$, $\underline{w} \in \mathcal{W}$ and $\Delta \triangleq \max_{\underline{w}\in\mathcal{W}, \underline{u}_{t-N}^t \in U^{N+1}} \|D(\underline{w}, \underline{u}_{t-N}^t)\|$, $\delta \triangleq \min_{\underline{w}\in\mathcal{W}, \underline{u}_{t-N}^t \in U^{N+1}} \|D(\underline{w}, \underline{u}_{t-N}^t)\|$. Let $r_\eta \triangleq \max_{\underline{\eta}_{t-N}, \ldots, \underline{\eta}_t \in E^{N+1}} \|\mathrm{col}(\underline{\eta}_{t-N}, \ldots, \underline{\eta}_t)\|$ and $\bar{k} \triangleq k\sqrt{N+1}$, where $k \in \Re^+$ is a suitable scalar such that $\left\|\frac{\partial \gamma(\underline{w},\underline{u})}{\partial \underline{w}}(\underline{w}', \underline{u}) - \frac{\partial \gamma(\underline{w},\underline{u})}{\partial \underline{w}}(\underline{w}'', \underline{u})\right\| \le k\, \|\underline{w}' - \underline{w}''\|\ \forall \underline{w}', \underline{w}'' \in \mathcal{W}, \forall \underline{u} \in U$. The following theorem guarantees, under suitable hypoteses, the boundness of the estimation error (for the proof, see [2]).

Theorem. *Suppose that Assumptions A and B are verified. Let* $\hat{\underline{e}}_t \triangleq \underline{w} - \hat{\underline{w}}_t^\circ$ *and consider the largest closed ball* $N(\hat{r}_e)$ *with radius* $\hat{r}_e$ *and center in the origin such that* $\hat{\underline{e}}_N \in N(\hat{r}_e)$. *If there exists a choice of* μ, *for which the inequality*

$$\left(\delta^4 - 8\bar{k}\Delta^2 r_\eta\right)\mu + \delta^6 > 0 \tag{3}$$

is satisfied, the second–order equation $(2\Delta\bar{k}\mu^2)\xi^2 + [\mu(\mu+\delta^2)^2 - (\mu+\delta^2)^3 + 4\bar{k}\Delta^2\mu r_\eta]\xi + 2\bar{k}\Delta^3 r_\eta^2 + \Delta r_\eta(\mu+\delta^2)^2 = 0$ *has the two real positive roots* ξ^- *and* ξ^+, *with* $\xi^- < \xi^+$.

Then, if the choice of μ *yields also the fulfilment of the inequality*

$$\mu^2 + 2\left(\delta^2 - \bar{k}\,\Delta\,\xi^+\right)\mu + \left(\delta^4 - 2\bar{k}\,\Delta^2\, r_\eta\right) > 0 \tag{4}$$

we have $\lim_{t\to+\infty} \|\hat{\underline{e}}_t\| \le \xi^-,\quad \forall\, \hat{r}_e < \xi^+$

Moreover, considering the sequence ξ_t *generated from the discrete-time system* $(\mu+\delta^2)^3\xi_t = (2\Delta\bar{k}\mu^2)\xi_{t-1}^2 + [\mu(\mu+\delta^2)^2 + 4\bar{k}\Delta^2\mu r_\eta]\xi_{t-1} + +2\bar{k}\Delta^3 r_\eta^2 + \Delta r_\eta(\mu+\delta^2)^2$, *where* $\xi_N \triangleq \|\hat{\underline{e}}_N\|$, *we have* $0 \le \|\hat{\underline{e}}_t\| \le \xi_t, \forall t \ge N$, *i.e., the sequence* ξ_t *constitutes an upper bound to the norm of the error dynamics.* □

For large r_η, (3) imposes an upper bound to μ, whereas (4) is very likely to impose a lower bound. In other words, the theorem guarantees the boundness of the estimation error $\forall \mu \in (\mu^-, \mu^+)$, where μ^- and μ^+ are suitable scalars. Assumption B is strictly related to an important property of networks. For

$\underline{H}(\underline{w}, \underline{u}_{t-N}^{t})$ to be an immersion, the Jacobian matrix $D(\underline{w}, \underline{u}_{t-N}^{t})$ must have rank n, $\forall \underline{u}_{t-N}^{t} \in U^{N+1}$. Since $D\left(\underline{w}, \underline{u}_{t-N}^{t}\right) \in \Re^{p(N+1)\times n}$, the bigger $p(N+1)$ is, the easier is for the above matrix to have rank n. According to this, we require that $p(N+1) \gg n$, i.e., we choose a very large window size, compared to the size of the weights vector. This is related to the estimation error of a network: since it is decreasing with P and, unlike the approximation error, increasing with n [4], in order to minimize it we would require $P \gg n$: this follows from $p(N+1) \gg n$, noticing that $P \geq N+1$ (the choice of N influences the condition on the rank of $D\left(\underline{w}, \underline{u}_{t-N}^{t}\right)$). As regards the injectivity of $\underline{H}$, we observe that the mapping $\underline{\gamma} : \mathcal{W} \times U \to Y$ is not injective, due to the invariance of $\underline{\gamma}(\underline{w}, \underline{u})$ to certain permutations of the weights [9]. Since we only require the estimator to find one of the weights vectors that minimize the cost (2) (considering as equivalent all those obtained through permutations), we can neglect the injectivity assumption.

As the robustness of the proposed algorithm is concerned, in [2] we have proved that, if the error in the estimate of the weights vector is suitably bounded, then the boundedness of the estimate is preserved.

2.3 d-step estimator

We now consider an estimator whose temporal window moves d stages at a time, where $1 \leq d \leq N$, and we define it *d-step estimator*. The following dynamical system is obtained by composition of (1) d-times with itself ($t = 0, 1, \ldots \bar{t}$):

$$\begin{cases} \underline{w}_{d(t+1)^{\circ}} = \underline{w}_{dt}^{\circ} \stackrel{\triangle}{=} \underline{w}^{\circ} \\ \underline{Y}_{dt} \stackrel{\triangle}{=} \begin{bmatrix} \underline{\gamma}(\underline{w}^{\circ}, \underline{u}_{dt}) \\ \underline{\gamma}\left(\underline{w}^{\circ}, \underline{u}_{dt+1}\right) \\ \vdots \\ \underline{\gamma}\left(\underline{w}^{\circ}, \underline{u}_{dt+N}\right) \end{bmatrix} + \begin{bmatrix} \underline{\eta}_{dt} \\ \underline{\eta}_{dt+1} \\ \vdots \\ \underline{\eta}_{dt+N} \end{bmatrix} = \underline{H}(\underline{w}^{\circ}, \underline{u}_{dt}^{dt+N}) + \underline{\eta}_{dt}^{dt+N} \end{cases} \tag{5}$$

Here, $\underline{Y}_{dt}$ plays the role of an "extended measurements", i.e., a set of $N+1$ measurements, and $\bar{t}$ is the biggest integer such that $P \leq d\bar{t}+N+1 < P+d$; if $d\bar{t}+N+1 > P$, we assume to use again the first $d\bar{t}+N+1-P$ measurements. The estimation cost becomes:

$$J_{d(t-N)+N} = \mu \left\| \underline{\hat{w}}_{d(t-N)+N} - \underline{\bar{w}}_{d(t-N)+N} \right\|^{2} + \left\| \underline{Y}_{d(t-N)} - \underline{H}\left(\underline{\hat{w}}_{d(t-N)+N}, \underline{u}_{d(t-N)}^{d(t-N)+N} \right) \right\|^{2}$$

The information set is now

$$\underline{I}_{d(t-N)+N}^{N} = \left\{ \underline{y}_{d(t-N)}^{d(t-N)+N}, \underline{u}_{d(t-N)}^{d(t-N)+N}, \underline{\bar{w}}_{d(t-N)+N-1} \right\}, t = N, N+1, \ldots, \bar{t}$$

and the optimal estimate $\underline{\hat{w}}_{d(t-N)+N}^{\circ}$ of $\underline{w}^{\circ}$, given $\underline{I}_{d(t-N)+N}^{N}$, is

$$\underline{\hat{w}}_{d(t-N)+N}^{\circ} \stackrel{\triangle}{=} \underset{\underline{\hat{w}}_{d(t-N)+N}}{\operatorname{argmin}} \; J_{d(t-N)+N}\left(\underline{\hat{w}}_{d(t-N)+N}, \underline{I}_{d(t-N)+N}^{N} \right)$$

where $\underline{\bar{w}}_{N}$ is the initial prediction. The estimator exploits an optimization window of size $N+1$, which slides d temporal stages at a time. From the point of view of the "extended system" (5), we are using only one "extended measurements" at a time. By noticing that, for system (5), the constants $\bar{k}, \delta, \Delta, r_\eta$ are the same as for system (1), we conclude that the convergence and robustness results above stated hold true also for the d-step estimator.

Given a fixed number of iterations, say l, the *one-step* estimator explores $N+l$ patterns of the data set, whereas *N-step* utilizes $Nl+1$ patterns (for the generic *d-step* estimator, $N+1+(l-1)d$). Note that the *N-step* estimator maximizes the number of patterns explored under the same number of iterations, giving rise to a batch training. This turns out to be useful whenever the data set is very large, almost infinite, and application of gradient methods over the entire epoch (i.e., steepest descent algorithms) becomes computationally unmanageable. In such a case, batch training overcomes the problem of too large data set size, dividing the P patterns into data batches of fixed length and training the network over every single data batch. In this situation, an useful feature of the *N-step* estimator is the presence of the prediction term in the cost which links the current estimate of the weights vector to the last one and, consequently, the current batch to the previous one by means of the prediction.

3 Simulation results and conclusions

The proposed algorithm, called WEST (Weights ESTimator), has been compared, for both classification and function approximation, with the following learning algorithms: BPX (error back-propagation, with momentum and adaptive learning rate), LM (Levenberg-Marquardt optimization [6]), EKF (iterative extended Kalman filter-based training). The simulations have been performed with the Neural Network Toolbox of MATLAB [5]. For BPX and LM we have used, respectively, the functions *trainbpx* and *trainlm* of MATLAB; EKF and WEST have been properly implemented.

As regards *function approximation*, we have trained a network with 30 hidden units to approximate the function $f(x, y) = \frac{\sqrt{x^2+y^2}}{1+\frac{1}{3}(x^2+y^2)}$ in the domain $[-10,10]\times[-10,10]$. The training set has been generated considering 2000 input/output pairs uniformly distributed in the domain; for the window size of WEST we have chosen $N = 249$. Only LM and WEST succeed in detecting the minimum of f; the behavior of the approximation of f through BPX, LM, EKF and WEST, not shown owing to the lack of space, is presented in [3]. Figure 1 shows the error during training (note that, to reduce the computational effort for EKF, the MSE has been calculated only every 200 patterns). The lowest value of the MSE on the test set is for LM (0.0007), but WEST is very near (0.0008); moreover, it provides an approximating function as good as the one of LM, but obtained with a reduced computational load and memory requirements (at each iteration, it deals with a Jacobian matrix 121×250, whereas LM with a 121×2000 one). BPX and EKF reach an MSE error of 0.0013 and 0.0011, respectively.

As regards *classification* problems, as testbed for our algorithm we have chosen a generalized 2-input XOR. The results are presented in [3]; the behavior of the error during training is analogous to that of the approximation problem.

To conclude, we summarize the most interesting features of the proposed training algorithm: - bounded network error; - robustness; - reduced computational burden; - no need for successive linearization. These properties

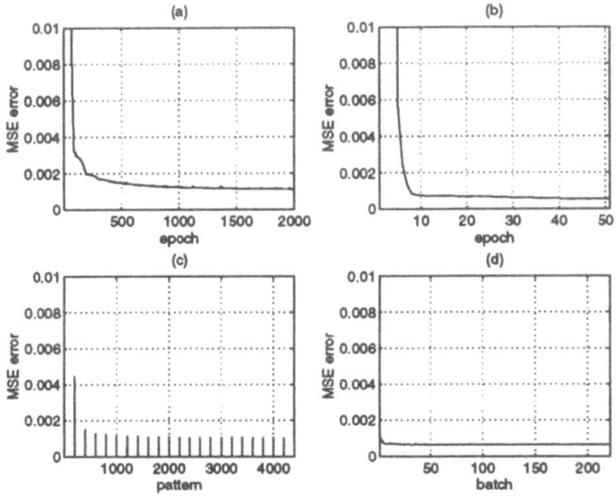

Figure 1: mean-squared error for BPX (a), LM (b), EKF (c), and WEST (d) training algorithms, in the function approximation problem.

make WEST a promising alternative to BP-based and EKF-based training algorithms.

References

[1] A. Alessandri, M. Maggiore, T. Parisini, and R. Zoppoli, Neural approximators for nonlinear sliding-window state observers. In *Proceedings of the 35th IEEE Conference on Decision and Control*, 1996, pp. 1461–1463.

[2] A. Alessandri, M. Maggiore, and M. Sanguineti. A new learning algorithm with bounded error for feedforward neural networks. Technical Report 97/1, DIST Tech. Report, 1997.

[3] A. Alessandri, M. Maggiore, and M. Sanguineti. Training feedforward neural networks through a parameter-estimation-based algorithm. In *Proceedings of NEURAP '98 Conference*, 1998, pp. 225-228.

[4] A. R. Barron. Approximation and estimation bounds for artificial neural networks. *Machine Learning*, 14:115–133, 1994.

[5] H. Demuth and M. Beale. *Neural Network Toolbox - For Use With MATLAB - User's Guide*. The Math Works, Inc., 1995.

[6] J. J. More. The Levenberg-Marquardt algorithm: Implementation and theory. In G. A. Watson, editor, *Numerical Analysis*, Lecture Notes in Mathematics 630, pages 105–116. Springer-Verlag, 1977.

[7] D. E. Rumelhart, G. E. Hinton, and R. J. Williams. Learning representations by back-propagating errors. *Nature*, 323:533–536, 1986.

[8] S. Singhal and L. Wu. Training multylayer perceptrons with the extended Kalman algorithm. In D. S. Touretzky, editor, *Advances in Neural Information Processing Systems 1*, pages 133–140, 1989.

[9] H. J. Sussmann. Uniqueness of the weights for minimal feedforward nets with a given input–output map. *Neural Networks*, 5:589–593, 1992.

Solving the Travelling Salesman Problem using the Kohonen Network Incorporating Explicit Statistics

B. John Oommen*
School of Comp. Science, Carleton University
Ottawa,CANADA
oommen@scs.carleton.ca

Necati Aras†
Türk Elektrik End. A. Ş.
Topkapı 34020
İstanbul, TÜRKİYE
naras@tee.com.tr

İ. Kuban Altınel†
Dept.of Industrial Eng.,
Boğaziçi University
İstanbul, TÜRKİYE
altinel@boun.edu.tr

Abstract

In this paper we introduce a new self-organizing neural network, the Kohonen Network Incorporating Explicit Statistics (KNIES) that is based on Kohonen's Self-Organizing Map (SOM). The primary difference between the SOM and the KNIES is the fact that every iteration in the training phase includes *two* distinct modules - the attracting module and the dispersing module. As a result of the newly introduced dispersing module the neurons maintain the overall statistical properties of the data points. Thus, although in SOM the neurons individually find their places both statistically and topologically, in KNIES they collectively maintain their mean to be the mean of the data points which they represent. The new scheme has been used to solve the Travelling Salesman Problem (TSP). Experimental results for problems taken from TSPLIB [13] indicate that it is a very accurate NN strategy for the TSP - probably the most accurate neural solutions available in the literature.

1 Introduction

In this paper we present a new Neural Network (NN) called The Kohonen Network Incorporating Explicit Statistics (KNIES). The primary difference between the Kohonen's Self-organizing map (SOM) and the KNIES is the fact that every iteration in the training phase includes two distinct modules - the attracting module and the dispersing module. In the attracting module a subset of the neurons migrate *towards* the data point that has been presented to

*Supported in part by the National Science and Engineering Research Council of Canada (NSERC).

†Partially supported by the Boğaziçi Research Fund Grant 97H301D, and Hewlett Packard.

the NN. This phase is essentially identical to the learning phase of the SOM. However, subsequent to this phase the rest of the neurons which have not been involved in the attracting module participate in a dispersing migration. Indeed, these neurons now move away from their current positions in a manner that ensures that the global statistical properties of the data points are mimiced by the neurons. Thus, although the neurons individually find their places both statistically and topologically *asymptotically*, in the KNIES they collectively maintain their mean to be the mean of the data points which they represent.

To prove the power of the KNIES we have used it to solve the well known Travelling Salesman Problem (TSP) which has been one of the oldest "hard nuts" of Operations Research and Mathematical Programming. It is known to be NP-hard [11]. Any algorithm devised to solve the TSP tries to answer the following question: Given a set of N cities and distances for each pair, what is the shortest tour that visits each city exactly once? There are many exact and heuristic algorithms in the literature for the TSP [10, 14], and to this collection of methods we now add one more, namely the KNIES, adapted to tackle the TSP which has been tested for instances obtained from TSPLIB problem set [13]. Indeed, it is probably the most accurate neural solution available in the literature.

2 The Kohonen Network Incorporating Explicit Statistics (KNIES)

2.1 Vector Quantization and the SOM

As opposed to the non-parametric methods which use all the data in the training and testing phases, the SOM compresses the information by representing it using a "small" set of vectors, **neurons**, called the **codebook** vectors in the SOM literature. These neurons are migrated in the **feature** domain so that they collectively represent the distribution under consideration.

We assume that we are given a set of N points $\mathcal{P} = \{X_i : 1 \leq i \leq N\}$ in a d-dimensional space. The primary aim of both VQ and the SOM is to represent these points $\mathcal{P}$ by M representative points, neurons, $\mathcal{Y} = \{Y_j : 1 \leq j \leq M\}$. In a typical *pattern recognition* setting it is generally assumed that $M \ll N$; as we shall see in our current application (the Travelling Salesman Problem (TSP)) this is not necessary. Indeed, in the TSP, it is not uncommon that in the learning phase $M \geq N$. Furthermore, although strictly speaking, we can represent $\mathcal{P}$ by M_N neurons (where M_N increases with N), in the interest of simplicity, in this paper we assume that the *final* number of representative neurons is a predetermined constant. In the case of the TSP the final number of neurons will be equal to the number of cities itself.

The set of neurons $\{Y_j : 1 \leq j \leq M\}$ are initially randomly assigned positions which are typically within the convex hull of the points they are to represent, $\mathcal{P}$. We now consider how the migration of the neurons is achieved.

In both VQ and the SOM the polarizing algorithm is repeatedly presented with a point X_i from $\mathcal{P}$. The neurons now try to incorporate the topological information present in X_i. This is done as follows. First of all the closest neuron to X_i, Y_{j*}, is determined. This vector and a group of neurons in its neighbourhood, B_{j*}, are now moved in the direction of X_i. The set B_{j*} is called the "Activation Bubble", and we shall presently specify how it is determined. The actual migration of the neurons is achieved by rendering the new Y_j to be a convex combination of the current Y_j and the data point X_i for all $j \in B_{j*}$. More explicitly, the updating algorithm is as follows :

$$Y_j(t+1) = \begin{cases} (1-\alpha(t))Y_j(t) + \alpha(t)X_i & \textit{if } j \in B_{j*} \\ Y_j(t) & \textit{otherwise} \end{cases} \tag{1}$$

where 't' is the discretized (synchronized) time index. This basic algorithm has two fundamental parameters, $\alpha(t)$ and the size of the bubble B_{j*}. $\alpha(t)$ is called the adaptation constant and satisfies $0 < \alpha(t) < 1$. The value of $\alpha(t)$ should be decremented to get good convergence as recommended by [9, 15].

The neighbourhood bubble B_{j*} is the quantity which makes VQ differ from the SOM. Indeed, if the size of the bubble is always set to be unity, only the closest neuron is migrated and this yields a VQ scheme. However, in the SOM, the nearest neuron and the neurons within a bubble of activation are also migrated, and it is this which permits the algorithm to be both *topology preserving* and *self-organizing*.

There are two ways of achieving the effect of the bubble size. The first method is by explicitly decreasing the size of the bubble, and this is often done by linearly decrementing the diameter of the bubble as :

$$W(t+1) = K_w \cdot W(t), \tag{2}$$

where K_w is a user-defined constant, and $[-W, W]$ specifies the activation bubble. An alternate method for reducing the effect of updating on the more distant nodes is by introducing a weighting function whose value decreases with the distance from the winning (or closest) neuron. Typical weighting functions are either of a Gaussian Window form (with a variance which steadily decreases), or can be of any kernel form used in non-parametric Parzen window methods [5, 7]. Although the SOM is described and implemented in such an elementary manner it has extremely elegant properties. A collection of these is found in [9, 15] and in the unabridged version of this paper [2].

2.2 Enhancing SOM using Global Statistical Information

In the effort to preserve the "non-parametric" form of the underlying algorithm both VQ and SOM completely ignore the statistical information that is already collectively resident in the data points. Thus we can conceive of enhanced SOM methods which, apart from only using the information in the data point presented, uses point-wise local information in conjunction with these pieces of statistical information.

Let us suppose that the mean of the set of points to be represented by the neurons in $\mathcal{P}$ is μ. The intention is that we maintain μ to be invariant for the entire training. The question which remains is merely one of determining how the neurons can be collectively updated so as to retain *their* mean to be also μ. To do this, as alluded to earlier, we shall decompose every training step into two phases, namely the **attracting** and **dispersing** phases.

When the data point P_i is presented to the network, the closest neuron and all the points within the activation bubble are migrated using an equation identical to (1). This constitutes the attracting module, and in this phase the neurons within the bubble use the local information present in the points as they appear individually and move towards them.

Observe that since we are partitioning the neurons as those which participate in the attracting phase and those which do not, in KNIES we assume that the activation bubble is maintained constant for a specified value of M. The question of decreasing the effect of the attraction as the neuron $j \in B_{j*}$ is more distant from j^* *must necessarily* be achieved by using a kernel function. For the sake of simplicity we assume that the kernel is the Gaussian kernel [5, 7] given by :

$$G(a,b) = K_g \cdot e^{\frac{-\|Y_a - Y_b\|^2}{2\sigma^2}}, \tag{3}$$

where K_g is a normalizing constant, σ^2 specifies the variance of the kernel which is progressively decremented (for example, linearly [3, 15]) to get the effect of reducing the activation bubble, and Y_a and Y_b are the coordinates of the points in question. In such a case the attraction phase can be completely described by equation (4) :

$$Y_j(t+1) = \begin{cases} Y_j(t) + G(j,j^*)(X_i(t) - Y_j(t)) & if\ j \in B_{j*} \\ Y_j(t) & otherwise. \end{cases} \tag{4}$$

In (4) X_i is the current point presented to the system and j^* is the winning neuron closest to X_i using the Euclidean norm.

Angéniol *et al.* [3] have recommended using a Gaussian kernel which is a function of the *indices* of the neurons — as opposed to being a function of their coordinates — and of the variance. The advantage of such a kernel function is that a single parameter σ^2, can take care of both the activation bubble and their intensity of attraction.

As a result of the migration of the neuron within the activation bubble we observe that the mean of all neurons moves away from the initially assigned value, μ. Thus, the dispersing phase now forces the neurons outside the activation bubble to migrate away from their current locations to render the mean of the neurons to continue to be invariant. This is done as follows. The total migration that has been effected as a result of the attraction module is the vector ΔY, where,

$$\Delta Y = \sum_{j \in B_{j*}} [Y_j(t+1) - Y_j(t)]. \tag{5}$$

The change in Y, ΔY, is now distributed among the neurons which did not participate in the attracting phase as :

$$Y_j(t+1) = Y_j(t) - \frac{1}{M - |B_{j*}|}\Delta Y, \quad if\ j \notin B_{j*}. \tag{6}$$

Although KNIES is *philosophically* superior to the traditional SOM, the additional effort used in maintaining the statistics is generally not useful for most problems unless the *quality* of the final solution of the underlying problem itself depends on the *intermediate distribution* of the neurons. One such problem is the TSP. In this case, as we shall see, a poor transient distribution of the neurons in the initial phase can cause the algorithm to yield a final TSP solution that is inferior to that which a superior transient distribution would yield. We shall use this property in next section to develop two new algorithms for the TSP.

3 Using KNIES to solve the TSP

3.1 KNIES_TSP

As we mentioned earlier, KNIES by itself would not be too powerful in solving problems for which the SOM suffices. However, in the TSP KNIES is a useful scheme because the quality of the final solutions depends on the transient stochastic distribution of the nodes on the ring, namely neurons. We shall make use of this as follows.

First of all we assume that the number of neurons at any time is $M(t)$, and that this number can increase or decrease by insertions or deletions. At any instant 't' we say that a neuron j located at Y_j is associated with a city $\Gamma_j(t)$ if :

$$\Gamma_j(t) = \min_i \|X_i - Y_j\| \qquad j = 1, \ldots, M. \tag{7}$$

Thus, at instant t, $\Gamma_j(t)$ is the city closest to the neuron j, and the intention is that it is represented by Y_j.

At every time instant a city X_i is presented to the network. The neuron j^* closest to X_i is now determined as the winner and it and all its neighbours in the activation bubble are moved toward X_i with a strength determined by an appropriate kernel function. In our experiments we have used the Gaussian kernel with a variance σ^2. However, rather than using the kernel associated with the coordinates of the nodes, we have used the kernel associated with the indices as described by Λ (see [3] for the details of the motivation for this) as follows :

$$\Lambda(j, j^*) = \frac{1}{\sqrt{2}} \exp\left(-\frac{d_{jj^*}}{\sigma^2}\right). \tag{8}$$

Here, d_{jj^*} is defined as $d_{jj^*} = \min\{|j - j^*|, M - |j - j^*|\}$ and σ (the standard deviation) is called the "gain parameter". The value of σ is decreased at the end of each complete pass by a factor K_σ (a user-defined constant). The decrement of σ has the dual effect of decreasing the the size of activation bubble and of decrementing $\alpha(t)$, the adaptation parameter of the SOM. The only parameters that are to be adjusted are the initial value of σ^2, and K_σ.

Subsequently, the rest of the neurons *outside* this bubble are moved in such a way that the mean of the neurons *currently on the ring* coincides with the mean of the cities represented by them. Notice that we do not make the mean converge towards the mean of all the cities - it is only made to move onto the mean of the cities, the set $\mathcal{P}$, represented by the *subset* $\{X_{\Gamma_j(t)} : j = 1, \ldots, M(t)\}$.

Observe now that the number of neurons $M(t)$ changes with time. Insertions and deletions are permitted in a manner analogous to the scheme by Angéniol *et al.* [3]. If a neuron wins the competition for two different cities in the same pass, a new neuron is created with the same coordinates as the winner neuron. This newly created neuron is inserted into the ring as the neighbour of the winner. Both neurons are inhibited until they disperse by the movements of the neighbouring units. In the same way, a neuron is deleted, if it does not win a competition during a fixed number of (say, three) complete passes.

The difference between our insertion/deletion processes and Angéniol *et al.*'s is quite subtle and is explained in detail in [2]. The pseudo-code for the algorithm is also included in [2].

3.2 KNIES_TSP_Global : A simplification of KNIES_TSP

KNIES_TSP can be computationally expensive. This is just because we try to maintain the mean of the neurons to be the mean of the cities they represent, and so, we are continuously posed with the problem of identifying the city associated with each neuron. Consequently, the subset of cities represented by the neuron, $\overline{\Gamma}$, has to be identified after the presentation of each neurons. A modification of KNIES_TSP (called KNIES_TSP_Global) approximates this by assuming that we attempt to always migrate towards the global mean of all the cities. The only difference between KNIES_TSP and KNIES_TSP_Global is that in the latter, the mean of the neurons on the ring always moves towards (not onto) the global mean of all the cities, instead of moving exactly onto the local mean only of the cities represented by the neurons currently on the ring.

This is accomplished by bookkeeping the amount by which the neurons inside the activation bubble are updated along the x and y coordinate axes. The total amount of updates along both coordinate axes which are computed by summing up the recorded quantities algebraically make up the components of a vector ΔY which represents the total change in the coordinates of the neurons in B_{j*}. This total change is equally distributed among the neurons which are not in B_{j*} as follows :

For all $j \notin B_{j*}$ **Do**

$$Y_j \leftarrow Y_j - \frac{1}{M(t)\cdot(M(t)-|B_{j*}|)}\Delta Y$$

EndFor

The above loop replaces their respective steps in KNIES_TSP to yield the procedure KNIES_TSP_Global. This procedure is computationally faster than KNIES_TSP since it does not have to identify $\overline{\Gamma}$ after the presentation of each city, which is done by identifying the nearest city to a neuron on the ring. It is also quite accurate even though it is theoretically a deviation from the principle motivating KNIES.

4 Computational Results

We have tested both algorithms described in the preceding section by using problem instances taken from TSPLIB [13]. The selected problems are geometric TSP instances which means that each city is represented by its coordinates on the two-dimensional Euclidean space and the distances among the cities are computed by using the Euclidean norm. Furthermore, the sizes of the problems range between 51 and 532.

In order to be able to compare the quality of the solutions obtained with our algorithms, all instances in the test bed were also solved by using the Pure Kohonen Network (PKN), the Guilty Net (GN) of Burke and Damany [4], and the approach of Angéniol, Vaubois and Le Texier (AVL) [3]. The neighbourhood function which is used in all the methods is based on the lateral distance between two nodes on the ring rather than their Euclidean distance. Namely, the distance between the winner neuron and any other neuron on the ring is the absolute value of the difference between the indices of the two neurons. Furthermore, in all methods — except the GN which does not have it — the parameter $\alpha(t)$, which is used to decrease σ^2 at every epoch, is kept fixed at 0.2 throughout the iterations.

Table 1 summarizes the results obtained for the problem instances in the test bed. The first column specifies the instance as it is referred in TSPLIB. The second column shows the optimal tour length for these instances. The numbers in the third to seventh column includes the solutions obtained by using PKN, GN, AVL, KNIES_TSP, (referred to as KL, L standing for "local", in the table) and KNIES_TSP_Global (referred to as KG in the table), respectively. These values are the best values recorded after running the algorithms for a large number of different parameter settings. The numbers in parentheses are the parameter values for which these results are attained.

With the exception of four cases, i.e., `lin105`, `pr107`, `pcb442`, and `rat195` (the details of which are given in [2]), the KNIES_TSP gives the best result for all instances. For the instance `eil51` both KNIES_TSP and KNIES_TSP_Global yield the same solution whereas for the remaining three instances the best solution is given by the KNIES_TSP_Global.

PKN and GN exhibited convergence problems. Similar observations have already been reported for both methods [3, 8]. This problem becomes more severe as the problem size increases. Therefore, for the instances `rat195`, `pcb442`, and `att532`no entries are included in the table.

Of course, it is impossible to make categorical statements about the various algorithms from the results presented here. However, to get an overall picture, we have computed the average deviations for PKN, GN, AVL, KL, and KG, which are respectively 4.5 %, 32.61 %, 4.84 %, 2.73 %, and 3.17 %. These figures do not consider instances `att532`, `pcb442`, and `rat195` for which the Pure Kohonen Network (PKN) and/or the Guilty Net (GN) did not converge for any value from our large parameter trial set. Based on these results, it is possible to draw the conclusion that KNIES_TSP (KL) performs better than the other methods. However, the performance of KNIES_TSP_Global (KG) is close to the KNIES_TSP's - which makes KG more attractiver than KL. More explicit experimental results and details of the actual final solution paths obtained by our algorithms can be found in [2].

Finally, we should mention that the performance of all methods become worse as the number of cities increases. For example, in the case of the instance `att532`, PKN and GN did not converge and AVL, KL, and KG yielded relative errors of 9.14 %, 6.74 %, 6.80 % respectively.

5 Conclusions

In this paper we have introduced a new self-organizing neural network, the Kohonen Network Incorporating Explicit Statistics (KNIES) based on the foundational principles of Kohonen's Self-Organizing Map (SOM). Unlike the SOM, in KNIES every iteration in the training phase includes two distinct modules - an attracting and dispersing module. In the attracting module a subset of the neurons migrate *towards* the data point that has been presented to the NN — just like in the SOM. However, subsequent to this phase the rest of the neurons not involved in the attracting module participate in a dispersing operation by migrating away from their current positions to ensure that the global statistical properties of the data points are mimiced. The new scheme has been adapted to solve the Travelling Salesman Problem (TSP). Experimental results for problems taken from TSPLIB [13] indicate that it is the most accurate NN strategy for the TSP reported.

Acknowledgement. The authors would like to acknowledge that the computational experiments of this research were realized on the hardware donated to the Department of Industrial Engineering at Boğaziçi University by Hewlett – Packard.

References

[1] Altınel, İ. K., B. J. Oommen and N. Aras, (1997), "Vector Quantization for Arbitrary Distance Function Estimation", *INFORMS Journal of Computing*, Vol. 9, 1997, pp. 439-451.

[2] N. Aras, B. J. Oommen and Altınel, İ. K., (1997), " The Kohonen Network Incorporating Explicit Statistics and Its Application to the Traveling Salesman Problem". Unabridged version of this paper.

[3] Angéniol, B., C. Vaubois, and J. Y. Le Texier, (1988), "Self-Organizing Feature Maps and the Travelling Salesman Problem," *Neural Networks* 1, pp. 289–293.

[4] Burke, L. I. and P. Damany, (1992), "The Guilty Net for the Travelling Salesman Problem," *Computers & Operations Research* 19, pp. 255–265.

[5] Duda, R. O. and P. E. Hart, (1973), *Pattern Classification and Scene Analysis*, Wiley, Chichester.

[6] Fort, J. C., (1988), "Solving a Combinatorial Problem via Self-Organizing Process : An Application of the Kohonen Algorithm to the travelling Salesman Problem, *Biological Cybernetics* 59, pp. 33–40.

[7] Fukunaga, K., (1990), *Introduction to Statistical Pattern Recognition, 2nd edition*, Academic Press, San Diego.

[8] Jeffreys, C. and T. Niznik, (1994), "Easing the Consious of the Guilty Net," *Computers and Operations Research* 21, pp. 961–968.

[9] Kohonen, T., (1995), *Self-Organizing Maps*, Springer-Verlag, Berlin.

[10] Lawler, E. L., J. K. Lenstra, A.H.G. Rinnooy Kan and D.B. Shmoys, (1985), *The Travelling Salesman Problem: A Guided Tour of Combinatorial Optimization*, Wiley, Chichester.

[11] Papadimitriou, C. H., (1978), "The Euclidean Travelling Salesman Problem is NP-Complete", *Theoretical Computer Science*, 4, pp. 237–244.

[12] Potvin, J.-Y., (1993), "The Travelling Salesman Problem: A Neural Network Perspective," *ORSA Journal on Computing* 5, pp. 328–348.

[13] Reinelt, G., (1991), "TSPLIB— A Travelling Salesman Problem Library", *ORSA Journal on Computing* 3, pp. 376–384.

[14] Reinelt, G., (1994), *The Travelling Salesman. Computational Solutions for TSP Applications*, Springer-Verlag, Berlin.

[15] Wong, Y., (1996), "A Comparative Study of the Kohonen Self-Organizing Map and the Elastic Net," *Computational Learning Theory and Natural Learning Systems*, 2, pp. 401–413.

Table 1: Comparison of the results obtained for different algorithms and for various TSP instances. In each case we report the best results and the values of the parameters associated with these results.

Instance	Opt. Value	PKN (M,σ^2,α)	GN (λ,μ)	AVL (σ^2,α)	KL $(M,\sigma^2,\alpha,\omega)$	KG $(M,\sigma^2,\alpha,\omega)$
att532	27686	No Con.	No. Con.	30217.6 (20,0.2)	29551.6 (200,30,0.2,0.15)	29569.8 (400,45,0.2,0.1)
bier127	118282	122211.7 (750,45,0.2)	155163.2 (20,0.5)	122673.9 (30,0.2)	121548.7 (100,15,0.2,0.1)	121923.7 (125,50,0.2,0.1)
eil51	426	443.9 (61,10,0.2)	470.7 (20,0.5)	443.5 (35,0.2)	438.2 (10,25,0.2,0.1)	438.2 (20,50,0.2,0.05)
eil76	538	571.2 (130,10,0.2)	614.3 (10,0.5)	571.3 (50,0.2)	564.8 (90,20,0.2,0.15)	567.5 (15,5,0.2,0.2)
eil101	629	688.7 (230,40,0.2)	771.9 (100,0.5)	671.4 (30,0.2)	658.3 (20,25,0.2,0.2)	664.4 (40,35,0.2,0.25)
kroA200	28568	30927.8 (450,25,0.2)	39370.2 (220,0.5)	30994.9 (50,0.2)	30200.8 (200,25,0.2,0.25)	30444.9 (160,10,0.2,0.05)
pr124	59030	59612.8 (250,10,0.2)	86853.4 (200,0.5)	62050.3 (5,0.2)	59075.7 (25,25,0.2,0.10)	59320.6 (125,10,0.2,0.05)
pr136	96772	103878.0 (450,20,0.2)	135887.7 (250,0.5)	103442.3 (45,0.2)	101156.8 (75,40,0.2,0.15)	101752.4 (30,35,0.2,0.05)
pr152	73682	74804.2 (400,25,0.2)	105230.1 (620,0.5)	74641.0 (15,0.5)	74395.5 (180,30,0.2,0.05)	74629.0 (60,10,0.2,0.2)
rd100	7910	8137.9 (160,15,0.2)	8731.2 (45,0.5)	8265.8 (15,0.2)	8075.7 (80,20,0.2,0.25)	8117.4 (60,10,0.2,0.05)
st70	675	692.8 (160,15,0.2)	755.7 (30,0.5)	693.3 (10,0.2)	685.2 (40,10,0.2,0.05)	690.7 (30,45,0.2,0.05)

Boosting Learning Algorithms for the Colorimetric Characterization of Color Printers

A. Artusi, P. Campadelli
Dipartimento di Scienze dell'Informazione
via Comelico 39/40, I-20135 Milano (Italy)
e-mail: campadelli@dsi.unimi.it
R. Schettini
Istituto di Tecnologie Informatiche Multimediali, CNR
Via Ampere 56, I-20131 Milano (Italy)
e-mail: centaura@itim.mi.cnr.it

Abstract

Our study examines whether the colorimetric characterization of a particular ink-jet printer, the HP DesignJet 755, can be faithfully achieved using learning algorithms trained on small sets of color samples. In particular our experiments were designed to analyze whether boosting, a general method for improving the accuracy of a learning algorithm, can improve the performance of multilayer perceptrons (MLPs) and multiple polynomial regression. The results show that the MPLs can not be effectively boosted, while multiple polynomial regression can. The accuracy obtained for both transformations from CIELAB to CMY and viceversa is good, but not sufficient for high quality color reproduction. To obtain better results the boosting algorithm must be modified.

Keywords: Boosting, regression, multilayer perceptrons, printers colorimetric characterization

1 Introduction

Faithful reproduction of color in CAD and multimedia systems requires many different processing steps, including device characterization to define the transformation between the native, device-dependent, n-dimensional color spaces (RGB, CMY, CMYB) and standard, device-independent, 3-dimensional colorimetric spaces, such as CIELAB, CIEXYZ. There are two main approaches to characterize color devices: modeling and sampling. While some scanners and displays can be simply and effectively characterized by modeling, as they use additive colors to achieve their results, this approach can not be adopted for printers, where characterization is based on subtractive color mixing, and requires extremely complex models that are difficult to construct [J]. For this reason color printers are commonly calibrated using a multidimensional look-up table [KNP]. But, since it takes a huge number of measurements to build large look-up tables, small look-up tables and interpolation functions are often employed. Multi-dimensional interpolation is simple only if the sampling is dense, that is, if a large look-up table has been built; otherwise complex interpolating functions are required. Considering that raster images for printers are today often as large as 5000x5000 pixels, quite powerful processors are required to keep the overall computation time within acceptable limits from the user's point of view.

Neural networks and polynomial approximations have also been experimented in color printer characterization [KA]. Feed-forward neural networks trained by back-propagation have produced promising results [ACS] [T] in terms of both color accuracy and computation costs. However this approach presents two disadvantages: a considerable number of color measurements are required (from several hundred to a few thousand examples have been used), and the maximum color error is in general still too large for high quality color reproduction. We have study if the colorimetric characterization of a particular ink-jet printer, the HP DesignJet 755, can be faithfully achieved using learning algorithms trained on small sets of color samples. Our experiments were designed to analyze in particular whether boosting, a general method for improving the accuracy of a learning algorithm, can improve the performance of MLPs and multiple polynomial regression. Our results show that the MPLs can not be effectively boosted while multiple polynomial regression can. The accuracy obtained for both the transformation from CIELAB to CMY and viceversa is good, but not sufficient for high quality color reproduction. To obtain better results the boosting algorithm must be modified.

Section two of the paper explains how the training and test sets were generated; section three describes the boosting algorithm, and section four reports some experimental results.

2 Training and test sets

Both the training and the test sets are formed by pairs of 3-dimensional vectors, one specifying the CMY coordinates of a printed color, the other specifying the corresponding CIELAB coordinates. To obtain a pair of vectors we printed a color, specified in the CMY space, in a square 4 cm wide at a resolution of 300 dpi. In three different points of the square we measured the radiant power reflected, with a spectrophotometer MINOLTA 2002. The mean of the three values was then used to obtain the CIELAB coordinates according to the CIE-1931 standard colorimetric observer [WS].

The creation of big data sets is a time-consuming activity and since printers are devices which may need to be characterized often (changes in paper, ink formulations and environmental conditions modify the printers behavior), characterization methods which use small training sets are highly desirable.

We built two training sets of 216 elements. Both of them were generated by highly non linear CMY sampling: Training set one, according to a method suggested in [M], and Training set two derived by an appropriate subsampling of a set of 729 colors, as suggested in [JMF]. The test set contains 760 elements obtained by random sampling of the CMY space.

3 Boosting

Boosting is a general method for improving the accuracy of a learning algorithm. The algorithm ADABOOST [FSa] for binary classification problems, has been shown to be very effective in experiments conducted by [DC] [FSb] [Q] and others. Two extensions of this algorithm have been proposed for regression problems: the algorithm ADABOOST-R [FSa] for functions $f : X \to [0,1]$ and the algorithm ADABOOST-RΔ [BCP] for functions $f : X \to [0,1]^m$.

Before describing ADABOOST-R Δ we must provide some preliminary definitions. Given a set X and $Y = [0,1]^m$, we let P be a probability distribution on XxY. A N−sample is a sequence $\langle(x_1, y_1), \dots, (x_N, y_N)\rangle$ with $(x_k, y_k) \in X \times Y$; we call $Samp_N$ the set of all the N−samples, and $Samp = \cup_N Samp_N$. Given a class of functions $H \subseteq \{f \mid f : X \to [0,1]\}$, a learning algorithm $\mathcal{A}$ on H is a function $\mathcal{A} : Samp \to H$.

The inputs of ADABOOST-R Δ are: a N−sample $S = \langle(x_1, y_1), \dots, (x_N, y_N)\rangle$, a distribution D on the elements of S, a learning algorithm $\mathcal{A}$, a number of iterations T, and a real number Δ. It works by repeatedly reweighting the examples in the training set, and then rerunning the algorithm $\mathcal{A}$ on these reweighted examples in order to effectively force $\mathcal{A}$ to concentrate on the hardest ones. At each cycle t, the algorithm computes a distribution D_t over the training examples and runs $\mathcal{A}$ on the examples weighted according to D_t obtaining a hypothesis h_t. After T iterations it produces a final hypothesis h_f which can be viewed as a weighted vote of the hypotheses $h_t \in H \quad (t = 1, T)$ made by $\mathcal{A}$. More precisely, $h_f = argmax_{y \in [0,1]^m} \sum \alpha_t HS(\Delta - ||h_t(x) - y||)$, where HS denotes the Heaviside function ($HS(x) =$ if $x \geq 0$ then 1 else 0), $|| \; ||$ is a norm on R^m, and α_t is the weight of the hypotheses h_t.

Even if $\mathcal{A}$ is moderately inaccurate, for a sufficiently large T the empirical error

$$\epsilon^{2\Delta} = \sum_{k=1}^{N} (HS(||h_f(x_k - y_k|| - 2\Delta) D(k))$$

made by the final hypothesis h_f on the sample S can be reduced close to 0.

Under weak conditions on H (that is, the Vapnik-Chervonenkis dimension [V] of H is finite), choosing elements of $X \times Y$ randomly and independently according to P, for sufficiently large samples, results in an empirical error close to the generalization error with high probability [V]. This suggests that a good learning algorithm would minimize the empirical error.

Algorithm ADABOOST-RΔ

Input: a N−sample $S = \langle(x_1, y_1), \dots, (x_N, y_N)\rangle$, a distribution D on S, a learning algorithm $\mathcal{A}$, an integer T, a real numberΔ.

Initialize the weight vector $w_i^1 = D(i)$ for $i = 1, \dots, N$

Do for $t = 1, 2, \ldots, T$

1. Set $\mathbf{p}^{(t)} = \frac{\mathbf{w}^{(t)}}{\sum_{i=1}^{N} w_i^{(t)}}$

2. Choose the sample $S^{(t)}$ from S randomly with distribution $\mathbf{p}^{(t)}$; call the learning algorithm $\mathcal{A}$, and obtain the hypothesis $h_t = \mathcal{A}_{S^{(t)}}$

3. Calculate the error $\epsilon_t = \sum_{i=1}^{N} p_i^{(t)} HS(||h_t(x_i) - y_i|| - \Delta)$; if $\epsilon_t > 1/2$ then $T = t - 1$ and abort loop

4. Calculate $\beta_t = \frac{\epsilon_t}{(1-\epsilon_t)}$

5. Set the new weights vector at $w_i^{(t+1)} = w_i^{(t)} \beta_t^{1-HS(||h_t(x_i)-y_i||-\Delta)}$

Output the hypothesis $h_f = argmax_{y \in [0,1]^m} \sum \alpha_t HS(\Delta - ||h_t(x_i) - y_i||)$ where $\alpha_t = \log \frac{1}{\beta_t}$

An upper bound to the error $\epsilon^{2\Delta} = \sum(HS(||h_f(x_k - y_k|| - 2\Delta)D(k))$ is set by following theorem [BCP] which shows that the empirical error goes to 0 exponentially fast with the number of combined predictions.

Theorem 3.1 *Suppose the learning algorithm $\mathcal{A}$, when called by ADABOOST-RΔ, generates hypotheses h_t with errors $\epsilon_1, \ldots, \epsilon_T < 1/2$, and let h_f be the final hypothesis output by ADABOOST-RΔ. Then*

$$\sum_{||h_f(x_k)-y_k||>2\Delta} D(k) \leq 2^T \prod_{t=1}^{T} \sqrt{\epsilon_t(1-\epsilon_t)} \quad .$$

4 Experimental results

Both the training sets described in section two were used to train multilayer perceptrons and multiple polynomial regression. All the parameters relevant to MLPs (network architecture, learning rate and momentum term) were selected in earlier experiments [A] in order to have a low mean error on both the training sets and the test set. To reduce the maximum error made by either MLPs or multiple polynomial regression, we applied ADABOOST-RΔ to both. The experiments were run on each mapping ($CMY \rightarrow CIELAB$ and $CIELAB \rightarrow CMY$) on both training sets; the algorithm performance was also evaluated on the test set of 760 colors. The results relative to MLPs with a fixed architecture are shown in Table 1 for the function $CMY \rightarrow CIELAB$ and in Table 2 for the function $CIELAB \rightarrow CMY$. In the columns for training error and test error the mean (left) and the maximum error (right) on the training and the test set are reported. Each row stands for one experiment and specifies the training set (Tr.1 or Tr.2) and the learning algorithm (backpropagation or boosted backpropagation). For a meaningful comparison we tried to run the boosted MLPs for a number of epochs comparable to that of the reference MLP,

Table 1: function $CMY \rightarrow CIELAB$

Tr.set	algorithm	tr. error	test error	epochs	T
Tr. 1	BP	0.737 1.996	1.866 6.904	19000	-
Tr. 1	BP + BOOST	1.453 3.595	2.085 6.830	17500	5
Tr. 2	BP	0.865 2.792	1.247 5.207	14500	-
Tr. 2	BP + BOOST	1.483 3.425	1.700 5.177	17500	5

Table 2: function $CIELAB \rightarrow CMY$

Tr.set	algorithm	tr. error	test error	epochs	T
Tr. 1	BP	2.339 10.286	8.692 30.212	18000	-
Tr. 1	BP + BOOST	3.368 11.338	9.031 35.592	17500	5
Tr. 2	BP	2.359 7.908	5.458 28.607	25000	-
Tr. 2	BP + BOOST	3.727 11.161	6.554 28.782	17500	5

but this was not always possible, since ADABOOST-RΔ may output from the Do loop (see step 3. of the algorithm).

Our results show that ADABOOST-RΔ is not able to improve the performance of MLPs. In fact, the maximum error decreases slightly only on the test set for the function $CMY \rightarrow CIELAB$, while it actually increases in all other cases. This suggests that backpropagation does not lend itself to boosting because of the high number of parameters that have to be set.

More encouraging results were obtained when the boosted algorithms were multivariate polynomials of different degree, whose coefficients were determined by multiple polynomial regression. We experimented with polynomials up to the fourth degree [A]. For limits of space we report only the results relative to the function $CMY \rightarrow CIELAB$ obtained with training set two. Since in the CIELAB space the distance between two colors is related to the perceptual distance, the error values give a clear idea of the quality of the learned transformation (a color difference of 1 in the CIELAB space corresponds approximatively to a barely noticeable perceptual difference, and a difference of less than 3 units is acceptable in most applications).

Table 3 is organized in the same way as the preceeding ones: each row refers to one experiment. Mulptiple polynomial regression is referred to as Regr.n, where n is the number of coefficients of the polynomial; Regr+BOOST means that ADABOOST-RΔ has been applied to multiple polynomial regression.

Looking at Table 3 we see that the mean and the maximum error decrease with the number of coefficients of the polynomial, and that boosting is effective in reducing the maximum error on both the training and the test set, with the exception of the fourth degree polynomial with 105 coefficients. The lowest maximum error has been obtained with a fourth degree polynomial with

Table 3: function $CMY \to CIELAB$

algorithm	tr. error	test error	T
Regr. 12	6.216 23.843	6.572 25.163	-
Regr. + BOOST	6.506 21.177	6.949 22.536	2
Regr. 21	5.130 18.531	5.653 19.700	-
Regr. + BOOST	5.337 17.81	5.713 19.239	3
Regr. 30	2.869 7.827	2.853 7.379	-
Regr. + BOOST	2.935 7.088	2.948 7.047	3
Regr. 39	2.465 8.118	2.635 6.827	-
Regr. + BOOST	2.651 6.987	2.814 6.313	4
Regr. 42	2.457 7.895	2.612 6.536	-
Regr. + BOOST	2.616 6.605	2.816 6.113	4
Regr. 60	1.647 4.050	2.022 5.690	-
Regr. + BOOST	1.675 3.834	2.037 5.529	4
Regr. 69	1.451 4.061	1.833 5.552	-
Regr. + BOOST	1.478 4.058	1.871 5.520	4
Regr. 87	1.254 3.065	1.637 5.058	-
Regr. + BOOST	1.296 2.701	1.703 5.034	11
Regr. 105	0.933 2.678	1.448 5.201	-
Regr. + BOOST	0.973 2.958	1.608 5.214	2

87 coefficients. The same general behaviour was registered for the function $CIELAB \to CMY$. Although satisfactory in most cases, a maximum error of 5.034 units in the CIELAB space is not good enough for high quality applications. However, ADABOOST-RΔ is capable of only a few iterations (eleven in this most favourable case), and this is far from the number required to radically reduce the empirical (and hopefully the generalization) error. This is a serious drawback. An interesting question is whether ADABOOST-RΔ could be modified to avoid the undesiderable output from the loop.

References

[A] A. Artusi.: Applicazione di algoritmi di apprendimento alla caratterizzazione colori-metrica di stampanti a colori. Tesi di laurea in informatica, Univ. degli studi di Milano, A.A 1996-1997.

[ACS] S. Albanese, P. Campadelli, R. Schettini.: Inkjet color printer calibration by back-propagation. Communication at the Workshop on Evaluation Criteria of Neural Net Efficiency in Industrial Applications, Vietri, November 1995.

[BCP] A.Bertoni, P. Campadelli, M. Parodi.: A boosting algorithm for regression. Proc. of ICANN97, Lecture Notes in Computer Science 1327 : 343-348, Springer, 1997.

[DC] H.Drucker, C. Cortes.: Boosting decision trees. In Advances in Neural Information Processing 8: 479-485, 1996.

[FSa] Freund, Y., Schapire, R.E.: A decision-theoretic generalization of on-line learning and an application to boosting. Journal of Computer and System Sciences 55: 119-139, 1997.

[FSb] Freund, Y., Schapire, R.E.: Experiments with a new boosting algorithm. Machine Learning: Proc. of Thirteenth Int. Conf.: 148-156, 1996.

[JMF] R. Jackson, L. MacDonald, K. Freeman.: Computer generated color J. Wiley and Sons Ltd, West Sessex, UK, 1994.

[J] T. Johnson.: Methods for characterizing colour printers. Displays 16(4): 193-202, 1996.

[KA] H.R. Kang, P.G. Anderson.: Neural network application to the color scanner and printer calibrations. Journal of Electronic Imaging 1: 25-135, 1992.

[KNP] J.M. Kasson, S.I. Nin, W. Plouffe, J.L. Hafner.: Performing color space conversions with three-dimensional linear interpolation. Journal of Electronic Imaging 4(3): 226-250, 1995.

[M] N. Moroni.: Barcellona HP Res. Labs, Personal Communication,1996.

[Q] J.R. Quinlan.: Bagging, boosting and C4.5. In Proceedings of the Thirteenth National Conference on Artificial Intelligence: 725-730, 1996.

[T] S. Tominaga.: A color mapping method for CMYK Printers. Proc. 4th IS and T and SID's Color Imaging Conference: Color Science, Systems and Applications, 1996.

[V] V.N. Vapnik.: Estimation of Dependences Based on Empirical Data. Springer-Verlag, 1982.

[WS] G. Wyszecki, W.S. Stiles.: Color Science. J. Wiley and Sons Ltd (1982).

Handwritten Character Recognition with an Analog Feature Extraction Module

G.M. Bo, D.D. Caviglia, and M. Valle
Department of Biophysical and Electronic Engineering
University of Genoa, Via all'Opera Pia 11/A, 16145 Genova, Italy
ph:+39 10 3532287, fax: +39 10 3532777, e_mail: gian_bo@dibe.unige.it

Abstract

The classification of handwritten digits through an analog feature extractor chip and a neural classifier is discussed in this paper. The chip implements a feature extraction algorithm onto analog circuits; it extracts a set of 112 features from the input character (32×24 binary pixel matrix). The features, coded by current signals, are given in input to a neural classifier which performs the recognition task. The chip validation results are reported: a set of handwritten digits have been classified by a neural network implemented by a software simulator. The resulting classification error rate has been successfully compared with the ones obtained by a high level model of the chip and to those obtained with other techniques reported in the literature.

1 Introduction

Among the different techniques for handwritten character recognition reported in the literature (see among others [1]), we consider in this paper those based on the following processing steps:

- the characters are pre-processed to extract a set of features;
- the features feed a classifier that performs the recognition task.

The reduction of information redundancy during the pre-processing step is the main advantage of this approach. An effective classification and reduced error rates can be achieved.

Since software implementations of feature extraction algorithms may require great computational power, dedicated hardware implementations have to be taken into account for real time target applications (i.e. post automation, check reading, etc.). In particular, analog VLSI implementations will be valuable if high processing speed and low power consumption are pursued.

In this paper the classification of handwritten digits through an analog VLSI feature extractor chip based in the algorithm discussed in [2] is presented. The chip extracts from handwritten characters, represented by 32×24 binary pixels matrices, a set of 112 features coded by current signals. Such features are given in input to a software implemented neural classifier (a Multi Layer Perceptron network, MLP [6]) performing the recognition task. The chip has been fabricated and succesfully tested.

This paper is organized as follows. In Section 2 a brief description of the feature extraction algorithm is given. In Section 3 the analog VLSI implementation of the feature extractor is presented. In Section 4 the classification results obtained from the recognition of handwritten digits from NIST 19 database are presented and discussed. The conclusion are drawn in Section 5.

2 The Feature Extraction Algorithm

The algorithm rationale [2] consists of extracting local information: the character matrix (i.e. 32×24 binary pixels) is partitioned into 16 sub-matrices of 9×7 pixels each, partially overlapped by one or two columns and/or rows. On each sub-matrix seven pattern recognition operators are applied to obtain a set of seven features: 112 features per character are thus obtained.

Let us consider a generic sub-matrix: the first feature, called smoothing (SM), evaluates the density of active pixels. It is defined as the weighted sum of pixel values in the sub-matrix (see [2] for details).

The other features evaluate to which extent the sub-matrix shapes match six major directions. In particular the horizontal, vertical, ±45 deg, and ±60 deg directions have been taken into account (HR, VR, P45, M45, P60, and M60 respectively). For each direction, a set of pixel paths have been defined: the sums of active pixels in each path are computed and an exponential function is applied to the sums. The exponential function is able to emphasize the most distinctive shape of the character. Finally, all the contributions are summed up to get the feature value [2].

3 Analog VLSI Implementation of the Feature Extractor

The feature extractor algorithm has been directly mapped onto analog circuits. The pixel information is coded by current signals, then the sum operator is easily implemented. A proper circuit based on a MOS transistor biased in the weak inversion region implements the exponential operator. More details about the analog VLSI implementation of the feature extraction algorithm can be found in [4].

The chip photograph and characteristics are shown in Fig. 1. One can see the modules corresponding to the 16 sub-matrices placed on a 4×4 array. The circuit has been fabricated through the EUROCHIP Service by using the ATMEL ES2 ECPD10 technology (a CMOS digital process with 1.0 μm channel length). The chip has 156 pads, and occupies an area of 7.4×9 mm^2. The measured average power consumption is 60 mW, and the character recognition throughput is of about 140 KChar/s.

In Fig. 2 the measured feature values obtained from the processing of three synthetic plus one real characters are reported: in each figure the character and the results of the processing are drawn. The feature values are coded by grey

Technology process	ES2 CMOS 1.0 μm
Character size	32×24 binary pixels
Input data size	32 bits
Throughput	140 KChar/s
Chip size	7.4×9 mm^2
Transistor count	28000
Power consumption	60 mW

Figure 1: The chip photograph and characteristics.

levels: the minimum and maximum values are coded by black and white levels respectively.

The character of Fig. 2.a has a horizontal shape on the top (sub-matrices 1 to 8) and a vertical shape on the bottom (sub-matrices 9 to 16): the SM, P45, M45, P60, and M60 have consequently small values. The HR feature has high values at the top, and the VR feature has high values at the bottom, according to the character shape.

Similar considerations can be pointed out regarding Fig. 2.b and Fig. 2.c. In Fig. 2.b the character has a +45 deg shape at the top, and a +60 deg shape at the bottom. In Fig. 2.c the character has a -45 deg shape on the top, and a -60 deg shape on the bottom. The results obtained from the feature extraction process are in accordance with the character shapes.

In Fig 2.d the features obtained from a real character are reported: the character is a "3" (NIST 19 database [3]). Again, the feature values are in accordance with the shape of the input character.

4 Validation of the Optical Recognition System

4.1 Validation Set-Up

The NIST 19 database [3] has been used to validate the feature extractor chip. It contains 402953 samples of segmented handwritten digits, each one represented by a matrix of 128×128 binary pixels. From this database we extracted 3000 characters divided into two subsets:

- 2000 characters (200 per class) from the Handwriting Sample Forms (HSF) HSF_1, used as training set;
- 1000 characters (100 per class) from HSF_4, used as test set [5].

The 3000 characters have been normalized as 32×24 binary pixel matrices and given in input to the chip. An 8-bits ADC converts the output analog

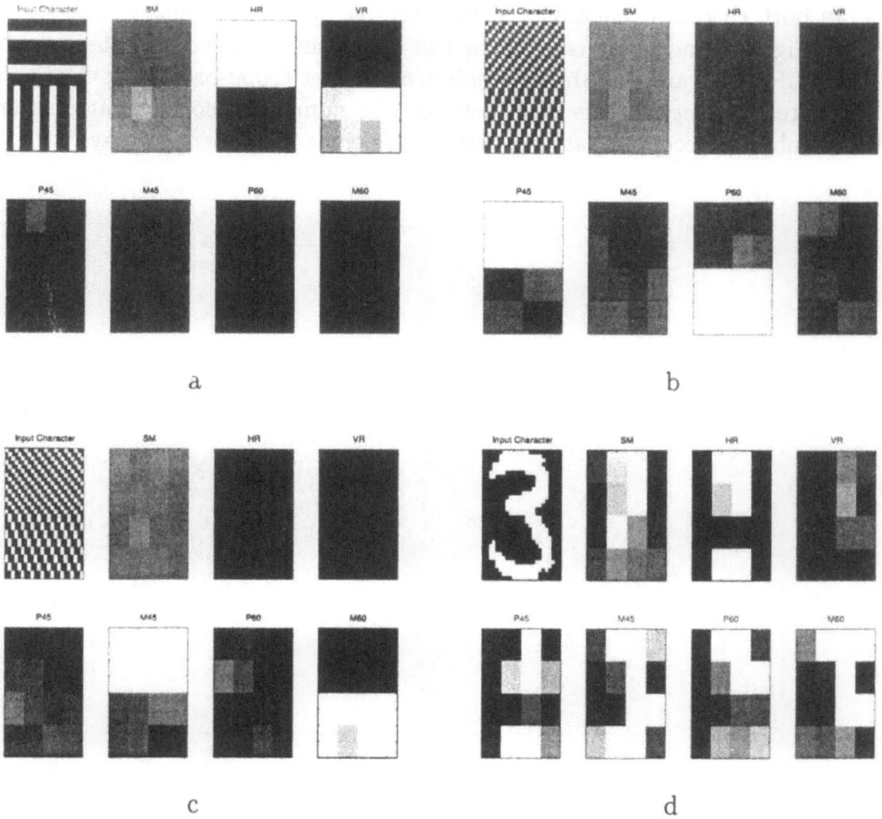

Figure 2: Chip measurement representations.

features (coded by currents) into digital values that vary in the range [-1,+1].

4.2 Simulation Results

The MLP network used for the classification task has 112 inputs (the number of features per character), 48 hidden neurons, and 10 output neurons (one per class).

The network has been trained by using the standard *by pattern* Back Propagation (BP) algorithm [6]. The classification task has been performed in two ways.

- With a rejection criterion: an input pattern will be classified if the difference between the maximum and second maximum of the network output values is greater than a predefined threshold, otherwise it will be rejected.
- Without a rejection criterion: all the patterns are classified.

In both cases, the classification is based on the highest output value.

In Fig. 3 the accuracy rate on the test set versus the rejection rate is shown. The MLP was trained with the whole training set (2000 patterns). The accuracy rate is defined as the ratio between the number of correct patterns and the number of accepted patterns (i.e. not rejected, classified anyway).

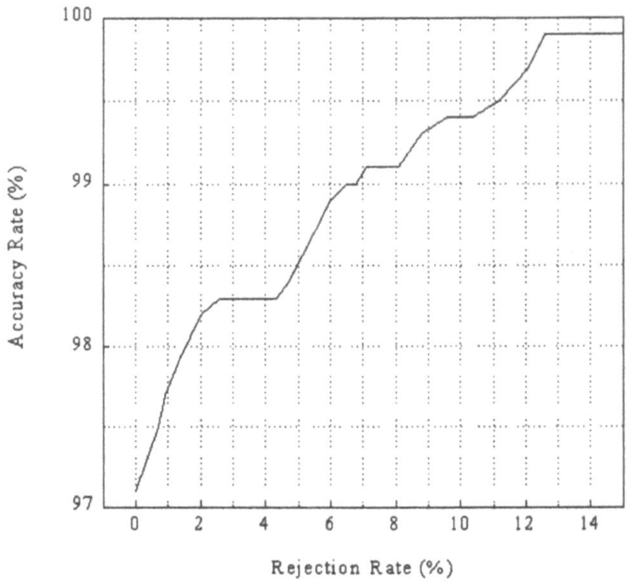

Figure 3: Accuracy rate vs rejection rate.

It is worth noting that an accuracy rate of 99% is obtained with a rejection rate of about 6.5% (93.5% of acceptance). An accuracy rate of 99.9% is obtained with a rejection rate of 13%.

These figures are satisfactory, taking into account even the small size of the training set and compare favourably to those discussed in [7].

5 Conclusion

The aim of our research is to develop an Optical Character Recognition system based on neural models and implemented by analog computational modules. In this paper the results obtained from the classification of handwritten digits through analog VLSI feature extractor chip have been presented and discussed. The chip extracts a set of 112 features based on local information from pixel matrices coding handwritten characters. The features are given in input to a software implemented neural classifier (a MLP network) which performs the recognition task.

The chip implementing the feature extractor module has been designed and fabricated: experimental results validate the chip functionality.

Acknowledgements

The authors wish to thank F. Orengo and G. Curatolo for their help in performing the chip measurements.

References

[1] S.N. Shrihari. *Recent Advances in Off-line Handwriting Recognition at CEDAR.* In Proc. of the Fifth Int. Workshop on Frontiers in Handwriting Recognition, IWFHR'96, pp. 1-15, 1996.

[2] D.D. Caviglia, M. Valle, A. Rossi, M. Vincentelli, G.M. Bo, P. Colangelo, P. Pedrazzi, and A.M. Colla. *Feature Extractor Circuit for Optical Character Recognition.* Electronics Letters, Vol. 30, No. 10, pp. 769-760, 1994.

[3] P.J. Grother. *NIST Special Database 19.* National Institute of Standards and Technology, 1995.

[4] G.M. Bo, D.D. Caviglia, and M. Valle. *An Analog VLSI Implementation of a Feature Extractor for Real Time Optical Character Reacognition* . IEEE Journal of Solid State Circuits, Vol. 33, No. 4, pp. 556-564, 1998.

[5] P.J. Grother. *Cross Validation Comparison of NIST OCR Databases.* In D.P. D'Amata Editor, volume 1906, SPIE, San Jose, 1993.

[6] J. Herzt, A. Krogh, and J.L. Palmer. *Introduction to the Theory of Neural Computation.* Addison-Wesley, 1989.

[7] S.B. Cho. *Neural-Netwok Classifiers for Recognizing Totally Unconstrained Handwritten Numerals.* IEEE Transacions on Neural Networks, Vol. 8, No. 1, pp. 43-53, 1997.

A High Performances and High Versatility Reconfigurable System for Fast Prototyping of Digital Neural Network Based on FPGA

Mario Costa, Davide Palmisano, Eros Pasero,
Luca Bovio Ferassa and Andrea DiLello
Politecnico di Torino
Dipartimento di Elettronica - Laboratorio di Neuronica
C.so Duca degli Abruzzi, 24 10129 TORINO (ITALY)

Abstract

In this paper we present the design and the realization of a system suitable to work as a fast prototyping platform for implementing several digital neural networks by means of FPGA each one with a different number of processing units and connectivity architecture. Thus we developed a system with a processing kernel working on neural algorithms in addition with a control unit and a communication unit that allow the managing and the interfacing both toward application field and a host computer for data downloading. To get the same performances varying the neural network implemented we developed a communication protocol that handle all the working operations. Its generality and adaptability at various working conditions allow an easy reconfiguration of the features of the neural network and therefore a saving in the time spent in testing and prototipation. Then we will show how a neural network can be implemented using up to four FPGA depending on its complexity. A case study will be presented to show the performances of the overall system.

1 Introduction

Since Artificial Neural Networks (ANN) have been introduced, their hardware implementation was interesting to solve problems in many fields, such as non linear processing, forecasting, classification and image processing. Though neural algorithms have an analog feature, their digital implementation had been investigated because of the reliability of digital VLSI fabrication technology. Moreover the last synthesis tools and testability techniques give designers good chance to get a working chip with the first run, avoiding the lost in time and money of a redesign. Performances and property of a neural system are highly dependent on its topology. As a function of the chosen architecture (one or multiple layer, locally or fully connected as a feed forward or feedback structure) and neurons and synapses number, the overall features and capabilities are completely different. Normally, in order to modify an hardware implemented

ANN adapting it to a different purpose, may be necessary to redesign it again at all. This is an extremely major drawback for the realization of a general purpose hardware implemented ANN that can be overcome only using multiplexed architectures with a reduced number of neurons or with configurable routing strategy [1]. In many cases this leads to a falling off of the performances respect to those of a special purpose ANN. Hence it seems clear the only way a user can easily modify the architecture of a digital ANN to fit it for several applications is to carry it out on a programmable logic system. At now Field Programmable Gate Array (FPGA) are the most interesting solutions. Usually it is essential to design interfacing circuits as well when a system must swap a data stream with another element such as a computer host [2]. Suppose, for example, an ANN for the control of a physical system. Signals from the transducers give an information about state variables and hence are the inputs of the network whereas the outputs are sent to the actuators. A changing of the physical system to be controlled takes to a new design of the interfacing circuit as well as of the ANN. A prototyping system which allows an efficient interface and reconfiguration of the function only by the manipulation the circuit working as ANN is a good improvement both for the costs reduction and saving in time spent for redesign and test.

2 Description of the system

The prototyping system block diagram can be represented as in fig 1. The guidelines for designing the whole system can be summarized as follow

- capability of building neural networks with different size. The only limit in the implementation of network topologies must come merely from the number of logic blocks available in the chosen FPGA. The best tradeoff between performances, complexity and costs can be obtained using four FPGA.
- development of a control unit implementing all the circuits that manage the working of the system regardless of the kind of neural network actually implemented.
- the minimum clock frequency for the system must be 25 MHz.
- parallelization of the processing
- external interfacing by a fast and commonly used bus to maximize the throughput and internal interfacing among memory and FPGA with a dedicated bus controlled by a suitable communication protocol

Here three main components can be distinguished. The Neural Block is made up of up to four FPGAs according to the user's requirements that can be arranged to implement various neural architectures by use of classical CAE tools, that is the user only need to define the structure by using of a schematic capture or behavioral description tools. The System Memory is used to store all processing data such as synaptic weights, neurons thresholds, input and output vectors and is concurrently accessed by the internal information stream and the external one carried on the external bus. The Control Unit handles

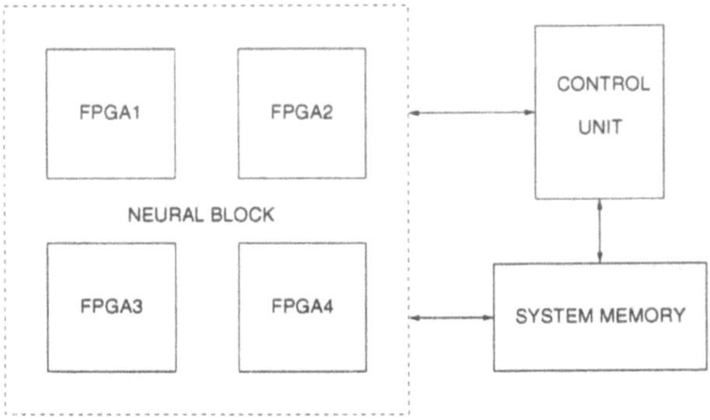

Figure 1: Block diagram of the system

the internal and external data flow using the communication protocol defined and must not be modified every time a new network is implemented.

The output data are sent to a block that provides a SCSI-2 interface toward off board. We chose to use this bus instead of VME [3] or PC-EISA bus because at now it is fast enough and allows a direct interfacing with a broad set of host computer regardless of the operating system running on them. In addition a bus named local bus is designed for transferring data from FPGAs to the System Memory. The System Memory must put up a different amount of data when the size and the topology of the network implemented are modified. It is important to notice that is quite hard to know a priori the required amount of memory space to store all data necessary for every network as well. Neural networks are often memory expensive systems, thus it is important to find a way to efficiently manage the on board memory to save both space and overall costs. We will examine now the working of the single FPGA, its interfacing with other elements and the way to make it working as a neural network.

2.1 The managing and the organization of the single FPGA

As previously presented, each FPGA contains a portion of the neural network as a function of the global complexity, the topology, the FPGA's occupation state and the number of input-output resources required. Each FPGA must share data with both the control unit and the system memory. For both these two kinds of interfacing the main problem is to find an optimization between the number of pins required and the amount of signal to be employed to grant a certain throughput. In fact, a complex communication protocol between FPGA and control unit allows better performances, but needs of a great number of signals and hence lots pins, as well as the exchange of data with the memory if wide addresses and data bus are used. Another important discrimination that must be accomplished among the input pins of the FPGA is about data input or weights input for the network. The simplest solution consisting in a fixed definition of a pin as a data input or weight input pin has the drawback to set the format of the input stream depending also on the width of the memory chips used. To get the best partition of the input-output pins of the single

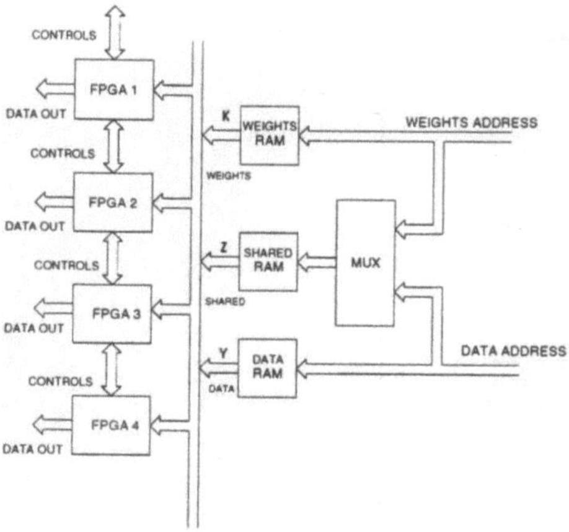

Figure 2: FPGA - memory datapath

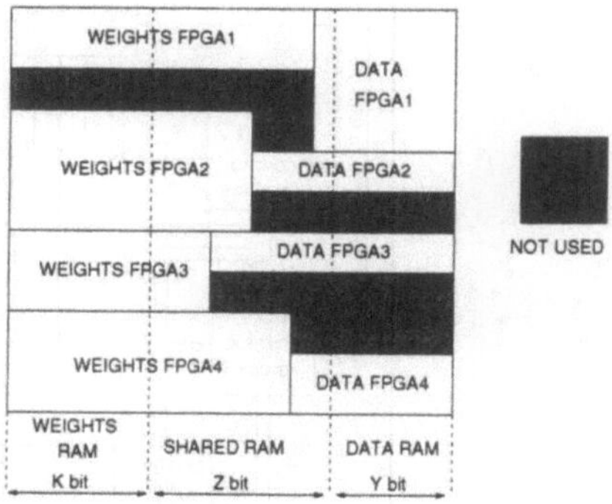

Figure 3: System memory organization

FPGA it is important to define some constraints on the data structures coming from a trade-off between resources utilization and system generality. They can be summarized as follows

- input data can be up to five bits
- weights can be from five to eight bits wide
- the same data information can be delivered up to four synapses once for cycle
- each FPGA requires eight control signals and a synchronization one
- the output bus is eight bit wide

It was derived a distribution of the pins based on three quantities named respectively **Y**, **K** and **Z** indicating the number of pins dedicated to input data, weights and shared data.

Another important element to be well defined is the data path between the system memory and the single FPGA. The following set of constraints raised for its developing

- minimization of RAM chips to be used on board
- reduction of the three state buffers
- minimum delay
- reduced set of control signal

The solution is showed in figure 2(a). The get a better throughput a particular timing in the reading cycle was created shifting the clock signals of the FPGA.

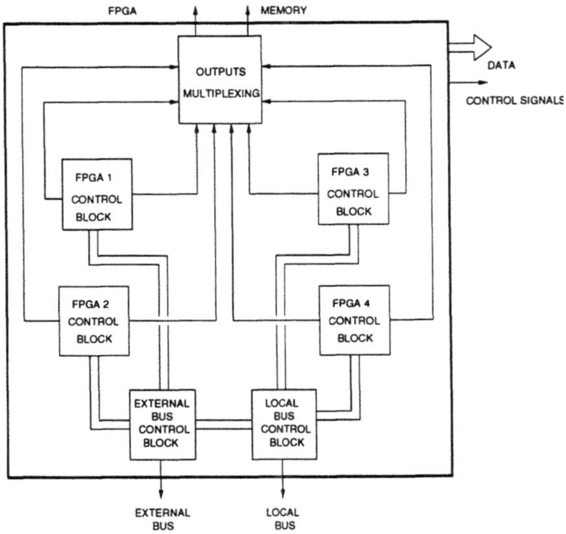

Figure 4: Control Unit

2.2 The system memory

The system memory can hosts both data and weights for the neural network. It is important hence to find a way to separate them to avoid misunderstanding or damaging accessing to the wrong item. There is a separation of the memory assigned to each FPGA in two sides which can be accessed at by two addresses named data address and weight address. This separation is not accomplished in the address space (that is a word can be a data or a weight according to its location) but in position of the word, that is some of the bits of a word can be data bit and the others weight bits. To save memory, a mechanism that downloads previously processed data overwriting new ones was developed with a control system to prevent damaging in still useful data. The system memory is hence structured as in figure 3(*b*).

3 Control unit

The Control Unit interacts with both the neural block implemented on the FPGA and the system memory monitoring the state during each processing step and carrying out its jobs. The items which the control unit bases the working on are its own control signal. The structure of this block can be seen in figure 4.

Every block named FPGA Control Block is constituted by two items demanded to handle data and weights, which are routed to the system memory by mean of a multiplexing network. A last item named *Operational Management Unit* is demanded to generate the control signals for the FPGA. It is synthesized as a synchronous machine and makes use of an its own subset of control signals. The External Bus Control Block is the most hard to be designed block of the control unit because it must operate almost entirely at the full clock frequency. Because of the requirement of a synchronous communication with the interface system, it must be able to introduce wait states if they are necessary. Moreover it must fetch and execute all the received instructions and generate a subset of signal related to the current operational state.

4 The control Protocol

The control protocol allows the user to define operations to be performed by the FPGA block by means of specific commands and moreover to control the information flow and the processing sequence giving some commands to the control unit before the beginning of the elaboration. The single command constitutes the parameter whose values the control unit determines from the manner the control must be executed. The control protocol acts in such a way to monitor and manage operations on the single FPGA, on the memory and on the local bus for the internal data exchange and the external data exchange by interfacing with the external bus chosen (in this case a SCSI-2 bus). Every kind of operations to be performed by the Control Unit is a particular subset of parameters and commands of the control protocol. The main feature of it is that it is a parametric protocol, that is each job is made up of a statement describing the task and a parameter for it (for example the number of memory accesses or the amount of bits to be processed in a certain sequence.

5 The design of the board

The cornerstone for the board design must be the interfacing system among memory, the FPGAs set and the control unit. Each component can access the memory only for one control unit clock period and must terminate in this time all its operations. It can gain a new memory access after a four clock cycles time. The memory is constituted by five blocks each one of 8 Kbyte and needs to have enable signals for the correct read-write cycles. The control unit was described in VHDL and synthesized to be implemented both with a XC4020 FPGA and with a ES2 ECPD10 ASIC in a pin compatible fashion depending on the user requirements. The main problem in the design of the control unit is its toggle rate, that force the designer to spend particular care in the FPGA design. An ASIC realization allows greater clock speed, but, on the other hand, is much more expensive both in design time and overall costs. To allow an easy interfacing with a SCSI-2 bus a commercial available controller was used to simplify the design.

6 Applications

We tested the system by building up a modular locally connected multy-layer perceptron [4] used for recognition of isolated hand-written digits trained with-the database provided by the National Institute of Standard and Technology. The general architecture of the MLP is based on partially connected modules, as it is shown in figure 5*a*, the number of which equals that of classes to be distinguish (in this case ten classes). In each module the unique output neuron is completely connected with the hidden layer, each neuron of which has

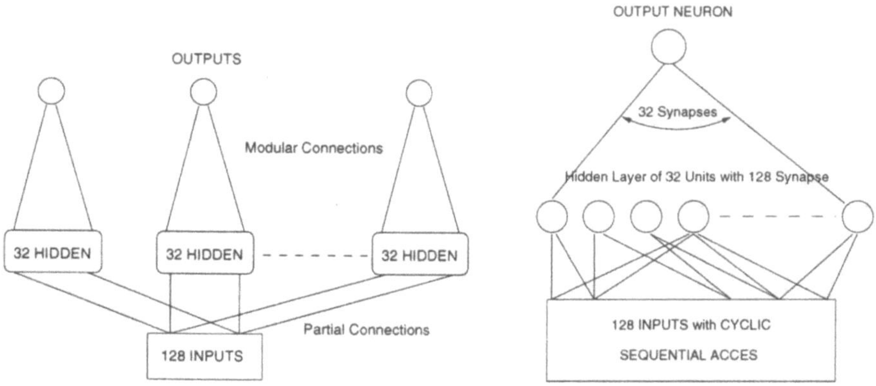

Figure 5: (*a*) Modular MLP implemented (*b*) Single module

access to only 28 input components (out of the total 128) in a cyclic sequential fashion. This solution gives two advantages to the designer: because of the fact each module covers the input vector an integer number of times, the same connection scheme may be preserved along the network and, second, a non degenerate network is obtained because different hidden neurons in the same module are connected with different subnets of the input vector. To avoid the use of the transfer function of the output neurons input pattern is classified by determining the maximum output activation by means of a winner-takes-all circuit.

Every FPGA is made to work alone thanks to the modular architecture independently from each other. Every FPGA consists of 2 hidden neurons with 7 synapses each and an output neuron with only 2 corresponding connections. The single module structure is depicted in figure 5*b* We got the best trade-off between performances and memory and FPGA sizes coding all the weights with five bits plus the sign and using only four bits for the transfer function of all the hidden neurons. When using this configuration an error rate at zero rejection rate of 3.7% arises, whereas an error rate of 4.3% comes from using 4+1 bits for weights and two for the internal transfer functions. In this conditions we got a recognition speed of 7812.5 digit/s for each FPGA, that means whether all FPGAs are employed, a total speed of 31250 digit/s or, equivalently, 290 milion MAC/s. To store all the weights set is necessary to provide 8800 bytes for each FPGA, but whether the modules in every components are identical this amount of memory can be reduces to one fourth.

References

[1] J. Cloutier, P. Y. Simard, "Hardware Implementation of the Backpropagation without Multiplication" in *Proceedings of* 4th *International Conference on Microelectronics for Neural Networks and Fuzzy Systems*, Turin, Italy, September 26-28 1994, pp 46-55

[2] G. Grossi, R. Posenato, M. Costa, D. Palmisano, E. Pasero," Fast Prototyping for Hardware Neural Networks", in *Proceedings of* 5th *International Conference on Artificial Neural Networks ICANN 95*, Paris, France, October 9-13 1995, vol. 1, pp. 411-416

[3] Adaptive Solutions," CNAPS System Architecture Manual", in *Technical Manual, CNAPS Core Series*, Adaptive Solutions Inc., 1400 N. W. Compton Drive, Suite 340, Beaverton, OR 97006, 1993

[4] M.Costa, E. Filippi, E. Pasero, "Relaxation of Hardware Requirements in Ensemble of Modular, non Degenerate Multy-Layer Perceptrons", in *Proceedings of* 14th *IMACS World Congress*, Atlanta, USA, July 1994, pp. 646-650

A L-based Energy Function for SE Prediction

Alexandra I. Cristea and Toshio Okamoto
AI lab., IS*S, The Grad. School of Info Systems
Univ. of Electro-Comm., Tokyo, Japan

Abstract

Neural Networks (NN), with their distributed parallel processing power, can be used as a tool to forecast stock exchange(SE) events, if these are seen as Time-Series(TS). In this paper we present a system for SE events prediction, based on an energy function that we deduce from the Lyapunov (L- also called *infinite*) norm. We focus on the mathematical deductions of the energy function and on the error minimization procedures. We present some comparative results of our method, the classical backpropagation method(BP), and the random walk generator.

1 Introduction

In this paper we discuss the mathematical basis of our SE events prediction tool and some of the results. In the whole work we study the TS of SE events, that, according to some authors([1,2]), can be forecasted with certain accuracy with NN, but are seen by others as chaotic events. We start with the assumption that SE data, even if not *fully deterministic*, are not *fully random*. This assumption is based on the fact that, as different agents act driven by their own private theories[1] it can be presumed that SE events are not influenced by a buy-sell pattern based on only one theory. We are making predictions in a *fixed window of time-values* and use as inputs *predictable data(PD) and other economical influence factors*, although a lot of the training takes place on simple PD. The data are some real-world data ([4]) and some user-designed series for testing and are divided into two subsets: a **training set**, that contains pairs of {*[old data], [new data]*}, that are fed into the system in order to teach it their correspondence, and a second set, the **test set**, of similar data, out of which only the *[old data]* are fed into the system, to check the system accuracy. We use values (prices on the SE market) over any desired period of days,scaled for a better representation and for a uniformization of the input: [0-min; 1-max], represented by the user-interface program on the 0y axis, while the 0x axis represents the time (fig.2).

2 Deduction of the energy function

We rewrite the TS problem in a standard *Energy-minimization* form. We use a L based development of an energy function(E), but while the L method is based on vector derivation, we used weights change, therefore matrix derivation, as follows.

Let r be the error vector of components $r_j = y_j - d_j; j = 1,..,n$, where $y_j = f\left(\sum_{i=1}^{m} w_{ij} x_i\right)$ is the NN's actual output, d_j the desired one, for neuron j in the output layer, x_i one of the inputs, $i = 1,..,m$, w_{ij} the weights.

[1] simplifying hypothesis (market is influenced by long term prediction windows: 25,50)

We define the maximum error radius:

$$w_0 \geq |r_j(w)|; j = 1, .., n; w_0 \geq 0, \tag{1}$$

the minimization of the error becomes equivalent to the minimization of w_0. Condition (1) becomes the restrictions: $g_{j1}(w) = w_0 + r_j(w) \geq 0; g_{j2}(w) = w_0 - r_j(w) \geq 0, w_0 \geq 0, j = 1, .., n$; where g_{j1}, g_{j2} are notations. A robust E can be defined by using quadratic penalties on the newly constructed restriction functions:

$$E(w) = \nu w_0 + \frac{k}{2} \sum_{j=1}^{n} \left\{ ([g_{j_1}(w)]_-)^2 + ([g_{j_2}(w)]_-)^2 \right\}, \tag{2}$$

with $[y]_- = min(0, y); \nu > 0, k > 0$, constants. The energy function depends only on the maximum error radius w_0, if the restrictions are satisfied, else the terms g_{j1}, g_{j2} also contribute to the error function. Let $[y]_- = y * S(y)$, where $S(y) = \{0, \text{ for } y \leq 0; 1, rest\}$ and $S_{j1} = S(w_0 + r_j)$; $S_{j2} = S(w_0 - r_j)$, then apply Gradient Descent(GD) method. The weight changing equations will look as follows:

$$\frac{dw_{i,j}}{dt} = \frac{-d}{dw_{ij}}(E(w)) = -kx_i f' \sum_{j=1}^{n} \{(w_0 + r_j) S_{j1} - (w_0 - r_j) S_{j2}\} \tag{3}$$

Similarly, the movement equation for the maximal error radius is:

$$\frac{dw_0}{dt} = \frac{-d}{dw_0}(E(w)) = -\nu - k \sum_{j=1}^{n} \{(w_0 + r_j) S_{j1} + (w_0 - r_j) S_{j2}\} \tag{4}$$

The basic idea of this approach is that w_0 not only controls all the other errors, but ensures convergence by it's monotonous decreasing. We can rewrite the components that are added in the sums of equations (3,4) :

for $\frac{dw}{dt}$ computation:

$$(w_0 + r_j(w))S_{j1} - (w_0 - r_j(w))S_{j2} \begin{cases} w_0 + r_j(w), & \text{for } r_j < -w_0 \\ 0, & r_j \in [-w_0, w_0] \\ -w_0 + r_j(w), & r_j > w_0 \end{cases} \tag{5}$$

for $\frac{dw_0}{dt}$ computation:

$$(w_0 + r_j(w))S_{j1} + (w_0 - r_j(w))S_{j2} \begin{cases} w_0 + r_j(w), & \text{for } r_j < -w_0 \\ 0, & r_j \in [-w_0, w_0] \\ w_0 - r_j(w), & r_j > w_0 \end{cases} \tag{6}$$

Both functions of r_j are linear ramps of absolute slope 1 and dead zone $[-w_0, w_0]$. The dead-zone linear ramp in (5) compares each error to w_0 and establishes how the weights should change in order to keep the errors inside the w_0 sphere. The function in (6) compares w_0 to the other errors, and tries to shrink the sphere, i.e., to adjust w_0 in order to become closer to the other errors. That is why w_0 can only become smaller, but each r_j must correct both deviations in negative and positive domain, so it can change both ways. Now that we have a weights' changing equation, let's consider the net design, and then make further mathematical deductions.

3 Error Backpropagation equations

The previously calculated weight changes can be used for the external layer. For the hidden layer, we need a BP error computation, for we cannot compute a direct error, as in the case of the external layer. The weight changes for the output layer are similar to the ones in (3, 4), with the only difference of letting x_i be h_k, the hidden layer values. With an analogy to the BP mechanism, the weights' ($v_{k,i}$) and biases' ($v_{0,i}$) changes before the hidden layer are:

$$\frac{dv_{i,k}}{dt} = -kx_i f'\left(\sum_{k=1}^{m+1} v_{k,i}h_k + v_{0,i}\right) * \sum_{j=1}^{n} w_{i,j}\left\{(w_0 + r_j)S_{j1} - (w_0 - r_j)S_{j2}\right\}, \quad (7)$$

$$\frac{dv_{0,i}}{dt} = -kf'\left(\sum_{k=1}^{m+1} v_{k,i}h_k + v_{0,i}\right) * \sum_{j=1}^{n} w_{i,j}\left\{(w_0 + r_j)S_{j1} - (w_0 - r_j)S_{j2}\right\}, \quad (8)$$

with $k = 1,..,m; i = 1,..,p$; $f(x) = \frac{1}{e\ \ +1}, \alpha \in \Re$, p the dimension of the hidden layer, and the rest of the notations the same as in chapt.2. We expect this model to be locally $H^{\inf}$ optimal in the sense of [3], as it is based on backpropagation and uses gradient-descent reasoning. The difference to the pure LMS based BP is the reduced sensitivity to minor errors by using the maximum error sphere and dead-zone delimiters for the movement equations.

4 The Network architecture

The designed net is based on L **GD NN**. The simplest net construction would be an 1 layer NN (1 input, 1 output layer) which isn't enough for the complexity of the analysed data. Previous results ([4]) showed that a 2-layer net is enough. Therefore, if the LGNN idea shows itself to be successful, it should at least be able to work on a net of a similar dimensional complexity.

Figure 1 shows the used net, containing a m-dimensional input layer, a hidden layer of dimension p and a 1-dimensional output layer. We consider 1 step

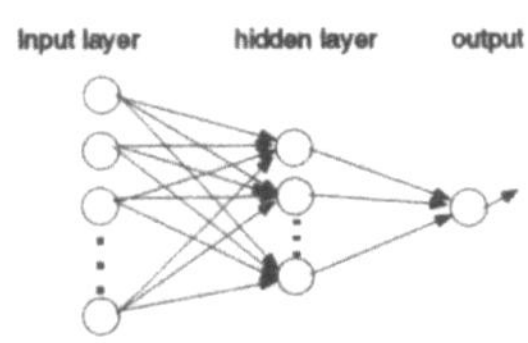

Figure 1: The net layout

ahead prediction, and for n step prediction (long-term prediction) we reinforce the system's output as a feedback to the input (from $x_1, x_2, ..., x_m$ we predict output y^t, from $x_2, x_3, ..., x_m, y^t$, y^{t+1}, a.s.o.)[2]. Evidently, the previously deducted weight changes (eq. 3, 4) cannot work for a 2-layer NN directly, so we have to take into consideration some error BP.

[2] t is the time variable

5 Implementation notes and details

The system can perform **training** (learning) through weight computing, **forecasting** with a 2-layer *feedforward Network* and serve also as an **user-interface**. These functions are implemented as independent programs. The communication between these processes is assured by the common resources, which are files, but can also be share memory resources or TCP/IP based internet information exchange resources. For a flexible design, the input files can be given to the system as fixed-format files, or user-designed files can be input through a user interface. A full help support is provided at each step by the designed program, both for current state explanations as for advice about possible steps to take by the system user. The **prediction error** is displayed in percentage to the maximal value(price) that occurred in the given time-period in a "Error dispersion" window (see fig.3).

6 Momentum and Convergence Step

An easy to implement method, related to *Momentum* convergence and *Weight Decay* ([5]), is the **step-changing**: if the same direction is taken for a given time-period, the step is increased: $step = 2 * step$; if there are oscillations around a value, decreased (to reach to the "valleys of lower potential"): $step = .5 * step$. This procedure requires less memory than the weight decay (where all the last changes of the weights have to be stored) and only an integer vector of directions, with values of $\{ -1, 0, +1 \}$. If the last weight-change had a negative sign, the new value of the direction vector will be [-1], if it was positive [+1], and for no change, [0]. This simple procedure has two consequences we observed: if the initial (random) weights are far from the desired ones, than this procedure tends to determine a shorter convergence time; else, the net has a slight tendency of oscillation till it finds the correct combination, because of the rougher approach of bigger steps for a short time period.

7 Results

7.1 Displays

System Learning Display: We present here a learning example of 10 input and 10 output data (fig.2). Therefore, out of 10 daily input SE values, 10 daily output values were to be learned. The weight matrix dimension had therefore a 10x10=100 elements dimension. The continuous, zig-zag line on the left side of the graphical chart are the past (or input) data, while on the right side, the desired outputs[3] and the prediction can be seen. The prediction is normally displayed by a dotted line, while the desired outputs are displayed by a continuous line[4] but here, with "0" error, the two outputs overlap, and therefore, a single line is visible. The program displays here the learning of the correspondence of 10 past prices with 10 future prices. The outputs are scaled from 0 to 1 (Oy axis), and the time (days) is represented on 0x.

Error display: is shown in a separate, 2-dimensional window (fig.2 upper right corner, fig.3).[5] The display of error was designed for a quick qualitative reading. Both the vertical and the horizontal axis represents errors (at different times). Errors can occur if the desired values and the predicted output values don't overlap. We represent the maximal error of the whole interval - exterior

[3] as in *Supervised Learning*

[4] just like the inputs

[5] displayed in percentage to the maximal value that occurred in the given time-period

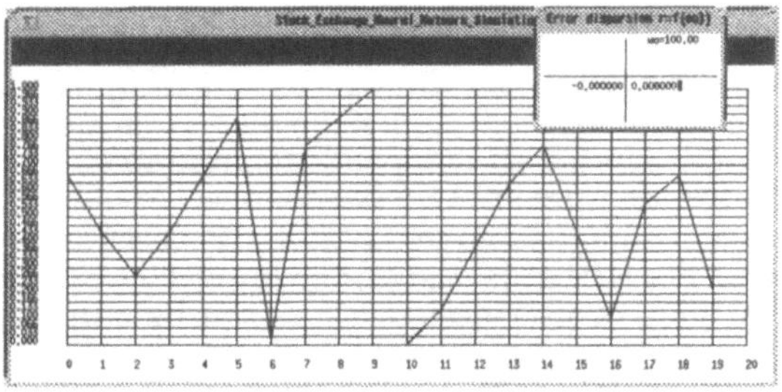

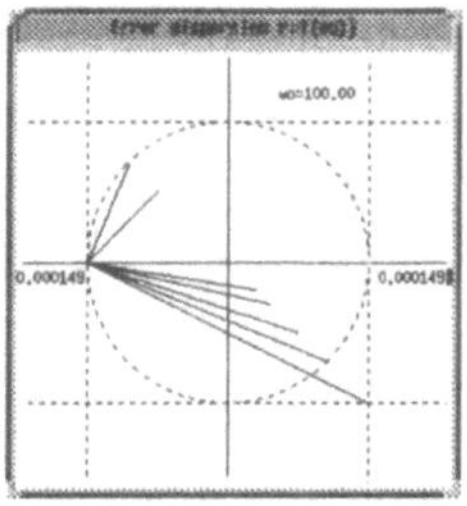

Figure 2: (left) A "perfectly" trained net
Figure 3: (right) Error display window

square - and the other errors (lower values) - lines starting in the left corner of the picture and ending at the intersection with the second diagonal.

These were the elements of the result display. We used them to view first the behaviour of a regular BP net, then the L NN training. There were little differences regarding the convergence. One of the similarities is that both methods tend to increase the weights during learning (if no momentum term is added). As a convergence difference we mention that the growth-rate is larger in the BP case, while our method seem to show a smoother pattern.

7.2 Adjustable step(AS) vs. constant step(CS)

We present here some results on the AS versus the CS. These computations were performed on a small quantity of irregular, user-design data (fig.4). The 0x-axis shows the number of input values, the 0y-axis shows the number of outputs. The convergence step number would be on a 3rd axis, which is not represented. The step on both axis is 1 day's time.

Out of 300 random starting points (see fig. 4 for used input-output combinations) that we used for training once with constant step, once with a variable step, there were 165 input-output combinations for which the adjustable step was better, and only 81 where the constant one showed better convergence. There were also 54 cases in which the two types had similar convergence times. The most frequently used starting point was the [20,10] (20 input values, 10 outputs).

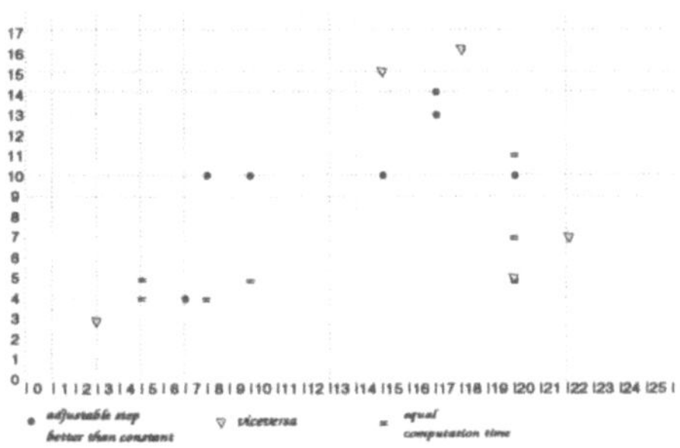

	correlation of set	perf. compared to random walk
BKP		
trained	0.951364	0.447806
test1	0.226734	0.853706
test2	0.308115	0.810235
modified Lyapunov		
trained	0.960783	0.356597
test1	0.480251	0.871593
test2	0.511318	0.795510

Figure 4: (left) Variable and constant step; the influence on the convergence.

Figure 5: (right) BP and L method compared with algebraic indicators.

7.3 Algebraic indicator comparison

We used two coefficients for comparison:

1) the **correlation** coefficient, that measures the linear correlation between the forecasted value y_j and the real value b_j (n - number of observations on validation set): $R = \frac{\sum_{=1} (b - \bar{b})(y - \bar{y})}{\sqrt{\sum_{=1} (b - \bar{b})^2}\sqrt{\sum_{=1} (y - \bar{y})^2}}$ with $\bar{x} = \frac{1}{n}\sum_{i=1}^{n} x_i$.

2) the **t test** or **Theill** coefficient, that measures the out-performance of the NN over random walk (the predictor that estimates the future value as $y_{j+1} = b_j + n_j$, i.e. the actual value plus a white noise): $T_r = \frac{\sqrt{\sum_{=1} (y - b\ )^2}}{\sqrt{\sum_{=1} (b - b_{-1}\ )^2}}$

If $T_r \leq 1$, the NN predictor is better than the random walk predictor.

Average results over a number of 100 learning experiments (with error margin of 0.1, and convergence time interval of 5-10 min.) are presented in fig. 5. We observed 2 methods: our method, versus the classical BP method. As in fig.5, we have tested these coefficients on the trained set of data, and also on two test sets. The Theill coefficient is best (lowest) in both cases, as expected, on the already trained set. The correlation of prediction and real value are also best for this case (closer to 1). Except for the performance measure of the test1 set, when compared to random walk, that has a higher value than the BP case (0.87.. when compared to BP's 0.85..), the rest of the values show that our method is superior: for the correlation coefficient, the values are closer to 1, so there is a higher correlation between prediction and real values, and for the comparison with random walk, the Theill coefficient has lower values for the method we proposed, when compared to the BP method.

8 Conclusions

We described in this paper an energy function construction, based on the L GD. We compared the $\bar{L}$ method with the classical BP method, and showed the differences between the two, and proved that the behaviour of our method is superior, by using indicator comparison. We also described a few features of the whole SE forecasting tool that was constructed, based on the TS behaviour of SE. For further work we intend to extend the system and apply it to other problems.

References

[1] T. Ankenbrand and M. Tomassini. ANN and GA, Springer (1995).

[2] B. Gas and R. Natowicz. ANN and GA, Springer (1993).

[3] B. Hassibi *et al.* Theor. Adv. in Neur.Comp., Kluwer (1994).

[4] D. Komo *et al.* Int.Smp.on Speech, Img.Proc.NN, IEEE (1994).

[5] A. Krogh and J.A. Hertz. Niels Bohr Inst., ftp (1992).

[6] T. Miyano and F. Girosi. ftp: publications.ai.mit.edu (1994).

[7] A. Roebel. *The Dyn. Pattern Select. Alg.*, TU Berlin (1994).

[8] P.D.Shawn and M.R. Davenport. IEEE Trans. on NN (1992).

Robust Processing of Visual Evoked Potentials

E.D. Di Claudio, G. Falvo, R. Parisi, R Perilli, and G. Orlandi

INFOCOM Dpt. - University of Rome "La Sapienza"
Mail address: Via Eudossiana 18, I-00184 Roma - Italy
Ph: +39-6-44585837 Fax: +39-6-4873300
Email: dic@infocom.ing.uniroma1.it

Abstract

Electrophysiology measurements attempt to get information about neural pathological conditions by measuring electrical cortical responses to appopriate stimuli.
Visual Evoked Potentials (VEPs) are the responses of occipital visual cortex to time-varying images captured by human eyes. They consist of a quasi-deterministic signal embedded in a strong background noise, produced by Electro-EncephaloGraphic (EEG) activity, sensors and amplifiers.
The VEP analysis is useful for early diagnosis and monitoring of diabetes, neural diseases such as multiple sclerosis, vascular diseases of the brain, check of the visual acuity in non-collaborative patients (children, old-aged people with degenerative neural diseases), and as a support for medico-legal investigations.
The signal-to-noise ratio is usually very low and a statistical averaging on multiple stimuli is required in order to recover an acceptable VEP waveform.
However the background noise is highly non stationary, clearly not Gaussian distributed and in some realization there is a strong quasi-periodic interference due to cerebral α-waves.
Moreover, the loss of attention and blinking may cause the lack or the weakening of the VEP signal in some trials. This impacts the reliability of standard (unweighted) average commonly used to reduce noise effects. So estimators that are resistant to the presence of *outliers* in the data set should be used for VEP preprocessing.
In this work, traditional robust estimators, such as the trimmed mean, and the median are compared with a modern neural version of the Iteratively Reweighted Least Squares (IRLS) which uses the recently developed Block Recursive Least Squares (BRLS) learning algorithm.
The same IRLS is used for the final deconvolution of brain spikes, that are the main target of VEP analysis.

1. Introduction

Electrophysiology measurements attempt to get information about neural pathological conditions by measuring electrical cortical responses to appropriate stimuli [1][2] and comparing them with a standard feature database. Visual Evoked Potentials (VEPs) are the responses of the occipital visual cortex to time-varying images captured by human eyes (see Fig. 1).

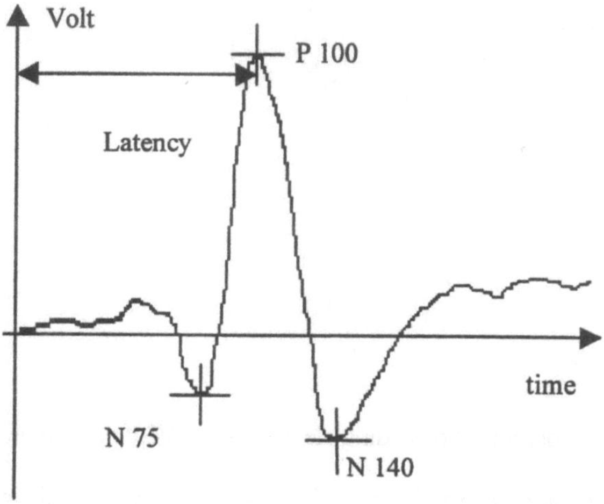

Fig. 1: Schematic of the VEP signal.

The VEP consists of a quasi-deterministic signal embedded in a strong background noise, produced by the Electro-EncephaloGraphic (EEG) activity [3], sensors and amplifiers. The frequency band of these evoked potentials ranges from 3 to 200 Hz. They are collected by applying electrodes to the head of the patient, as shown in Fig. 2. A sudden change in the visual pattern stimulates retinal photoreceptors, originating an electrical potential, which is partially processed within the retina by several neural cell types and finally transmitted through the visual pathway to the occipital cerebral cortex, where it is collected by the electrode(s). The types of visual patterns used for stimulation are relatively standardized in medical literature and practice [1][2].
The VEP analysis is useful for early diagnosis and monitoring of diabetes, neural diseases such as multiple sclerosis, vascular diseases of the brain, check of the visual acuity in non-collaborative patients (children, old-aged people with degenerative neural diseases), and as a support for medico-legal investigations [4]. In particular, diseases impact the latency and/or the peak amplitude of the main positive pulse, called the P100 point because the normal latency from stimulus to response is about 100 ms. The recorded VEP waveform is characterized by the amplitude and the latency of the P100 peak, as well as of the negative peaks, usually referred at 75 ms (N75), and 140 ms (N140) [2].

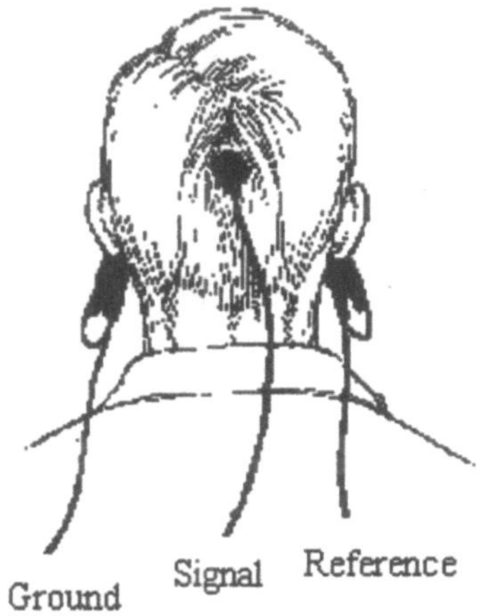

Fig. 2: Position of recording electrodes for VEP measurement.

The signal-to-noise ratio is usually very low and a form of statistical averaging on multiple stimuli is required in order to recover an acceptable VEP waveform [4][5][6][7].

We have found that the background noise is highly non stationary, clearly not Gaussian distributed, and in some realization there is a strong quasi-periodic interference due to cerebral waves [8]. The loss of attention and blinking may cause the lack or the weakening of the VEP signal in some trials. This diminishes the reliability of standard (unweighted) average commonly used to reduce noise effects. So estimators that are resistant to the presence of *outliers* in the data set should be used for VEP preprocessing. In this work, traditional robust estimators, such as the *trimmed mean*, and the *median* are compared with a modern neural version of the *Iteratively Reweighted Least Squares* (IRLS), which uses the recently developed Block Recursive Least Squares (BRLS) learning algorithm [9][10].

The same IRLS is used for the final deconvolution of brain spikes [8], that are the main target of VEP analysis. In fact, the VEP waveform appears as the result of passing the neuron spike train originated by the stimulus through a dispersive distributed channel, which models the visual pathway.

2. The VEP Signal Model

The VEP waveform is usually modelled as a quasi-deterministic signal $\mathbf{s}(t)$, buried in a zero-mean random noise $\mathbf{n}(t)$. So the observed signal $\mathbf{x}(t)$, uniformly sampled with period T, obeys the discrete-time equation:

$$\mathbf{x}(kT) = \mathbf{s}(kT) + \mathbf{n}(kT) \quad , k=1, 2, \ldots, N; \tag{1}$$

$$E\{\mathbf{s}(kT)\} = \mathbf{p}(k); \tag{2}$$
$$E\{\mathbf{n}(kT)\} = 0, \tag{3}$$

where the symbol $E\{.\}$ indicates the statistical expectation operation [11]. The sampling rate chosen for the experiments was 2 KHz. The noise is usually considered independent from the VEP signal, but from personal experience this assumption might be a relatively crude model simplification at low spatial resolutions.

The noise is clearly not Gaussian, and has a distribution characterized by moderately long tails, as shown by the typical histogram depicted in Fig. 3. It is also markedly not stationary.

M independent realizations of the VEP, or *snapshots*, each of length N, are sequentially collected (Fig. 4) and stored columnwise in the (M by N) raw data matrix **X**.

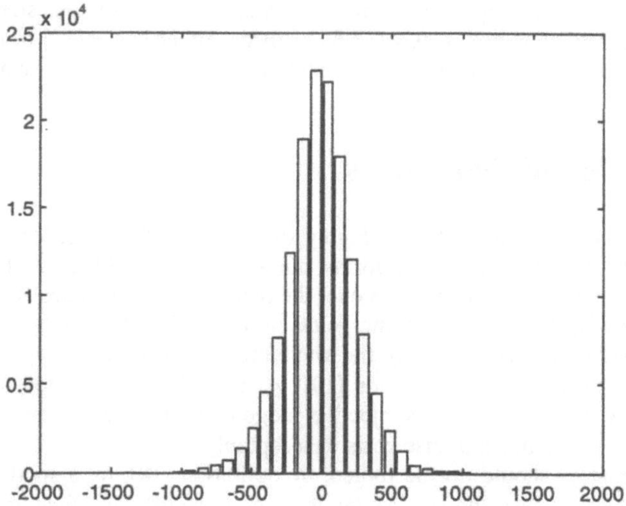

Fig. 3: Histogram of a typical EEG noise realization.

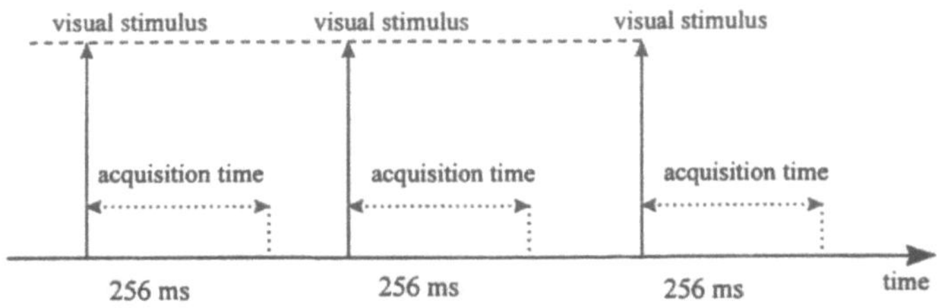

Fig. 4: Collection of independent VEP realizations.

The basic analysis procedure for extracting the VEP signal from crude realizations stored in matrix **X** consists in the following steps:
Step 1: Remove from each column of matrix **X** the DC offset originated by the analog-to-digital (A/D) conversion;
Step 2: Estimate the VEP {**p**(k), k=1, 2,..., N} by averaging the rows of matrix **X** over the M realizations.
Step 3: Filter and downsample the estimate obtained from the previous step to reduce the noise. If necessary, an autoregressive (AR) analysis, either linear [11] or non-linear [8], can be performed for deconvolving the exciting train of spikes originated by cortex neurons.

3. Robust Mean Estimators

The presence in each data batch of highly disturbed VEP realizations (*gross-errors*, according to [10]) and missing waveforms, due to blinking and loss of attention, makes the simple arithmetic mean of samples over the realization (row) space an unreliable tool for estimating **p**(k). In fact, the missing points may introduce bias [11], while high noise peaks (*outliers*) may corrupt locally the waveform, even if they are present only in a single realization on several hundreds [10]! A manual data selection stage could be inserted, but it will rely on the subjective judgement and the personal ability of the operator, will waste time, and raise costs unacceptably.
A robust estimators should not be fooled, by definition [10], by a small percentage of outliers and missing data, though mantaining a *good statistical efficiency*, ideally close to the Cramer-Rao bound [11] for uncontaminated data.
Two classical robust estimators for the mean of a stochastic process (with a symmetric distribution) are the *sample median* [10][11], and the *sample trimmed mean* [10]. In the trimmed mean approach, the M samples are first ranked. Then the q ($q < M/2$) smallest and the q larger samples are discarded from the data set. The trimmed mean is finally

evaluated as the arithmetic mean of the (M-2q) remaining samples. The trimming factor t, given by the formula $t = 2q/M$, is usually chosen between 0.2 and 0.75 [10]. The trimmed mean is an unbiased estimator of the mean of a random variable characterized by a symmetric distribution, and remains near-optimal also for Gaussian signals, while the sample median is less statistically efficient [10], but it can cope with a larger number of outliers than the trimmed mean (up to the 50% of the data set, instead of the 50q%).
The third estimator used was the classical IRLS algorithm for robust linear regression [10]. P basis functions $\{f_i(x), i=1, 2,..., P\}$ are evaluated on the same set of Q points $Z=\{x_j, j=1, 2,..., Q\}$. The PQ samples are ordinately stored in the (Q by P) system matrix **A**, so that $\mathbf{A}_{ij} = f_i(x_j)$. The (Q by 1) vector **b** contains the noisy measurements of the target function $y(x_j)$, sampled at the same points in Z. The IRLS seek for the (P by 1) weight vector **w** which minimizes the functional:

$$J(\mathbf{w}) = \sum_{i=1}^{Q} \rho\left(\left|\frac{\mathbf{b}_i - \sum_{j=1}^{P} \mathbf{A}_{ij}\mathbf{w}_j}{\beta \cdot \sigma}\right|\right). \tag{4}$$

The selected function $\rho(x)$ should be monotone and positive on the real axis, upper and lower bounded by some constants L_1 and L_2, and retain most of information from good samples. The normalizing parameter σ is a robust estimate of the standard deviation of fitting errors (tipically the median of the absoute values of residuals [10]); β is a scaling factor, usually chosen between 1.2 and 1.5, by the Chebychev inequality [11]. Clearly the IRLS applies to the robust mean estimation problem, if it solves the overdetermined problem:

$$\begin{bmatrix} 1 \\ \vdots \\ 1 \end{bmatrix} m \cong \mathbf{b}, \tag{5}$$

for the regression parameter m.

4. The IRLS by the BRLS Algorithm

Equation (5) can be solved directly by the BRLS fast learning algorithm developed for neural networks, first formalized in [9]. It is worth noting that formulas (4) and (5) could be referred to a single-layer ADALINE neural network [12].
However the convergence of the algorithm to the local minimum is not guaranteed if σ is adaptively computed from data [10]. The experimental evidence suggests that practical convergence is typically reached within 10-20 iterations. The best use of the

IRLS-by-BRLS algorithm in this application was however reached in the nonlinear AR modelling outlined in Step 3 after data preprocessing. In fact, most linear prediction equations [8][11] in VEP estimates exhibited very small errors, except for the few ones excitated by neuron spikes. Conventional Linear Least Squares Estimators (LLSE), that assume identically distributed residuals, failed to resolve the spiky nature of the excitation.

5. Experimental Results

Three different estimators (*trimmed mean*, IRLS and *median*) have been compared to the traditional *mean* in terms of performance and robustness with several patients not affected by diseases. The same estimators were used both for removing the DC offset and estimating the VEP. Results are summarized by the performance index (PI) reported in Table 1.

The PI is related to the VEP signal-to-noise ratio after averaging, under the simplifying hypotheses of statistical (near-)independence ampong the EEG noise and the VEP. The procedure for evaluating the PI can be summarized as follows:

1) the EEG spectrum is first estimated from a lot of snapshots (300 in our case), taken from the patient waiting in a completely dark room;
2) the sample average noise power P_N in the band 5-40 Hz is computed;
3) the VEP is estimated by one of the averaging methods presented above, and its power P_S is computed in the same 5-40 Hz band;
4) the PI is finally obtained as the ratio between P_S and P_N.

Due to the presence of spurious quasi-periodic waves in some noise snapshots, the background EEG spectrum is also estimated by a fast and statistically robust procedure, which possesses several interesting features:

1) each noise snapshots (472 samples long in our case) is zero padded up to 1024 samples, and transformed in the frequency domain by the use of the FFT algorithm;
2) the median of the squared modulus of FFT bins, taken over the realization space, is considered as the (scaled) estimate of the spectrum at the frequency bin.

Performance index	Angular pattern resolution (cycles/deg.)				
Estimator	**0.5**	**1.5**	**4**	**8**	**11**
Trimmed mean	**152.5**	**140.8**	**202.6**	**64.7**	**16.7**
IRLS	**137.7**	**126.3**	**173.3**	**55.0**	**14.9**
Sample mean	**99.8**	**96.6**	**148.3**	**47.5**	**11.7**
Sample median	**129.5**	**115.7**	**180.3**	**56.6**	**14.2**

Table 1: Sample linear PI for different VEP estimators and visual pattern resolutions over the bandwidth 5-40 Hz (N=472, M=100 for VEP and M=300 for EEG noise).

It is worth noting that the PI decreases abruptly when increasing the pattern resolution over 4 cycles/deg. The relative reduction of the PI at low spatial resolutions (0.5

cycles/deg) seems to be an effect of a statistical dependence between VEP and EEG noise. The classical mean suffers clearly from the missing data problem (about the 5% of the data set). Robust estimators exhibit a better performance than the sample mean, with only a slight advantage for the trimmed mean on this data set.
The raw VEP waveform, obtained by the previous step, was filtered and downsampled by a factor five (400 Hz). After compensating for statistical effects of the EEG spectrum, independently measured on each patient, the weighted AR coefficients and residuals were computed, confirming theoretical considerations. A sample realization is shown in Fig. 5.

6. Conclusions

Robust neural tools have been developed for the VEP analysis, demonstrating a performance comparable with the best estimators availables. The general purpose IRLS-by-BRLS algorithm was shown to be flexible both as VEP estimator and as robust AR predictor. Further work will be addressed to adaptive processing of prediction residuals to better estimate the latency of the P100 peak of the VEP, and investigate about the causes of amplitude losses and delay changes observed in many pathologies.

7. Acknowledgement

This work was supported in part by the Italian Ministry of Scientific and Technological Research.

8. References

[1] *Visual Electrodiagnostic* – A guide to procedures, ISCEV (International Society for Clinical Electrophysiology of Vision), 1996.

[2] J.R. Heckenlively, G.B. Arden, *Principles and Practice of Clinical Electrophysiology of Vision*, Mosby Year Book, 1991.

[3] P.L. Nunez, R.D. Katznelson, *Electric Fields in Brain – The Neurophysics of EEG*, Oxford University Press, 1981.

[4] A. Remonds, "A Didactic Review of Methods and Applications of EEG Data Processing," *International Course of EEG Data Processing* – Paris 1977, EEG Informatics by Elsevier Scientific Publishing Company, 1977.

[5] V.D. Constantinescu, "Revealing the Distinction Between Perception and Cognition Through Intra-individual Variability of Visual Evoked Responses," International Journal of Bio Medical Computing Vol. 40, pp. 169-178, Elsevier Science, 1996.

[6] S. Panzeri, G. Biella, E.T. Rolls, W.E. Skoggs, A. Treves, "Speed, Noise, Information and the Graded Nature of Neural Responses." *Computation in Neural System*, No. 7, pp. 365-370, 1996.

[7] P. Gurzi, G. Biella, A. Spalvieri, "Estimate of Mutual Information Carried by Neuronal Responses From Small Data Samples," *Computation in Neural System*, No. 7, pp. 717-725, 1996.

[8] A. Uncini, M. Marchesi, G. Orlandi, F. Piazza, "An Adaptive Neural Network Filter for Evoked Potentials," *IEEE Transactions on Biomedical Engineering*, 1990.

[9] R.Parisi, E.D. Di Claudio, G. Orlandi and B.D. Rao, "A generalized learning paradigm exploiting the structure of feedforward neural networks," *IEEE Transactions on Neural Networks*, Vol.7, No.6, pp. 1450-1460, November 1996.

[10] P.J. Huber, *Robust Statistics*, John Wiley, New York, 1981.

[11] A. Papoulis, *Probability Random Variables and Stochastic Processes*, Mc Graw-Hill, New York, 3rd Edition, 1991.

[12] S.Haykin, *Neural Networks-A Comprehensive Foundation*, IEEE Press, 1994.

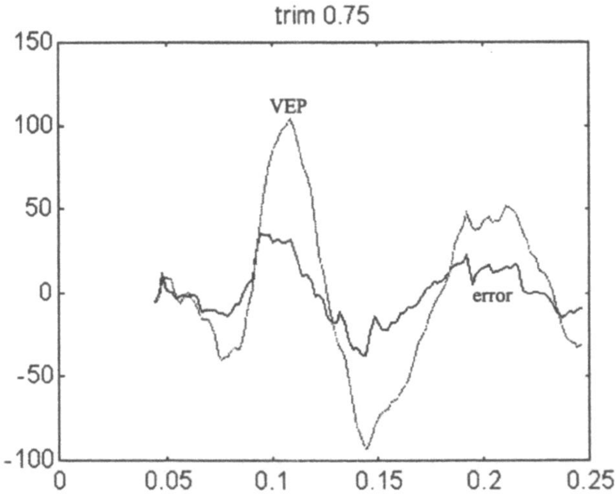

Fig. 5: Sample plot of VEP and residuals vs. time [s], 8 cycles per degree.

Convergence Properties of the Circular Cascade Correlation algorithm

Gian Paolo Drago
Istituto per i Circuiti Elettronici - CNR
via De Marini, 6 - 16149 Genova, Italy

Sandro Ridella
DIBE - Universita' di Genova
via Opera Pia, 11A - 16145 Genova, Italy

Abstract

Circular Cascade Correlation (CCC) is an improvement of the well known Cascade Correlation algorithm: the modification consists in adding a new input, whose value is the mean of the squared original inputs. This addition allows hidden units to detect closed subspaces of the input space besides the usual open ones and it increases the representation capability of the network with a moderate growth of its complexity. A convergence theorem is proved, which shows that the number of hidden units needed by CCC to learn a finite training set is equal or lower than its cardinality.
Keywords: Neural Network Computing, Learning, Adaptive Architecture, Growing Neural Networks, Cascade Correlation.

1 Introduction

The well-known Cascade Correlation (CC) learning is one of the most important incremental training algorithms for Feedforward Neural Networks [1] [2] [3]. This algorithm starts with a minimal network and incrementally let it grow until desired performance is reached. It works both with analog and digital inputs and only requires the training of perceptron-like sub-structures, avoiding learning of multilayer structures. In this paper we suggest the use of the Circular Backpropagation (CBP) [4] inside the Cascade Corralation (CC): the modification we propose will be called Circular Cascade Correlation (CCC). The new algorithm retains the above described advantages and needs no modification in the software implementation except for the addition of an input to the data file, whose value is the mean of the squared original inputs. It can be shown that this addition allows a perceptron to detect closed subspaces (besides the usual hemispaces) of the input space. This suggestion and the advantages of this input encoding have been discussed previously [4]. In particular it has been shown that the CBP model contains as particular cases both the traditional MLP sigmoidal and the radial basis function activation.

In this paper the advantages of using CCC are discussed: in particular a convergence theorem is proved, which allows to show that CCC can learn every finite training set using a number of hidden units no greater than the cardinality of the training set itself. This convergence is much faster than the one proved in [1] for traditional CC learning.

2 The Circular Back Propagation

We will now briefly recall the main modification to BP needed to obtain CBP. The usual argument of the activation function for BP

$$\chi = \sum_{i=0}^{n_i} w_i x_i \tag{1}$$

becomes

$$\chi = g[\sum_{i=1}^{n_i} (x_i - c_i)^2 + \theta] = \sum_{i=0}^{n_i+1} w_i x_i \tag{2}$$

where
n_i is the number if inputs
w_i is the weight connected to the input i-th
x_i is the input i-th, x_0 is the bias and x_{n_i+1} is the new CBP input:

$$x_{n_i+1} = \sum_{i=1}^{n_i} x_i^2 \tag{3}$$

The relationships between (g, c_i, θ) and (w_i) are straightforward to be obtained and they may be found in [4]. In eq.2 g is a gain term used for adjusting the smoothness of the activation function, c_i is the i-th component of the vector representing the center of the hypersphere, while θ is proportional to the square of its radius. Eq.2 includes as special case eq.1: this means that the hypersphere degenerate into an hyperplane. The quantity of expression 2 is the argument of a sigmoidal function

$$y = sgm(\chi) = \frac{1}{1 + e^{-\chi}} \tag{4}$$

We recall that using expression 2 into eq.4 let one obtain as particular case the RBF [4]. We stress that using CBP it is possible to find an hypersphere such as the activation function given by eq.4 has a value equal 1 inside and 0 outside.

In order to show an example where adding the new input is particularly advantageous, let us consider a circular region inside a square one. Suppose that the points inside the circular region are of a different class than those outside this region. Clearly, a perceptron is unable to separate the two classes, since they are not linearly separable. The addition of the CBP term allows separation.

More about the CBP properties in particular with regard to its generalisation capabilities, such as the Vapnik-Chervonenkis dimension may be found in [4].

3 The Circular Cascade Correlation Learning Algorithm

The output of the CCC network, containing n_h hidden units, is given by:

$$y_{n_h} = \sum_{j=1}^{n_h} v_j \{sgm(\sum_{i=0}^{n_i+1} w_{ji} x_i)\} \tag{5}$$

where
v_j is the weight connecting the j-th hidden neuron to the output
w_{ji} is the weight connecting the the input i-th to the hidden neuron j-th

As a first step, when the $(n_h + 1)$-th hidden unit is added, its input-side weights are found by maximizing:

$$R = |\sum_{l=1}^{n_p} \delta_l z_l| \tag{6}$$

where

$$\delta_l = t(l) - y_{n_h}(l) \tag{7}$$

n_p is the number of training patterns
$t(l)$ is the target for the pattern l
$y_{n_h}(l)$ is the value of eq.5 for the pattern l
z_l is the output of the new hidden unit using eq.2 and eq.4

This step is similar to the Fahlman's one except that in eq.6 the average of the terms δ_l and z_l are not taken into account: the reason of this modification will be justified in the following, where the convergence properties of this algorithm will be discussed. Moreover some caution can be taken in maximazing expression 6, particularly when classification problems are considered, in order to avoid that patterns previously correctly classified become misclassified. Both theory and practice indicate that it may be not convenient to search for a global maximum of expression 6.

The second step of the algorithm follows [3]. The new hidden unit becomes a permanent feature-detector in the network. The output training consists in finding the weights connecting the old hidden units and the new hidden unit to the output unit, through minimisation of the neural network mean square error (MSE):

$$C = \frac{1}{n_p}\sum_{l=1}^{n_p}(t(l) - y_{n_h+1}(l))^2 \tag{8}$$

where $y_{n_h+1}(l)$ is the value of the eq. 5 for the pattern l, taking into account for the new hidden.

4 Circular Cascade Correlation Convergence Properties

In [1] we have proved a theorem which guarantees CC convergence, which means that by increasing the number of hidden units sufficiently, the cost function can always be kept under a prescribed rate. We will now prove the following theorem:

THEOREM: given a training set containing n_p patterns, the CCC needs no more than n_p hidden units to represent the training set with MSE equal to zero.

The main advantage of this result with respect to that presented in[1] lies in the certainty to reach a zero error with a finite number of hidden units. This is due to the local representation capability of CBP, which is able to correct errors in some regions of the input space while avoiding to introduce errors elsewhere.

PROOF: the proof of the theorem starts considering that at the beginning of the algorithm the errors as defined in eq.7 coincide with the targets. It is easy to show that the value of expression 6 is certainly larger than the maximum among the absolute values of the targets of the training set, which corresponds to a pattern that will be called from now on maximum absolute valued target (MAVT) pattern. This can be shown by using a CBP unit which contains inside the MAVT pattern, while leaving outside all the others. The term z_l in expression 6 takes value 1 inside the hypersphere and 0 outside. The weight connecting the new hidden unit and the output must take the value of the target of the MAVT pattern. Consequently the error of eq. 7 related to the MAVT pattern becomes zero: the training algorithm guarantees that it remains zero up to the end of the learning algorithm. The procedure can be iterated: n_p hidden units are sufficient to have a network whose MSE is zero on all the patterns of the training set.

5 Computer simulation

Two classification problems are considered:

1. Generalised XOR this consists of two inputs $x_1, x_2 \in [-1, +1]$; 400 samples positioned on a regular lattice have been considered; the target is $+1(-1)$ if $x_1 x_2 \geq 0 (\leq 0)$.

2. Twospirals (Wieland, as cited in [3]): this consists of 194 samples located on two spirals, each of which belongs to a different class.

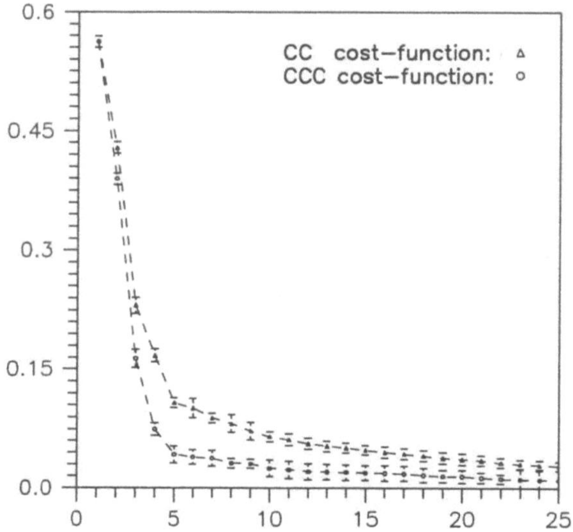

Figure 1: The MSE cost function versus the number of hidden units n_h for the Generalized XOR test. CC and CCC are compared. Each point shows the mean value of a set of 10 trials.

In Fig.1 and Fig.2, the abscissa is the number of hidden units, while the ordinate is the Mean Square Error, both for the CC and the CCC. In both the cases and for each considered number of hidden units, ten trials were performed: the best, the mean and the worst result have been plotted. Looking at the classification error, as defined in [3], CCC was able to solve exactly the Generalized Xor and the Two-spirals tests using 11 and 24 hidden units respectively.

Let us briefly comment these results. The strong advantage of CCC with respect to CC in the continuous generalised XOR test can be explained by remembering that it has been shown by [5] that the exact solution can only be obtained by a BP using at least two hidden layers. CBP can solve exactly the problem with one hidden layer containing 4 hidden units [4]: this confirm that one layer CBP can represent input-output relationships which needs two BP hidden layers [4].

The Two-spirals test is solved by CBP with only one hidden layer containing 7 hidden units [4], while in [3] no classification error was obtained by CC using 12 hidden units. From these numbers one can compute the numbers of weights of the two networks: CC needs 117 weights, while CCC needs 121 weights. The classification performance of CC and CCC are rougly equivalent, but CCC gives a better MSE then CC.

6 Conclusions

In this paper we have shown that the use of CBP in a growing neural network, like CC, allows training hidden units to correct errors in restricted regions of the input space. As a consequence, CCC is able to reach a predefined mean square error with a lower number of hidden units with respect to CC.

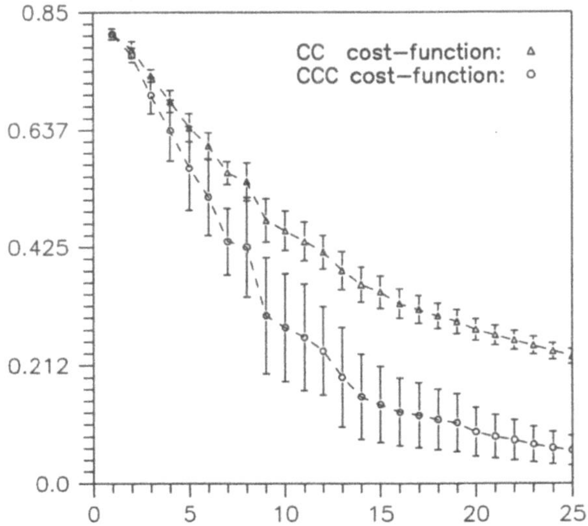

Figure 2: The MSE cost function versus the number of hidden units n_h for the Two-spirals test. CC and CCC are compared. Each point shows the mean value of a set of 10 trials.

References

[1] Drago G.P. and Ridella S. 1994, Convergence Properties of Cascade Correlation in Function Approximation, *Neural Computing & Applications*, **2**, 142-147.

[2] Drago G.P. and Ridella S. 1996, On the convergence of a growing topology neural algorithms, *Neurocomputing*, Vol.12, n.2-3, 1996, 171-185.

[3] Fahlman S.E. and Lebiere C. 1990, The Cascade-Correlation Architecture, advances in Information Processing Systems 2, Touretzky, D. (Ed.), 524-532.

[4] Ridella S., Rovetta S., Zunino R., Circular Back Propagation Networks, *IEEE Trans. on Neural Networks*, Vol. 8, n. 1, January 1997, 84-97.

[5] Zwietering P.J., Aarts E.H.L. and Wessel J., Exact classification with Two-layered perceptrons, *International Journal Neural Systems*, Vol.3, no.2 (1992), 143-156.

Neural Learning and Weight Flow on Stiefel Manifold

Simone Fiori, Aurelio Uncini, and Francesco Piazza *
Dipartimento di Elettronica e Automatica,
Università di Ancona, Ancona – Italy
{simone,aurel,upf}@eealab.unian.it

Abstract

The aim of this paper is to present a new class of learning models for linear as well as non-linear neural layers called *Orthonormal Strongly-Constrained* (SOC or Stiefel). They allow to solve orthonormal problems where orthonormal matrices are involved. After general properties of the learning rules belonging to this new class are shown, examples derived independently or by reviewing learning theories known from the literature are presented and discussed.

1 Introduction – Stiefel learning

Learning without supervision in a neural layer means adapting a weight matrix so that a predefined target is achieved. The columns of the weight matrix individuate directions in the weight-space, thus learning a weight matrix means also discovering *interesting* directions [6]. This way, an index is established over the set of all possible directions, so that any 'interesting' direction is quantitatively defined as the one maximizing the index.

In the present work we deal with the special case of the constrained search where the constraint is that of orthonormality of the columns of the neural layer's weight-matrix representing the state-variable. The associated learning (searching) algorithms are called *Orthonormal Strongly-Constrained* (SOC or Stiefel) and they allow to solve the orthonormal problems. Orthonormal problems (ONP) arise in several contexts, as those of the Optimal Linear Compression (Principal Component/Subspace Analysis, [4, 10, 11]), Whitening [7], Blind Separation of Sources [1, 4, 7, 8], DOA estimation [1]. In a ONP, the target of the adaptation rule for a neural network is to learn an orthonormal weight matrix related in some a way to the input signal. Since it is a-priori known that the final state must belong to the subset $\mathcal{N}$ of the whole space of searching containing orthonormal matrices, the evolution of the weight matrix may be strongly bounded to *always* belong to $\mathcal{N}$.

To denote an orthonormal matrix the set $\mathcal{N}_r^{p\times q}$ of the *r–orthonormal* $p \times q$ *matrices* as $\mathcal{N}_r^{p\times q} \stackrel{\text{def}}{=} \{\mathbf{W} \in \mathcal{R}^{p\times q} | \mathbf{W}^T\mathbf{W} = r^2\mathbf{I}\}$ is defined. By definition r

*This work was supported by the Italian MURST.

must be non-null and $q \leq p$. The set $\mathcal{N}_r^{p\times q}$ is known as a *Stiefel manifold*. Formally we can then define a class of learning algorithms as:

$$\mathcal{C} \stackrel{\text{def}}{=} \{\mathcal{A}_J(\mathbf{W}) | \forall t \in \mathcal{T} : \mathbf{W}^T(t)\mathbf{W}(t) = w^2(t)\mathbf{I}\} \ , \tag{1}$$

where $\mathcal{A}_J(\mathbf{W})$ is a generic searching-algorithm with state-variable $\mathbf{W}$, $\mathcal{T}$ is a time interval the algorithm runs within, and $w(t)$ is a function differentiable at least once and non-null. J is the objective function whose extrema are searched. Each algorithm contained in $\mathcal{C}$ is called SOC (Orthonormal Strongly-Constrained) or Stiefel. Any algorithm in $\mathcal{C}$ is characterized by a fundamental property: Differentiating both members of $\mathbf{W}^T(t)\mathbf{W}(t) = w^2(t)\mathbf{I}$ with respect to the time yields:

$$\frac{d(\mathbf{W}^T\mathbf{W})}{dt} = \dot{\mathbf{W}}^T\mathbf{W} + \mathbf{W}^T\dot{\mathbf{W}} = 2\dot{w}w\mathbf{I} \ ,$$

which can be rewritten as:

$$\dot{\mathbf{W}}^T\mathbf{W} + \mathbf{W}^T\dot{\mathbf{W}} = \pi\mathbf{I} \ , \ \text{with } 2\dot{w}w = \pi \ , \tag{2}$$

where $\dot{\mathbf{W}} = d\mathbf{W}/dt$. Hereafter, examples of SOC learning rules are presented and discussed.

2 A Xu's Principal Subspace algorithm

As a first example of an algorithm belonging to $\mathcal{C}$, here the one proposed by L. Xu in [11] is recalled. It allows the extraction of the Principal Subspace of fixed dimension from a multivariate random signal $\mathbf{x}$ with covariance matrix $\mathbf{C}_x$ having distinct and positive eigenvalues.

2.1 Xu's Subspace Rule is SOC

Xu proposed as an objective function the following criterion:

$$\varrho(\mathbf{W}) \stackrel{\text{def}}{=} \frac{\varphi[\det(\mathbf{W}^T\mathbf{C}_x\mathbf{W})]}{\psi[\det(\mathbf{W}^T\mathbf{W})]} \ , \tag{3}$$

where $\varphi(z)$ and $\psi(z)$ are positive monotonically increasing functions defined for $z > 0$, and $\det(\cdot)$ returns the *determinant* of the matrix contained within. Xu proved [11] that under some conditions the above function maximizes when the columns of $\mathbf{W}$ span a Principal Subspace of $\mathbf{x}$.

The Gradient Steepest Ascent (GSA) algorithm $\mathcal{A}_J(\mathbf{W})$ based on the above objective function is SOC. This property is proved in the following result.

Theorem 1. *If for the system* $\dot{\mathbf{W}} = \partial\varrho/\partial\mathbf{W}$ *an initial condition* $\mathbf{W}(0) \in \mathcal{N}^{p\times q}$ *is assumed, then for all* $t > 0$ *the flow* $\mathbf{W}(t) \in \mathcal{N}^{p\times q}$.

Proof. By handling the plain expression of the gradient $\partial \varrho / \partial \mathbf{W}$ and defining the functions:

$$\beta(\mathbf{W}) \stackrel{\text{def}}{=} \frac{\varphi'[\det(\mathbf{W}^T \mathbf{C}_x \mathbf{W})] \det(\mathbf{W}^T \mathbf{C}_x \mathbf{W})}{\psi[\det(\mathbf{W}^T \mathbf{W})]} ,$$

$$\alpha(\mathbf{W}) \stackrel{\text{def}}{=} \frac{\psi' \varphi \cdot \det(\mathbf{W}^T \mathbf{W})}{\psi \varphi' \cdot \det(\mathbf{W}^T \mathbf{C}_x \mathbf{W})} ,$$

the system $\dot{\mathbf{W}} = \partial \varrho / \partial \mathbf{W}$ reads:

$$\beta \frac{d\mathbf{W}}{dt} = \mathbf{C}_x \mathbf{W} (\mathbf{W}^T \mathbf{C}_x \mathbf{W})^{-1} - \alpha \mathbf{W} (\mathbf{W}^T \mathbf{W})^{-1} . \tag{4}$$

Pre-multiplying the preceding equation by $\mathbf{W}^T$ yields:

$$\beta \mathbf{W}^T \frac{d\mathbf{W}}{dt} = (\mathbf{W}^T \mathbf{C}_x \mathbf{W})(\mathbf{W}^T \mathbf{C}_x \mathbf{W})^{-1} - \alpha \cdot (\mathbf{W}^T \mathbf{W})(\mathbf{W}^T \mathbf{W})^{-1} . \tag{5}$$

Moreover $\beta \mathbf{W}^T \dot{\mathbf{W}} = (1-\alpha)\mathbf{I} \Rightarrow \beta \dot{\mathbf{W}}^T \mathbf{W} = (1-\alpha)\mathbf{I}$, and their hand-by-hand addition gives:

$$\mathbf{W}^T \frac{d\mathbf{W}}{dt} + \frac{d\mathbf{W}^T}{dt} \mathbf{W} = \frac{d(\mathbf{W}^T \mathbf{W})}{dt} = \frac{2(1-\alpha)}{\beta} \mathbf{I} , \tag{6}$$

therefore the SOC strong property $\mathbf{W}^T(t)\mathbf{W}(t) = w^2(t)\mathbf{I}$ holds for all $t \geq 0$, whereby the thesis follows. □

The quantity $w(t)$ is a real function of the time and of the initial value $w^2(0) = \frac{1}{q}\mathrm{tr}[\mathbf{W}^\mathrm{T}(0)\mathbf{W}(0)]$. Its evolution is governed by the function π_X given by $\beta \pi_X = 2(1-\alpha)$. It would be interesting to note that with the choice $\varphi(z) = \log(z)$, the preceding expression simplifies, since $\beta[\mathbf{W}(t)] = 1/\psi(w^{2q}(t))$. Likely, the simplest form of α is obtained when $\psi(z) = z$, indeed in that case $\alpha(\mathbf{W}) = \det(\mathbf{W}^T \mathbf{C}_x \mathbf{W})$.

2.2 A 'learning control law' $\bar{\alpha} = \bar{\alpha}(t)$

From equation (6) (see also [11]), it is known that at the equilibrium the property $\alpha(\mathbf{W}) = 1$ must hold. We now propose to replace in the algorithm (4) the function $\alpha(\mathbf{W})$ with an arbitrariy function of the time only, $\bar{\alpha}(t)$, chosen so that at least condition:

$$\lim_{t \to t_\star} \bar{\alpha}(t) \; = \; 1 , \tag{7}$$

holds true, and such that the second term in (4) is weighted in a proper way. The function $\bar{\alpha}(t)$ is called here *learning control law.* Time $t_\star$ may not be bounded.

Actually, by replacing the true expression of the parameter α with an arbitrarily chosen time-function we do not change the structure of the algorithm, but simply force the temporal evolution of the variables to a non-self-controlled course, and control the speed the state $\mathbf{W}(t)$ travels with along the Stiefel manifold $\mathcal{N}^{p\times q}$. As a consequence of this assumption the system (4) becomes:

$$\beta\frac{d\mathbf{W}}{dt} = \mathbf{C}_x\mathbf{W}(\mathbf{W}^T\mathbf{C}_x\mathbf{W})^{-1} - \bar{\alpha}(t)\mathbf{W}(\mathbf{W}^T\mathbf{W})^{-1} \ .$$

The main consequence of the earlier assumption is that, as $\bar{\alpha}(t)$ is a known function of the parameter t only, the scalar differential equation in (2) can be easily tackled (at least in the intentions) by solving the following integral equation:

$$\int_{w_0}^{w(t)} \frac{u\,du}{\psi(u^{2q})} = \int_0^t [1 - \bar{\alpha}(\theta)]d\theta \ , \tag{8}$$

under the usual condition of the orthonormal initial state for the algorithm, providing that $\varphi(x) = \log(x)$.

3 A 2nd-order differential learning equation

As a second example of algorithms belonging to $\mathcal{C}$, consider the following second order non-linear differential learning rule for the neural layer described by the input-output relation $\mathbf{y} = \mathbf{G}(\mathbf{W}^T\mathbf{x} + \mathbf{w}_0)$:

$$\mathbf{W}^T\ddot{\mathbf{W}} - \ddot{\mathbf{W}}^T\mathbf{W} = 2\sigma\rho\mathbf{W}^T(\mathbf{H} - \mathbf{H}^T)\mathbf{W} \ , \tag{9}$$

where $\ddot{\mathbf{W}} = d^2\mathbf{W}/dt^2$, σ and ρ are positive constants and $\mathbf{H}$ is an arbitrarily variable $p \times p$ matrix, which controls the learning. $\mathbf{G}(\cdot)$ is a non-linear diagonal activation operator and $\mathbf{w}_0$ is a biasing vector arbitrarily adapted.

Let us prove that algorithm (9) is SOC if $\mathbf{W}(0) \in \mathcal{N}_{w_0}^{p\times q}$. To start with, it is useful to define the following coupled systems:

$$\dot{\mathbf{W}} = \sigma\mathbf{P}\mathbf{W} \text{ with } \mathbf{W}^T(0)\mathbf{W}(0) = w_0^2\mathbf{I} \ , \tag{10}$$

$$\dot{\mathbf{P}} = \rho(\mathbf{H} - \mathbf{H}^T) \text{ with } \mathbf{P}^T(0) = -\mathbf{P}(0) \ . \tag{11}$$

Due to the structure of the sub-system (11), it is straightforward to see that for all $t \geq 0$ the property $\mathbf{P}^T = -\mathbf{P}$ (skew-symmetry) holds true. Besides, the following result holds:

Theorem 2. *Consider the dynamical system $d\mathbf{W}/dt = \sigma\mathbf{P}\mathbf{W}$ where $\mathbf{P}$ is skew-symmetric and σ is a real-valued scalar, if $\mathbf{W}(0) \in \mathcal{N}_{w_0}^{p\times q}$ for some w_0 real constant, then the flow $\mathbf{W}(t)$ is Stiefel.*

Proof. The matrix $\mathbf{P}$ is always skew-symmetric, that means $\mathbf{P}^T = -\mathbf{P}$. With some algebra it can be found that $d(\mathbf{W}^T\mathbf{W})/dt = \sigma\mathbf{W}^T(\mathbf{P}^T - \mathbf{P})\mathbf{W} = \mathbf{0}$, therefore $\mathbf{W}^T(t)\mathbf{W}(t) = \mathbf{W}^T(0)\mathbf{W}(0) = w_0^2\mathbf{I}$. □

From the above Theorem it follows that the algorithm (10) together with (11) is a SOC one. It remains to show that systems (10)+(11) and (9) are equivalent.

Differentiating equation (10) with respect to the time yields $\ddot{\mathbf{W}} = \sigma\dot{\mathbf{P}}\mathbf{W} + \sigma\mathbf{P}\dot{\mathbf{W}}$. Replacing equations (10) and (11) in the preceding formula gives:

$$\mathbf{W}^T\ddot{\mathbf{W}} = \sigma\rho\mathbf{W}^T(\mathbf{H} - \mathbf{H}^T)\mathbf{W} + \sigma^2\mathbf{W}^T\mathbf{P}^2\dot{\mathbf{W}} \ .$$

Finally, by subtracting $(\mathbf{W}^T\ddot{\mathbf{W}})^T$ from $\mathbf{W}^T\ddot{\mathbf{W}}$ (observe that $\mathbf{P}^2$ is symmetric), the rule (9) arises.

3.1 Mechanics-type control

In [3] it was shown how to interpret the (10) and (11) as equations describing the dynamics of a rigid system of masses moving in an abstract space under a potential energy field and a viscous braking effect, providing that:

$$\mathbf{H} = -\left(\kappa\frac{\partial U}{\partial \mathbf{W}} + \mu\mathbf{P}\mathbf{W}\right)\mathbf{W}^T \ , \tag{12}$$

where U is a function bounded below to be minimized under the restriction $\mathbf{W} \in \mathcal{N}^{p\times q}_{w_0}$ and $\mu \geq 0$, $\kappa > 0$.

It would be interesting to study the existence of special equilibrium points for the second-order system (10)+(11) when the control matrix (12) is used, in those cases when $U = U(\mathbf{W})$ only.

Theorem 3. *Let $\mathcal{S}$ be the dynamic system (10)+(11) where the control matrix is assumed as in (12), the initial state is chosen so that $\mathbf{W}(0) \in \mathcal{N}^{p\times q}_{w_0}$ and $\mathbf{P}(0)$ is skew-symmetric. Define the matrix function $\mathbf{F}(\mathbf{W}) \stackrel{\text{def}}{=} -\kappa\frac{\partial U}{\partial \mathbf{W}}$. A matrix $\mathbf{W}_\star$ is a stationary point of $\mathcal{S}$ if $\mathbf{F}^T(\mathbf{W}_\star)\mathbf{W}_\star$ is diagonal and $\mathbf{P}(\mathbf{W}_\star) = \mathbf{0}$.*

Proof. Using the Lagrangian function:

$$J_1(\mathbf{W}) \stackrel{\text{def}}{=} J(\mathbf{W}) + \tfrac{1}{2}\mathrm{tr}[(\mathbf{W}^T\mathbf{W} - w^2\mathbf{I})\mathbf{E}] \ ,$$

where $J = \kappa U$ and $\mathbf{E}$ is a diagonal matrix containing Lagrange multipliers, recalling that at each instant $\mathbf{W}^T\mathbf{W} = w_0^2\mathbf{I}$, it can be proved that the standard Kuhn-Tucker condition for the existence of an (orthonormal) extreme of U reads $-\mathbf{F} + \mathbf{W}\mathbf{E} = \mathbf{0}$, where $w_0^2\mathbf{E} = \mathbf{W}^T\mathbf{F}$, therefore $w_0^2\mathbf{F} = \mathbf{W}(\mathbf{W}^T\mathbf{F})$. This equation implies:

$$w_0^2(\mathbf{F}\mathbf{W}^T - \mathbf{W}\mathbf{F}^T) = \mathbf{W}(\mathbf{W}^T\mathbf{F} - \mathbf{F}^T\mathbf{W})\mathbf{W}^T \ .$$

By hypothesis $\mathbf{P}(\mathbf{W}_\star) = \mathbf{0}$ and $\mathbf{D}_\star \stackrel{\text{def}}{=} \mathbf{F}^T(\mathbf{W}_\star)\mathbf{W}_\star = \mathbf{D}_\star^T$, thus from equation (12) it follows that matrix $\mathbf{H}_\star = \mathbf{F}(\mathbf{W}_\star)\mathbf{W}_\star^T$ is symmetric. □

Another fundamental feature of the system (10)+(11)+(12) is its asymptotic stability. Note that system (9) is represented by the pair of coupled systems (10) and (11), whose state-matrices $\mathbf{P}$ and $\mathbf{W}$ form an unique extended state $\mathbf{X} = (\mathbf{P}, \mathbf{W})$.

Theorem 4. *Be U a real-valued, bounded below function of $\mathbf{W}$, with a minimum in $\mathbf{W}_\star$. Then the equilibrium state $\bar{\mathbf{X}} = (\mathbf{0}, \mathbf{W}_\star)$ of the system $\mathcal{S}$ obtained by gathering (10)+(11)+(12) is asymptotically stable.*

Proof. Due to space limitations, the proof, based on the Lyapunov stability theory, is here omitted. The proof of a slightly less general result may be found in [5]. □

Note that function $U(\mathbf{W})$ may have more than one minimum (local minima) corresponding to local maxima of $-U(\mathbf{W})$. Besides, note that *the local extreme for which $\mathbf{P} = \mathbf{0}$ depends not only upon $\mathbf{W}(0)$, but also on $\mathbf{P}(0)$.*

4 Relationships with other theories

In this part, relationships among the SOC theory and other learning theories known from the literature are briefly discussed.

4.1 The Moreau-Pesquet's theory

The main theoretical instrument for developing a Principal Component Analysis theory in connection with the SOC one, is the definition of *semicontrast* functions specialized for orthonormal mixtures. Said $\mathcal{S}_d$ the set of *uncorrelated* random source vectors with n entries, and $\mathcal{M}_d$ the set of random vectors $\mathbf{x} = \mathbf{R}\mathbf{s}$ where $\mathbf{s} \in \mathcal{S}_d$ and $\mathbf{R} \in \mathcal{N} \stackrel{\text{def}}{=} \mathcal{N}_1^{n \times n}$. Denote with $\mathcal{P}$ the subset of $\mathcal{N}$ containing matrices $\mathbf{R}$ satisfying the *independence property*: $\mathbf{R} = \mathbf{D}\mathbf{Q}$ where $\mathbf{D} = \mathrm{diag}(\pm 1, \ldots, \pm 1)$ and $\mathbf{Q}$ is a permutation matrix. Then, the concept of *semicontrast* is defined [8] as follows:

Definition 5. *A semicontrast on $\mathcal{M}_d$ is a multivariate mapping φ from the set $\mathcal{M}_d$ to $\mathcal{R}$ which satisfies the following three requirements:*

M1. $\forall \mathbf{s} \in \mathcal{M}_d,\ \forall \mathbf{R} \in \mathcal{P},\ \varphi(\mathbf{R}\mathbf{s}) = \varphi(\mathbf{s})$;

M2. $\forall \mathbf{s} \in \mathcal{S}_d,\ \forall \mathbf{R} \in \mathcal{N},\ \varphi(\mathbf{R}\mathbf{s}) \leq \varphi(\mathbf{s})$;

M3. $\forall \mathbf{s} \in \mathcal{S}_d,\ \forall \mathbf{R} \in \mathcal{N},\ \varphi(\mathbf{R}\mathbf{s}) = \varphi(\mathbf{s}) \Leftrightarrow \mathbf{R}\mathbf{s} \in \mathcal{S}_d$.

Suppose that a neural network described by $\mathbf{y} = \mathbf{W}^T\mathbf{x}$ is used for decorrelating random signals drawn by orthogonal mixtures of uncorrelated source waves. An example of semicontrast function evaluated on $\mathbf{y}$ is given in [8] $\varphi_{\mathbf{y}}^f \stackrel{\text{def}}{=} \sum_{i=1}^n f(\sigma_{y_i}^2)$, where f is a strictly convex function from $\mathcal{R}_+$ to $\mathcal{R}$, and the $\sigma_{y_i}^2$ are the variances of the y_i. This function was already used by Watanabe with $f(z) = z\log(z)$ [8].

It is worth to note that a function φ^f depends uniquely upon the matrix variable $\mathbf{W}$, and requirements M1 – M2 – M3 ensures that it has a unique *maximum* (but possible sign and permutation). This example allows to show that in order to perform decorrelation a function $U \propto -\varphi$ should be used in (12), where φ is a semicontrast. Since U depends uniquely on the variable $\mathbf{W}$, the obtained system is autonomous, hence Theorems 3 and 4 may be used.

4.2 The Deco-Brauer's theory

In [2] Deco and Brauer introduced the concept of *volume-conserving* neural networks. A volume-conserving 'square' architecture performs a mapping $\mathbf{y} = \mathbf{S}(\mathbf{x})$ between a set of input patterns and a set of output ones characterized by the property $\det\left(\frac{\partial \mathbf{S}}{\partial \mathbf{x}}\right) = 1$. For a linear network with weight matrix $\mathbf{W}$ the condition above becomes $\det(\mathbf{W}) = 1$. The same condition ensures the bijectivity of the mapping. Volume conservation allows for example the preservation of the entropy $H(\mathbf{x})$ when the input signal is mapped into the output signal whose entropy is $H(\mathbf{y})$, that means $H(\mathbf{x}) = H(\mathbf{y})$, in fact:

$$H(\mathbf{y}) = H(\mathbf{x}) + E[\log|\det(\partial \mathbf{S}/\partial \mathbf{x})|] \,.$$

Symbol $E[\cdot]$ represents the *mathematical Expectation* operator. A SOC algorithm with matrix $\mathbf{W} \in \mathcal{N}_1^{n \times n}$ for a linear network is always volume-conserving, indeed $\mathbf{W}^T\mathbf{W} = \mathbf{I}$ implies $\det(\mathbf{W}) = \pm 1$.

4.3 The Laheld-Cardoso's 'PFS' approach

System (10)+(11) allows *equivariant* square blind source separation by orthonormal mixtures. Strictly speaking, 'equivariant' [7] means that the performance of the separation is independent of the mixing matrix, especially of its conditional number (eigenvalues' spread). This means that a *parameter free separator* (PFS) can manage well-conditioned problems as well as ill-conditioned ones. Formally, let $\mathbf{A}$ be the orthonormal square mixing matrix of the mixing model $\mathbf{x} = \mathbf{A}\mathbf{s}$, where s contains the source signals, and $\mathbf{T} \stackrel{\text{def}}{=} \mathbf{W}\mathbf{A}$ the source-to-output transference matrix, where $\mathbf{W}$ is the weight matrix of the neural net $\mathbf{y} = \mathbf{W}\mathbf{x}$. It is easy to check that:

Property 6. *If the control matrix* $\mathbf{H}$ *depends uniquely on the output* $\mathbf{y} = \mathbf{W}\mathbf{x}$ *then the system (10)+(11) is PFS.*

Proof. In this case equations rewrite:

$$\dot{\mathbf{T}} = \sigma \mathbf{P}\mathbf{T} \;,\; \dot{\mathbf{P}} = \rho[\mathbf{H}(\mathbf{T}\mathbf{s}) - \mathbf{H}^T(\mathbf{T}\mathbf{s})] \;, \tag{13}$$

thus the system's behavior depends on $\mathbf{T}$ and $\mathbf{s}$ only. □

Changing $\mathbf{A}$ is equivalent to change the starting point $\mathbf{T}(0)$ (and $\mathbf{P}(0)$) of the algorithm. For knowing details about the PFS approach, readers please refer to [7] and references therein.

5 Re-derivation of the Oja's Subspace Rule

The Oja's Principal Subspace Rule [9] has been studied extensively for many years. Probably, the deepest and most complete and lighting results about it have been obtained by Yan, Helmke and Moore in [12]. Here we aim to re-derive the same subspace rule by the Xu's algorithm discussed above.

Start by post-multiplying gradient $\partial\varrho/\partial\mathbf{W}$ in (4) by $\mathbf{W}^T\mathbf{C}_x\mathbf{W}$. The obtained system is described by the following differential equation:

$$\frac{d\mathbf{W}}{dt} = \frac{1}{\beta}[\mathbf{I} - \alpha\mathbf{W}(\mathbf{W}^T\mathbf{W})^{-1}\mathbf{W}^T]\mathbf{C}_x\mathbf{W} . \tag{14}$$

The new algorithm (14) no longer allows the orthonormal evolution of the state-matrix $\mathbf{W}$. This consequence can be easily shown by observing that:

$$\mathbf{W}^T\frac{d\mathbf{W}}{dt} = \frac{1}{\beta(\mathbf{W})}[1 - \alpha(\mathbf{W})](\mathbf{W}^T\mathbf{C}_x\mathbf{W}) ,$$

whereby it is easily seen that:

$$\frac{d(\mathbf{W}^T\mathbf{W})}{dt} = 2\frac{1-\alpha}{\beta}(\mathbf{W}^T\mathbf{C}_x\mathbf{W}) .$$

Even if for $t = 0$ the $\mathbf{W}^T\mathbf{W} = w_0^2\mathbf{I}$ holds true, for $t > 0$ the product matrix $\mathbf{W}^T\mathbf{C}_x\mathbf{W}$ does not result a scaled identity nor a diagonal. Thus rule (14) is not interesting to our concerns.

We now propose a useful modification of the system (14) that leads to a surprising result. Firstly, rewrite equation (14) in a different way, that is:

$$\frac{d\mathbf{W}}{dt} = \frac{\alpha}{\beta}[\alpha^{-1}\mathbf{C}_x - \mathbf{W}(\mathbf{W}^T\mathbf{W})^{-1}\mathbf{W}^T\mathbf{C}_x]\mathbf{W} \stackrel{\text{def}}{=} \hat{\gamma}\mathbf{Z}\mathbf{W} , \tag{15}$$

where $\mathbf{Z}(\mathbf{W}) \stackrel{\text{def}}{=} \alpha^{-1}(\mathbf{W})\mathbf{C}_x - \mathbf{W}(\mathbf{W}^T\mathbf{W})^{-1}\mathbf{W}^T\mathbf{C}_x$ and $\hat{\gamma}(\mathbf{W}) \stackrel{\text{def}}{=} \alpha(\mathbf{W})/\beta(\mathbf{W})$. The *new* system defined by:

$$\frac{d\mathbf{W}}{dt} = \sigma(\mathbf{Z} - \mathbf{Z}^T)\mathbf{W} = \sigma\mathbf{P}\mathbf{W} , \tag{16}$$

is Stiefel, moreover, recalling that $\mathbf{C}_x$ is symmetric, there holds:

$$\begin{aligned}\mathbf{P} &= \alpha^{-1}\mathbf{C}_x - \mathbf{W}(\mathbf{W}^T\mathbf{W})^{-1}\mathbf{W}^T\mathbf{C}_x - \alpha^{-1}\mathbf{C}_x + \mathbf{C}_x\mathbf{W}(\mathbf{W}^T\mathbf{W})^{-1}\mathbf{W}^T \\ &= \mathbf{C}_x\mathbf{W}(\mathbf{W}^T\mathbf{W})^{-1}\mathbf{W}^T - \mathbf{W}(\mathbf{W}^T\mathbf{W})^{-1}\mathbf{W}^T\mathbf{C}_x .\end{aligned}$$

If for the algorithm an initial condition $\mathbf{W}(0)$ such that $\mathbf{W}^T(0)\mathbf{W}(0) = w_0^2\mathbf{I}$ is assumed for some real constant w_0, it always remains $\mathbf{W}^T\mathbf{W} = w_0^2\mathbf{I}$. Replacing this result in the equation (16), its properties do not change, since $\mathbf{P}$ still remains skew-symmetric, therefore the above equation reads $w_0^2\mathbf{P} = \mathbf{C}_x\mathbf{W}\mathbf{W}^T - \mathbf{W}\mathbf{W}^T\mathbf{C}_x$. Post-multiplying both members of the latter equation by $\mathbf{W}$ yields:

$$w_0^2\mathbf{P}\mathbf{W} = \mathbf{C}_x\mathbf{W}\mathbf{W}^T\mathbf{W} - \mathbf{W}\mathbf{W}^T\mathbf{C}_x\mathbf{W} = (w_0^2\mathbf{I} - \mathbf{W}\mathbf{W}^T)\mathbf{C}_x\mathbf{W} , \tag{17}$$

therefore the *equivalent* form of the system (16) is:

$$\frac{d\mathbf{W}}{dt} = \frac{\sigma}{w_0^2}(w_0^2\mathbf{I} - \mathbf{W}\mathbf{W}^T)\mathbf{C}_x\mathbf{W} . \tag{18}$$

This closely recalls the continuous-time version of the well-known Oja's Principal Subspace estimation rule [12]. In turn, learning rule (18), attempts to maximize a simple quadratic cost function under the constraint of orthonormality. In fact, it is well-known that the Oja's rule (18) arises from the GSA maximization of the function:

$$T(\mathbf{W}) = \mathrm{tr}[\mathbf{W}^T\mathbf{C}_x\mathbf{W}] + \tfrac{1}{2}\mathrm{tr}[(\mathbf{W}^T\mathbf{W} - w_0^2\mathbf{I})\mathbf{E}] \ ,$$

where $\mathbf{E}$ is a diagonal matrix containing the Lagrange multipliers for the constraints. The initial state has to belong to $\mathcal{N}^{p\times q}$.

References

[1] P. Comon, *Independent Component Analysis, A New Concept ?*, Signal Processing, Vol. 36, pp. 287 - 314, 1994

[2] G. Deco and W. Brauer, *Nonlinear Higher-Order Statistical Decorrelation by Volume-Conserving Neural Architectures*, Neural Networks, Vol. 8, No. 4, pp. 525 - 535, 1995

[3] S. Fiori, *Unsupervised Neural Artificial Learning Models for Blind Signal Processing*, M.Sc. Dissertation, Dept. of Electronics and Automatics, Univ. of Ancona (Italy), July 1996 (in italian)

[4] S. Fiori, A. Uncini and F. Piazza, *Mechanical Learning Applied to Principal Component Analysis and Source Separation*, Artificial Neural Networks, pp. 571 - 577, 1997, Springer-Verlag

[5] S. Fiori and F. Piazza, *Orthonormal Strongly-Constrained Neural Learning*, Proc. of IJCNN, pp. 2332 - 2337, 1998

[6] C. Fyfe, *A General Exploratory Projection Pursuit Network*, Neural Processing Letters, Vol. 2, No. 3, 1995

[7] J.F. Cardoso and B. Laheld, *Equivariant Adaptive Source Separation*, IEEE Trans. on Signal Processing, Vol. 44, No. 12, pp. 3017 - 3030, Dec. 1996

[8] E. Moreau and J.C. Pesquet, *Independence/Decorrelation Measures with Application to Optimized Orthonormal Representations*, Proc. of ICASSP, pp. 3425 - 3428, 1997

[9] E. Oja, *Neural Networks, Principal Components, and Subspaces*, International Journal of Neural System, Vol. 1, pp. 61 - 68, 1989

[10] F. Palmieri, J. Zhu and C. Chang, *Anti-Hebbian Learning in Topologically Constrained Linear Networks: A Tutorial*, IEEE Trans. Neural Networks, Vol. 4, No. 5, pp. 748 - 761, 1993

[11] L. Xu, *Theories for Unsupervised Learning: PCA and Its Nonlinear Extension*, Proc. of IJCNN, pp. 1252 - 1257, 1994

[12] W.-Y. Yan, U. Helmke, and J.B. Moore, *Global Analysis of Oja's Flow for Neural Networks*, IEEE Trans. Neural Networks, Vol. 5, No. 5, pp. 674 - 683, 1994

From Behavior to VHDL:
A CAD Environment for SPNNs

W.Fornaciari, V.Piuri, F.Salice
Politecnico di Milano, Dipartimento di Elettronica, P.zza L. Da Vinci 32,
20133 Milano (MI), Italy.

Abstract

This paper presents an automatic design flow for digital special purpose feed-forward multi-layer neural networks (SPNNs). The target architecture is constituted by a collection of basic elements called *Pseudo-Neurons* (PNs) connected in a *pipelined-tree* manner, to suit efficient VLSI implementation and, due to an internal pipelined multiprocessing, achieving low latency and good throughput. The related CAD environment implements three successive stages: the weights discretization (*WD*), the architectural synthesis (*AS*), and the VHDL model generation (*VHDL_G*).

1. Motivations and CAD environment description

This paper is focused on *special purpose neurocomputers*. The aim is to present a design methodology allowing to cope with the constraints of an efficient VLSI implementations of *feed forward* ANNs without on-line learning, while assuring *a priori* no degradation in the expected behavior.

The first stage within the proposed design flow depicted in fig.1, transforms the weights set derived by a theoretical analysis into an equivalent discretized one. Successively, the *AS* module maps weights and network topology onto an optimized digital structure (*PNs tree structure*) characterized by both a pipelined evaluation of the neuron computation and an efficient schema for bus sharing and local synchronization. Finally, the obtained macro-blocks system implementation is translated into a VHDL description suitable for synthesis by *VHDL_G*.

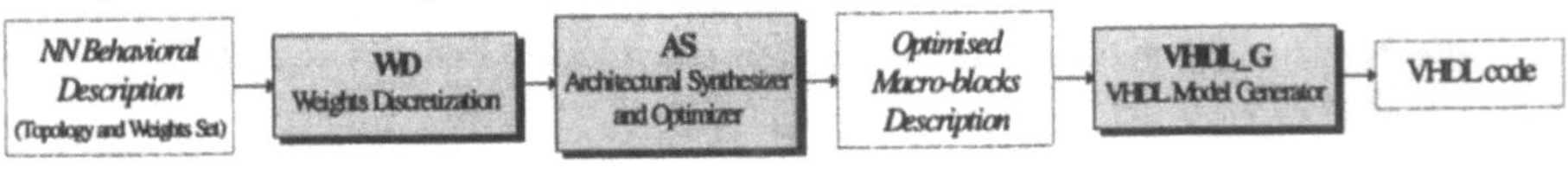

Figure 1. The proposed design flow for SPNNs.

The paper is organized as follows: section 2 presents the design methodology allowing to cope with the finite-precision constraints while assuring *a priori* no degradation in the predict behavior. Section 3 presents the design step to deal with performance and silicon constraints. Section 4 (*VHDL_G*), discusses the translation towards the VHDL code. Concluding remarks are drawn in section 5.

2. Weights Discretizator - *WD*

The proposed analysis allows to identify a class of Neural Network (NN) implementations such that, once the learning error ε is fixed, it is guaranteed that the perturbation introduced by a discrete finite-precision arithmetic, will not degrade the performance over the acceptable bound ε.

Let us consider a multi-layers neural network classifier, characterized by a learning error ε and by an approximation of the sigmoidal activation function of the output neurons (at layer *n*) realized via a one-step hard-limiter (fig.2). Given a certain learning error ε, the corresponding weighted sum $Net[1\text{-}\varepsilon]$ (fig.2) represents the maximum perturbation at which the network is *insensitive*. In fact, even if an error *Eq* induced by the quantization of weights and non-linear functions is added to the nominal output, the network still produces a correct value. In [1] is demonstrated that the following constraint must be fulfilled to guarantee the same original performance:

$$|Eq| < \ln\left(\frac{1-e}{e}\right) = |Net[1-\varepsilon]| = |Net[\varepsilon]| = Eq_{max} \qquad (1)$$

$$Net_{n,i} = \sum_{j=1}^{k} w_{n,i,j}\, in_{n,j} = W_{n,i}\, I_{n,i}^{T}$$

$$Output_{n,i} = nlf\,(Net_{n,i})$$

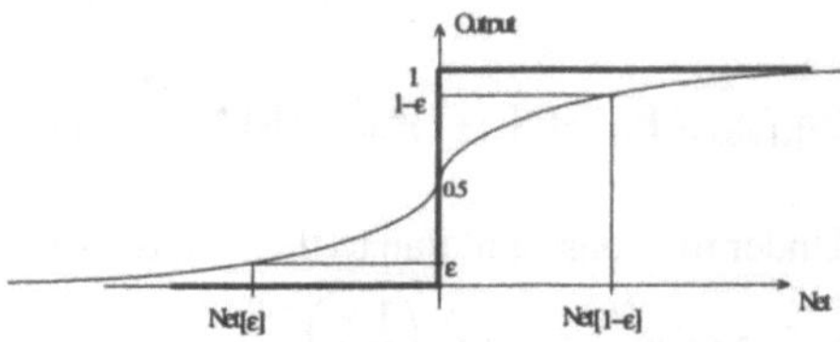

Figure 2. The neuron model (left) and the sigmoidal activation function of the neurons composing the output layer (right). *Output*$_{n,i}$ pertains the *i*-th neuron at output layer *n* which is obtained by computing the weighted sum *Net*$_{n,i}$ of its inputs *in*$_{n,j}$ and by filtering the result via the activation function *nlf*; $w_{n,i,j}$ is the synaptic weight associated with the connection between neuron *i* at layer *n* and *j* at layer *n-1*.

The proposed methodology to properly discretize the weights set while ensuring the insensitivity is based on the following definitions:

- $I_{l,iN}$ and $W_{l,iN}$ are the *non-discretized* input vector and weights vector of neuron *i* at layer *l*, respectively. $I_{l,iD}$ and $W_{l,iD}$ are the discretized input and weights vectors, respectively; $\Delta W_{l,iD}$ the weights discretization vectors of neuron *i* at layer *l* such that $W_{l,iD}=W_{l,iN}+\Delta W_{l,iD}$;
- δ is the module of the maximum error due to the discretization of the activation function (e.g., in case of non linear functions realized as *sum of steps* or *look up table*, δ is equal to half of the maximum step high [2]);
- $Eq_{l,i}$ is the quantization error representing the perturbation of the theoretical value (neuron *i-th* at layer *l-th*) due to the discretization introduced by the digital implementation.

Since all hardware components (multiplier, memory, adders, ...) typically composing a neuron have to be dimensioned in size according to the worst case, there is no significant convenience in having different weight lengths for the same neuron. As a consequence, each of the quantization terms $|\Delta w_{l,i,jD}|$ can be replaced by a common constraint on the bit widths:

$$|\Delta w_{l,i,jD}| \le |\Delta w_{l,i,jD} \text{ on } m_{l,i} \text{ bits}| \qquad (2)$$

Furthermore, since neurons belonging to the same layer process the same number of weights, it is convenient (in terms of design complexity and modularity) to have both the same weight lengths and the same input discretization for the whole layer, so that:

$$|\Delta w_{l,i,jD} \text{ on } m_{l,i} \text{ bits}| \le |\Delta w_{l,i,jD} \text{ on } m_l \text{ bits}| \text{ and } |\Delta in_{l\text{-}max}| \le |\Delta in_{l\text{-}max} \text{ on } h_l \text{ bits}| \qquad (3)$$

To obtain better approximations, the method used to compute finite precision representation of the weights is based on the *rounding operator* [5]. By calling 2^{μ_l} the bit weight associated with the m_l*-th* bit of $w_{l,i,j}$ and 2^{η_l} the bit weight associated with the h_l*-th* bit of inputs at layer l, the following relations are obtained for the rounding errors on weights (ε_{wr}) and inputs (ε_{inr}):

$|\Delta w_{l,i,jD}| \leq |\Delta w_{l,i}$ on m_l bits$|_{max} = \varepsilon_{wr} = 2^{\mu_l - 1}$ $\quad |\Delta in_{l\text{-}max}| \leq |\Delta in_{l\text{-}max}$ on h_l bits$|_{max} = \varepsilon_{inr} = 2^{\eta_l - 1}$

By using the expression of the maximum error, in [1] is demonstrated the validity of the following iterative equations for the different layers, the first hidden and the following others, respectively:

$$Eq_{1,i\text{-}max} \leq E_{1,i} = (N_0+1)*2^{\mu_1-1} + 2^{\eta_0-1}*\left(\sum_{j=1}^{N_0} |w_{1,i,jN}| + N_0*2^{\mu_1-1}\right) \tag{4b}$$

$$Eq_{l,i\text{-}max} \leq E_{l,i} = (N_{l-1}+1)*2^{\mu_l-1} + MAX_{j=1}^{N_{l-1}} \left(th\left(\left|\frac{E_{l-1,i}}{4}\right|\right) + \delta_{l-1,j}\right)*\left(\sum_{j=1}^{N_{l-1}} |w_{l,i,jN}| + N_{l-1}*2^{\mu_l-1}\right) \tag{4a}$$

Under the constraint due to the insensitivity requirement:

$$E_n = MAX_{j=1}^{N_n} E_{n,j} \leq \ln\left(\frac{1-\varepsilon}{\varepsilon}\right) \tag{5}$$

In the following m_l represents bits number (for each layer l) to be adopted for weights finite precision representation (the LSB weight of m_l is μ_l) and h_0 indicates the bits number on which the inputs are encoded (the LSB weight of h_0 is η_0).

Eq.4a,b and eq.5 define a range of possible architectural parameters identifying a class of implementation insensitive neural networks (class CΔ). The actual use of such a result, once the learning error ε is known, is in the direction of computing *Eqn* via eq.5 and carrying out a iterative backward application of eq. 4a,b from the output to the input layer. A global comprehensive equation will be obtained, whose vector X of unknowns comprises: the number of bits, h_0 (equivalently, h_0), on which the net input is encoded (if it is not a design constraint), the number of bits m_l (equivalently, m_l), for each layer l, to be adopted for weights representation, and the activation function approximation error $\delta_{l,i}$.

As a consequence of fixing m_l and $\delta_{l,i}$, the parameters characterizing the implementation such as the communication buses and the dimensional characteristics of any processing element (i.e., registers and arithmetic units), are established. The solutions can be computed by exhaustively exploring the space spanning on $X=\{\mu_n,\ldots, \mu_2, \mu_1, \delta_{n-1}..\delta_2, \delta_1, \eta_0\}$. The optimal solution among the selected candidates composing the class CΔ will be chosen according to some criteria, such as, a cost function gathering silicon cost and performance, as shown in section 3.

3. Architectural Synthetizer: *AS*

The adopted architecture suitable for multi layer feed forward NN, is based on the decomposition of the neuron functionality onto a number of processing elements, called Pseudo Neurons (PNs) [3]. Each PN receives its input values from both the outside world (or previous neural network *layer*) and the preceding PN (this last one, named Linear Input -LI-, is a part of the neuron global computation) and forwards a partial sum computed on its subset of inputs added to its LI (fig.3a,b). The memory necessary to store the synaptic weight value is partitioned in some memories local to each PN. With this approach, neurons with any number of inputs can be obtained by connecting the proper number of PNs.

The digital architecture of the feed-forward multi-layer NN considered is specified in terms of internal structure of each neuron, contents of the memories where the synaptic weights will be stored, paths of the data/control signals and, modality for the input/output of signal from/to the external environment.

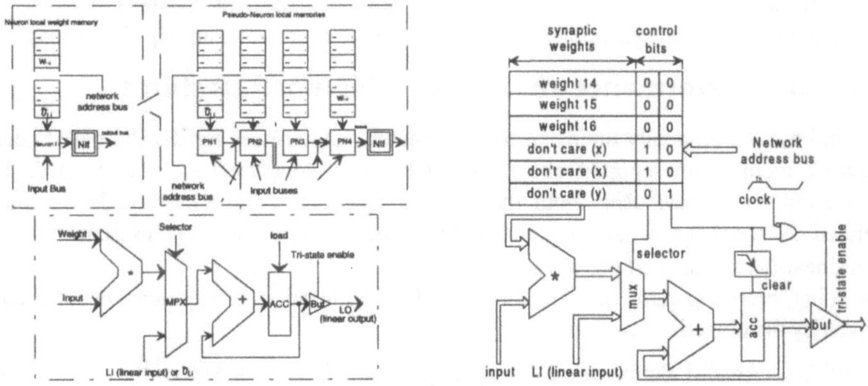

Figure 3. Blocks-schema of one neuron. Decomposition of one neuron and internal structure of a PN.

The whole structure is based on a mapping of the neural network topology onto the corresponding set of digital processing elements. In [3] has been shown that if all PNs compute the same number of synaptic products, the synchronization is simplified and no additional *wait cycles* for inter-layer data communication must be introduced. For this reasons each PN is associated with a local memory of the same size; furthermore, a proper encoding of memory words is performed to support both a *local* generation of the synchronization signals and an efficient *mux/demux* of the data (the control unit is distributed). The placement of the synaptic weights inside the PNs local memories and the positioning of some delay cells used for demultiplexing purposes, can be automatically obtained [3].

To obtain a data flow exploiting the computational characteristics of a PN, the internal architecture of each neuron is organized as a multi-level structure (a *tree*) (fig.4).

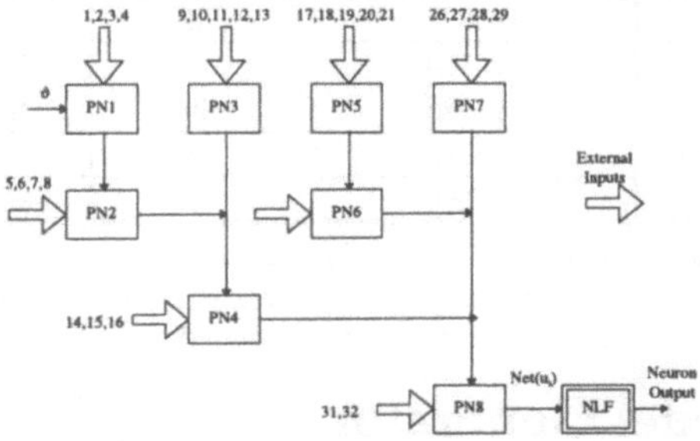

Figure 4 . Internal architecture of a neuron: tree structure of a 32-inputs neuron.

The *latency* of the propose structure is $PNw + \lceil \log_2 (K_l) \rceil$ where PNw is the number of words in each PN memory and K_l is the number of PNs per neuron at layer l.

In [4] is argued a structured methodology to obtain a formal representation for the complete architecture given the network topology and the architectural parameter PNw. The final representation is summarized in the following information:

- Allocation of *synaptic weights* into the PN memories;
- Definition of *delay cells* required to obtain synchronization both inside the layer and between adjacent layers.

From these information it is possible to build the complete digital architecture: positioning the delay cells inside the network and allocating the synaptic (and synchronisation) weights in the PN memories allow to map interconnections onto the buses and to guarantee synchronization both inside the neuron and among all neurons.

3.1 Cost-performance analysis of ANNs architectures

To find out the optimally tuning of the architectural parameters for the PN-based neural network architecture reported above, a suitable goal function has been defined. The evaluation is obtained using both the network gate count and the throughput-latency, defined as functions of the number of bits used for both weights discretization and non-linearity implementation, whose value is defined by *WD* [1].
By using such a parametric cost function, the optimization process can be efficiently driven by the objective function:

$$\text{ObFu} = \text{Area}^{\alpha} * \text{Performance}^{1-\alpha} = f(m, h_0, d_1, \ldots d_n, \text{PNw})^{\alpha} * g(\text{PNw})^{1-\alpha} \quad (6)$$

which relates the architectural parameters (m, h_0, d_1, ...d_n, PNw) to both area and performance. The parameter α can be chosen to represent the designer's goals e.g., latency vs silicon cost.

In tab.1 a tradeoff analysis performed by considering different values of α is shown. Only the results concerning the best ObFu for a given α have been reported, together with the corresponding precision required for implementing the activation function and representing the weights. As far as the requirements of the application are established, the designer's task will consist simply of selecting the optimal implementation among the ones corresponding to the rows of tab.1. It has to be pointed out that, in addition to the optimization of a given implementation, the proposed approach can be used to compare different design solutions, i.e. to discover their different impacts on the real behavior.

α	ObFu	Area (NgNN)	Latency	PNw	m	h
0	16	161960	16	3	6	6
0.2	86	43837	18	6	6	6
0.4	381	24104	24	10	7	5
0.6	1386	14985	39	18	8	4
0.8	3608	8234	133	66	8	4
1	8234	8234	133	66	8	4

Table 1. Example of optimization for a 3 layers neural network whose sub-space of solutions is Q = {{mi, hi},...}={{6,6}, {6,7}, {7,5}, {8,4}, {6,8}}. Area is expressed in terms of gates. h is the number of bits used to represent the activation function output.

4. VHDL Model Generator: *VHDL_G*

The implementation of digital neural networks conform to the proposed design methodology requires to use both the parametric pseudo-neural macro-blocks and the corresponding interconnection structure.
As previously reported, each PN is composed by a multiplier, a multiplexer, an adder, one accumulator, a control unit and, if the PN is the last of neuron, a step-wise non linear function. To efficiently implement the connections between PNs, the output of each element shares the interconnection BUS with other PNs by using a tri-state buffer.
The multiplier generates the product between an input and its corresponding weight. Since the Baugh-Wooley architecture is characterized by an regular structure and its

performance is similar to other cellular multiplier presented in [5], it is has been chosen as target architecture.

Given two inputs, with *m* and *n* bits respectively, the multiplier is constituted by m*(n-1)+3 full-adder and the computational time is (m+n)*Δ, where Δ represents the computational time of a single full-adder. By representing with I_m and F_m integer and decimal parts of the input, with I_n and F_n the integer and decimal components of the weight and with N_{l-1} the number of weights that any neuron belonging to the layer *l*-th process, the adder and accumulator maximum number of bits evaluates:

$I_{add\ acc} = I_n + I_m + \lceil \log_2(N_{l-1}+1) \rceil$ and $F_{add\text{-}acc} = F_n + F_m$.

To select the proper operation (the sum between the accumulated value and the product *weight*input*, the sum between the linear input value (LI) and, the accumulated value and activation of the tri-state buffer), two additional signals (*S0, S1*) are introduced: (0,0) for product evaluation, (1,0) for LI or threshold and (0,1) for the tri-state buffer activation.

In order to obtain a local synchronization each weight is associated with the corresponding set of control signals. It is remarkable that the synchronization weights, represented as *x* and *y* in fig.3b, are *don't care* so that the implementation cost of the combinatorial network of the PN control unit can be significantly reduced. Furthermore, the adopted memory encoding allows to reduce the connectivity requirements since it is avoided the broadcasting to all PNs, of the signal controlling the transmission.

5. Concluding remarks

The paper presented a design methodology tailored to design feed-forward, multi-layer neural networks for specific applications (special purpose neurocomputer), moving from the abstract definition of the network behavior down to the architectural synthesis. The work addressed the problem of achieving *realistic* (cost-effective) digital architectures through a relaxation of the design constraints. Furthermore, the system can be easily integrated in industrial environments, since the final output is automatically provided in terms of a set of VHDL models. The ongoing effort is in the direction of integrating the presented set of CAD tools in an exiting framework for Hw/Sw co-design of embedded systems, providing the user a CAD environment allowing to explore alternative designs to tradeoff both the Hw vs Sw domains as well as the third *neural* dimension.

6. References

[1] A.Bàsaglia, W.Fornaciari, F.Salice, "Behavior-driven minimal implementation of digital ANNs", Proc. of IEEE ICNN'95 Int. Conf. on Neural Networks, pp 1644-1649, Perth (Australia), Nov-Dec., 1995.

[2] P.Murtagh, A.C. Tsoi, "Implementation issues of sigmoid fuction and its derivative for VLSI digital neural networks," IEE Proc.-E, vol.139, n.3, May, 1992.

[3] W.Fornaciari and, F.Salice "A structured approach for automatic design of PN-based digital neural networks" in Proc. of IEEE-IJCNN'92, Neijing (China), November 1992.

[4] W.Fornaciari, F.Salice, "A New Architecture for the automatic Design of Custom Digital Neural Networks", Transaction on Very Large Scale Integration (VLSI) Systems, vol.3, n.4, Dec. 1995.

[5] K Hwang, "Computer Arithmetic, Principles, Architecture and Design", John Wiley & Sons, New York, 1979

Incremental Min-Max Network.
Part 2: Continuous and Discrete Spaces

F.M. Frattale Mascioli, M. Biancavilla, & A. Rizzi
INFO-COM Dpt., University of Rome "La Sapienza"
Via Eudossiana 18, 00184 Rome, Italy
mascioli@infocom.ing.uniroma1.it

Abstract

Neuro-fuzzy networks based on Simpson's Min-Max model are a very important and active research topic. In the present paper an incremental approach for hybrid data is discussed, with the aim of obtaining a better classification when data in the input space are completely or partially discrete. Experiments show promising results.

1 Introduction

The data available for training a neural network are often represented by both discrete and continuous values [1]. The simultaneous utilization of them is often hampered by the necessity of reducing the data set under a homogeneous form for reason of compatibility. This reduction is dangerous since it introduces unwanted constraints of arbitrary nature. In the present work, we face the problem of dealing with these different input data types in a more natural way in the context of incremental Min-Max (IMM) nets, introduced in [2]. IMM nets were proposed for overcoming some serious problems affecting previous versions of Min-Max algorithm, including the original one. Two modifications were proposed in [2]; in the present case we will consider only one of them based on the saturation of the parameter θ of Min-Max learning algorithm.

Discrete inputs are often available in classification problems and contain valuable information. The problem of handling continuous and discrete data without introducing arbitrary and undesired constraints is complicate and would require a deep investigation. We present in the following the result of a preliminary analysis of the problem based only on some aspects. In spite of this partial sight of the context of interest, the proposed method already yields satisfactory results. This will be illustrated in Sect. 4, where few real hybrid benchmarks will be considered and the gain obtained by considering hybrid data will be stressed. The nature of discrete data is not precisely defined. This aspect of the problem will be considered in Sect. 2. The definition of discrete input is strictly connected to the way of handling it. We will introduce at this regard the notion of *template* and *discrete space*. The hybrid input is therefore represented in a hybrid space with the consequence of extending the notion of Min-Max hyperbox. The application of the previous notion results in an extended IMM neural net which will be discussed in Sect. 3.

2 Discrete input data

In actual situations, useful information is introduced by expressing important properties of the context under examination by means of qualitative attributes. They correspond to variables characterized by linguistic values belonging to finite sets of very low cardinality. Some examples are:

- Green/red/yellow/etc....;
- North/South/East/West;
- Italian/English/French.

This type of discrete attributes are usually replaced with numeric values (depending on a given codification procedure) in order to render the input data handleable by traditional neural networks. This operation, however, is very critical, since it can distort the information associated to the above qualitative attributes. A simple example at this regard will clarify this point. Let us consider linguistic attributes as {green, red, yellow}. The usual way to deal with them consists with translating their linguistic information into a numeric form, assigning an arbitrary value to each determination. For instance, a possible choice is:

$$\text{green} \longrightarrow 1$$
$$\text{red} \longrightarrow 2$$
$$\text{yellow} \longrightarrow 3$$

The result of the translation is to have available numbers in place of linguistic values. An undesired effect of this codification is the introduction of a synthetic ordering relationship, inducing in the set of discrete attributes a norm. Since by definition true discrete value sets are not characterized by any type of norm, this additional information may mislead the training algorithm. It is important to remark that, in some cases, the norm induced in the discrete value space is acceptable (for example: {very low, low, medium, high, very high} and {small, big}).

On the basis of the previous discussion we will define as discrete, a variable which:

- presents a finite number of determination;
- is defined in a domain without norm;
- does not provide for an order.

The translation of the determinations of a discrete attribute under a numeric form is undertaken by recurring to a binary string. In order to avoid the previous inconvenience of introducing an order, the simplest way of creating the string is to associate to each determination one bit of the string. The value "1" will denote the presence of the attribute at hand; the value "-1" indicates its absence.

Referring to the previous example the proposed representation is:

$$\text{green} \longrightarrow -1\ -1\ \ 1$$
$$\text{red} \longrightarrow -1\ \ 1\ -1$$
$$\text{yellow} \longrightarrow 1\ -1\ -1$$

In the present work we are interested in considering hybrid input data in the context of Min-Max models. Consequently, it is necessary to extend the basic notion of hyperbox used by Simpson in the original algorithm with regard to any continuous input data. From the previous discussion, we conclude that the input data both span a continuous space and include strings of binary values. When we cluster these data, we obtain the usual hyperboxes with regard to the continuous data and simultaneously we get sets of strings with regard for the discrete data. In order to deal with the latter, we introduce the notion of *template*. The template is itself a string of numeric elements taking values in the set {+1, -1, 0}. The values of its components depend on the corresponding values of all the strings belonging to it. Namely, if the k-th component of all the strings of the set is equal to the same value "1" ("-1"), then the k-th component of the corresponding template is equal to "1" ("-1"); on the contrary, it is set to zero. For instance, in the case of a set of two strings with 4 components, we have:

$$\begin{matrix} 1 & -1 & 1 & -1 \\ 1 & 1 & -1 & -1 \end{matrix} \longrightarrow \begin{matrix} 1 & 0 & 0 & -1 \end{matrix}$$

It is evident that the number of zeros of a template grows with the cardinality of the set of binary strings it represents. Moreover, we note that the template is a sort of hyperbox in a discrete space where we can think to represent the discrete data. The presence of a zero in the template means the maximum size of the template for the corresponding dimension. Consequently, we define as cardinality of the template the number of zeros it contains. It is useful to visualize the ensemble of discrete and continuous variables in a suitable hybrid space, obtained by replica of the basic continuous space. The latter is a unitary hypercube of the N-dimensional continuous space, where N is the number of continuous inputs. The reason for having a region with the shape of a unitary hypercube is the required normalization in the interval [0, 1] of the continuous inputs in the case of Min-Max models. The hybrid space is constituted by an ensemble of M N-dimensional hypercubes, where M is the number of attributes of the discrete variables. Fig. 1 clarifies the previous definition of hybrid space in the case of N=2 and M=3 (one discrete variable with 3 attributes).

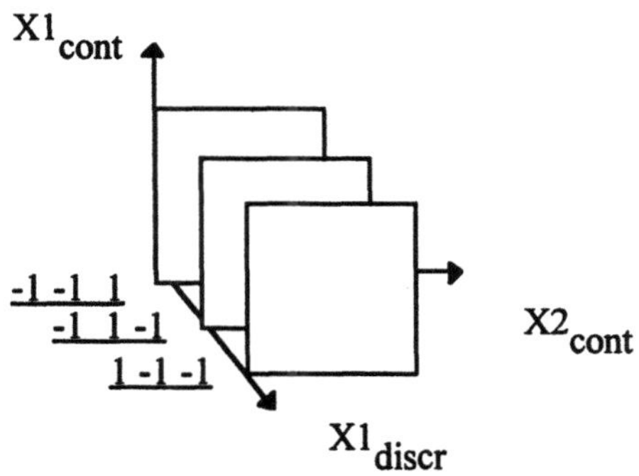

Fig.1: hybrid space in the case of two continuous variables and a discrete variable with 3 determinations. The resulting input space is composed by 3 unitary squares in the 3-dimensional space.

A cluster in the hybrid space is constituted by a hyperbox Γ in the continuous space (hypercube) and by a template with regard to discrete variables. It can be visualized, as said, by as many replica of Γ as there are zeros in the template. Consequently, we can associate it a "volume" given by the cardinality of the template times the volume of Γ. The number of zeros φ of the templates tends to grow excessively with the consequence of rendering the clusters too large and less useful for the classification job. This problem can be remedied by optimizing φ during training, as it will be discussed in the following section. It is interesting to note that φ plays in the discrete space the same role of the parameter θ in the original Min-Max algorithm.

3 Hybrid IMM neural classifier

The extension of the IMM neural classifier proposed in [2] for continuous input data is based on the options proposed in the previous section for handling discrete variables. The version of IMM net taken into consideration is that based on saturation of the parameter θ (θ=1) [2]. The basic iterative learning process developed in the continuous case is still valid in the present hybrid case. Each iteration consists of 4 steps. They will be described in detail in the following, with the modifications introduced with respect to [2] for handling the discrete data.

Step 1: Expansion. In the correspondence to the first iteration, all the training patterns are considered. The task to be pursued is to form one cluster per class. Each pattern consists of a continuous and discrete part. The former is represented by a point in the N-dimensional unitary hypercube. It is associated to a hyperbox (in the following HB) of the corresponding class by enlarging as necessary its sides [2]. The latter, which coincides with a binary string (as said in Sect. 2) is associated to the discrete part of the HB already considered. Its template is enlarged following the rule introduced in Sect. 2. The step is concluded when all the patterns to be processed are considered.

Step 2: Overlap test. This test is necessary in order to meet the basic requirement of the IMM learning algorithm, that is covering the input space with pure HBs. The latter are defined as the HBs containing patterns belonging to a unique class. During the learning process, we will have besides the *pure* clusters, also *mixed* clusters, i. e. those containing patterns of two or more different classes. The purpose of step 2 is to determine the overlaps between HBs obtained in step 1. Firstly, we have to define the overlap between two hybrid HBs. With regard to the continuous part, we apply the same definition used in [2]. Two continuous HBs are overlapped if each projection of them on the coordinate axes produces two segments which are superimposed. Two discrete HBs, i. e. two templates, are overlapped if there is not any component which takes on the values "+1" ("-1") for the other. Two examples clarify this definition.

Example 1: T1={-1 0 0 0 1} T2={1 0 1 1 0 -1}

In this case there is not overlap, since the last component of the two templates takes on 1 and -1 respectively.

Example 2: T1={1 -1 0 0 1 0} T2={1 0 -1 0 0 1}

In this case no component of the previous type is present and consequently the two templates are overlapped. Finally, we can define what we mean for overlap between two hybrid HBs. They are said overlapped if and only if simultaneously overlap in both continuous an discrete spaces. Step 2 is concluded when all the HB obtained in step 1 are compared. The final result is represented by two lists: the pure list (non overlapped HBs) and the mixed list (overlapped HBs).

Step 3: Contraction. The purpose of this step is to eliminate overlaps. Consequently, at its end the mixed list is empty. The HBs belonging to the mixed list are reduced and their overlap regions are added to a suitable list (overlap list).

The contracted HBs are added to the pure list, since they contain patterns of only one class. The contraction operation can be carried out in the discrete space in several different alternatives, which introduce suitable parameters to be optimized. On the contrary, it does not present any inconvenience in the continuous case, as seen in [2]. For this reason, we adopt in the present case the simply strategy of operating only in the continuous space. This can be done, since it is sufficient for eliminating the overlap according to the previous definition. The contraction is therefore carried out by applying the procedure followed in [2] with regards to the continuous part of the HBs involved and leaving unaltered their templates.

Step 4: Merging. The purpose of the merging step is to reduce the complexity of the net without impairing its performance. It is obtained by merging the HBs of the same class belonging to the pure list. It is important to undertake this step by avoiding to create new overlaps with HBs of different classes.

The iterative learning pass terminates when the overlap list is empty. If this is not the case, the iteration continues by applying step 1 to each cluster of the overlap list. Consequently, the learning process carries out a sort of progressive "zoom", reducing the portion of the input space to be examined after each iteration.

The iterative learning process yields a fuzzy neural Min-Max model at each iteration. Each net is characterized by a complexity measured by the number N_{Cl} of pure HBs obtained and by the number N_{mis} of misclassifications, which measures its performance. On the basis of learning theory, the optimal architecture of the net corresponds to the minimum of a function constituted by the weighted combination of the two previous measures, i. e.

$$F=\lambda*N_{Cl}+(1-\lambda)*N_{mis}$$

The value of λ is empirically determined. A value of 0.5 yields acceptable results and is used as default. The function F depends on iteration number and on parameter φ. Its minimum determines the optimal net, i. e. that obtained in the corresponding iteration.

Let R be the number of clusters. The resulting fuzzy neural network is shown in Fig. 2.

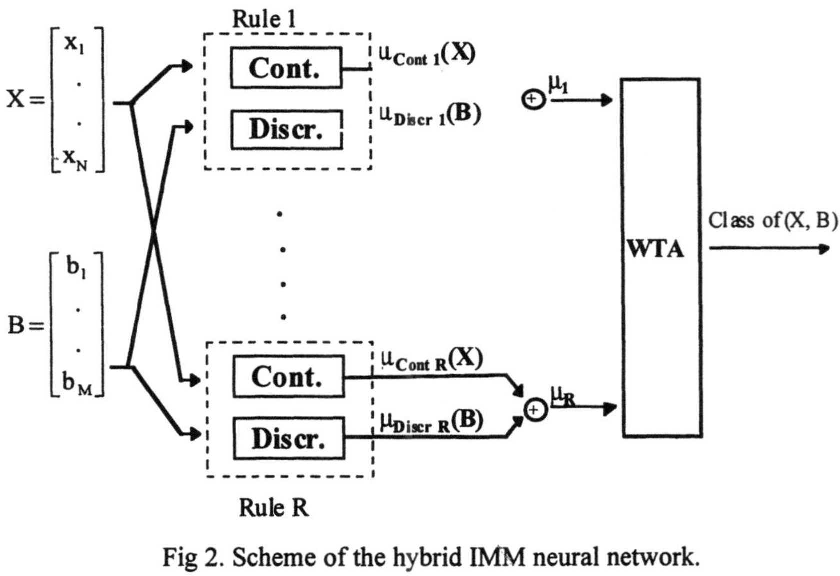

Fig 2. Scheme of the hybrid IMM neural network.

It consists of one hidden layer composed by R couples of neurons, where R is the number of rules determined by the learning process: each rule corresponds to a cluster. The neurons are associated in couples due to the hybrid nature of the HB. Their activation function coincides with the membership function defined in the following. Their outputs are added; the result is the output of the couple. The final layer of the net is a winner-takes-all element which selects the couple with maximum output. The class of this couple represents the class of the input, yielding the result of the classification. The operation of the net is affected by the use of the membership function considered. In the continuous case we use the function considered in [2]. In the discrete case a deep investigation on several classic benchmarks has allowed to select the function described below and to use the sum of the memberships of the continuous and discrete parts as representative of the hybrid HB. The membership function used in the discrete case is based on a reasonable definition of distance L between the input binary string B and the template of the discrete part of the HB under consideration. The definition of L, we propose to use, is the number of components which are different between B and the template, without considering the zero components of the latter. The membership function is

$$\mu_{Discr}(B) = 1 - \alpha \cdot L$$

The value of parameter α is chosen with a criterion similar to that followed in the continuous case, i. e.

$$1 - \alpha \cdot \frac{M}{2} = 0.1$$

where M is the cardinality of B.

4 Simulation tests and conclusions

The main goal of the simulation tests was to ascertain the capacity of the proposed net to take advantage of the discrete information. For this reason the results obtained with and without discrete information have been considered. The comparison of the results obtained in the two cases has proved the effectiveness of our net. Moreover, the results compare favorably with respect to those cited in the technical literature. In the following, for reasons of space, we report the results concerning 3 real problems: two of them are classical benchmarks and the remaining one regards the diagnosis of osteoporosis illness. The data sets of the first two problems (horses, credit cards) are available from [3]. The data set regarding the third one was suppied by Istituto Radiologico Policlinico Umberto I (Rome, Italy). The results are summarized in Tabs 1, 2, 3.

Credit cards benchmark (Tab. 1):

number of data in training set = 518
number of data in test set =172
number of variables = 51
B = 45

Including parts	Error in test set (%)	Neurons
Hybrid	13	76
Continuous	27	190

Tab. 1: comparison between continuous and hybrid credit cards data set

Horse data set (Tab. 2):

number of data in training set = 273
number of data in test set =91
number of variable = 58
B = 44

Including parts	Error in test set (%)	Neurons
Hybrid	22	71
Continuous	35	102

Tab. 2: comparison between continuous and hybrid horse data set

Osteoporosis (Tab. 3):

More critical results are shown in the third table using osteoporosis data set:
number of data in training set = 397
number of data in test set =132
number of variable = 60
B = 27

Including parts	Error in test set (%)	Number of rules
Hybrid	33	220
Continuous	42	131

Tab. 3: comparison between continuous and hybrid osteoporosis data set

Work in progress regards mainly the attempt of improving the clustering technique in the hybrid space, taking into account recent works regarding binary clustering [4,5], with the aim of better integrate continuous and discrete membership functions. Finally, in order to overcome the excessive number of input variables (both continuous and discrete) in many real applications, an input selection method can be incorporated in the proposed IMM classifier, as suggested in [6].

References

[1] A. Likas, K. Blekas, and A. Stafylopatis. Application of the fuzzy min-max neural network classifier to problems with continuous and discrete attributes. NNSP'94, Ermioni (Greece), pp. 163-170, 1994.

[2] A. Rizzi, M. Biancavilla, and F. M. Frattale Mascioli. Incremental Min-Max Network - Part 1: Continuous Spaces. WIRN Vietri 1998.

[3] L. Prechelt. Proben1 - A Set of Neural Network Benchmark Problems and Benchmarking Rules. Technical Report 21/94, September 30, 1994. File name: proben1.tar.gz Anonymous FTP from machine ftp.ira.uka.de in directory /pub/neuron.

[4] M. Muselli. Hamming clustering: improving generalization in binary neural networks. ICANN'94, Sorrento, Italy, pp. 1083-1086, 1994.

[5] M. Muselli. On sequential construction of binary neural networks. *IEEE Trans. on Neural Networks*, Vol. 6, pp. 678-690, 1995.

[6] M. Sugeno, T. Yasukawa. A fuzzy-logic-based approach to qualitative modeling. *IEEE Trans. on Fuzzy Systems*, Vol. 1, No. 1, pp. 7-31, 1993.

AN ADAPTABLE BOOLEAN NEURAL NET OBTAINED BY CONNECTING SEVEN NODE ASSSEMBLIES

F. E. Lauria, R. Prevete, M. Milo, S. Visco
Dipartimento di Scienze Fisiche dell' Università di Napoli
& Istituto Nazionale di Fisica della Materia
Mostra D' Oltremare, Pad. 19, I-80125 NAPLES (Italy) (EU)
lauria@na.infn.it
http://www.na.infn.it/Gener/cyber/nnet.htm

Abstract

We present the seven node Boolean neural subnet playing the assembly role so allowing us to implement an Hebbian rule in the Boolean net obtained by connecting them.

1 Introduction

We discuss a new sufficient condition, see [1 and 2 cap. 4], for the implementation of a Hebbian rule in a Boolean neural net. Using the Caianiello formalism, [4], we have proposed a sufficient condition based on the special purpose subnets we call the assemblies, see [2]. Here, we present a different assembly model implemented in as a seven node Boolean neural subnet.
Let us consider the time as a discrete variable, its smallest indivisible unit being tau, or τ. For us a Boolean net is a net composed of linear threshold units, ie, the net nodes compute the linear combination of their inputs and compare the result with a threshold T, see [4]. If the result is no smaller than zero, the node output assumes the value 1. Otherwise the node output is 0. Following [2], we call "coupling coefficient" the weight, $C_{h,k,d}$, associated to the directed connection between the node k and the node h, in the order, with a $d\tau$ delay. We call "facilitatory" a connection impinging onto the node h if its associated C is positive. We call "inhibitory" a connection impinging onto the node h if its associated C is negative and in absolute value greater than the sum of T plus all the facilitatory connections impinging onto the node h.
A delay dependent coupling coefficient can be either time independent, as the one just introduced, or time dependent, $C_{h,k,d,t}$. Let us call "Hebbian" a connection associated to a time dependent coupling coefficient $C_{h,k,d,t} > 0$, "non-Hebbian" the connection associated to the time independent $C_{h,k,d}$. For each h and d, let us represent with $C_{h,k',d,t}$ the time dependent coupling coefficients associated to the

Hebbian connections impinging upon the node h and stemming out of the nodes k', with a $d\tau$ delay. We call "Hebbian-in", "Hebbian-out", a node receiving, sending, at least one Hebbian connection. The presence in a net of a Hebbian-in node implies the presence of a Hebbian-out node and of a Hebbian connection, and vice versa. Obviously, as a node can receive, and send, both Hebbian and non-Hebbian connections, it can be at the same time Hebbian-out, Hebbian-in and non-Hebbian.

Let us call "Hebbian parameters" the positive constants a, b, e. Then, the $C_{h,k',d,t}$ value is computed every τ by applying the "Hebbian rule":

$$
C_{h,k',d,t+1} = \begin{cases} C_{h,k',d,t} + a & \text{if } f[h,t] = f[k', t - d*\tau] = 1 \text{ and } \Sigma_{k'} C_{h,k',d,t} * f[k', t - d*\tau] < T, \\ C_{h,k',d} & \text{if } f[h,t] = f[k', t - d*\tau] = 1 \text{ and } \Sigma_{k'} C_{h,k',d,t} * f[k', t - d*\tau] \geq T, \\ C_{h,k',d,t} - b & \text{if } f[h,t] = 1 - f[k', t - d*\tau] = 1 \text{ and } 0 \leq C_{h,k',d,t} - b, \\ 0 & \text{if } f[h,t] = 1 - f[k', t - d*\tau] = 1 \text{ and } C_{h,k',d,t} - b < 0, \\ C_{h,k',d,t} - e & \text{if } f[h,t] = 0 \text{ and } 0 \leq C_{h,k',d,t} - e, \\ 0 & \text{if } f[h,t] = 0 \text{ and } C_{h,k',d,t} - e < 0. \end{cases}
$$

In other words, if both the node h and the node k are active then the coupling coefficient associated to the connection between them is increased, as long as the sum of the coupling coefficients associated to the connections stemming from the presently active nodes k' and impinging onto the node h is smaller than the node threshold T. Otherwise, namely for the second clause as soon as the latter sum is no smaller than the node threshold T, from the next τ on, the coupling coefficients $C_{h,k',d,t}$ associated to the connections between the presently active Hebbian-out nodes and the node h become “frozen”, ie, become the time independent constants $C_{h,k',d}$. If the node h, the node k or both are inactive then the coupling coefficient associated to the connection between them is decreased, until it becomes zero.

As we are considering Boolean neural nets with absolute inhibition, but see also the discussion in [2, cap. 4], it is meaningless to consider time dependent negative coupling coefficients and to extend the Hebbian rule to them.

To implement the Hebbian rule in a n node Boolean neural net so to obtain a net trainable to compute a Boolean function f, we propose to consider n smaller nets, the assemblies, connected so to obtain a larger Adaptable Boolean Neural Net, or ABNN, learning to compute f, see also [2, cap. 5].

Besides the normal assemblies, the ABNN contains a finite number of special purpose assemblies, the primitives, with a reduced number of Hebbian-out nodes, ie, with a reduced number of output Hebbian connections. The net output non-Hebbian connections stem from the primitives to activate the system effectors, so generating the elementary actions seen by the net environment.

As confirmed by our simulations, see [1, 2] and the works there quoted, a net organized in assemblies and primitives learns, unassisted in a finite number of training sessions, to control the execution of any finite sequence of elementary actions. That is, if the set of elementary actions is universal, the net can be trained to control the computation of any finite algorithm.

2 The Seven node assembly

The assemblies presented in [2] can be universally trained, in the sense just introduced. However, a net organized in assemblies, it is a waste of resource. In fact, the number of nodes per assembly increases with the number of assemblies, as each assembly must contain a sufficient number of nodes to hold the identifier, namely the name, of another assembly, or of a primitive, to work properly. A slimmer wiring diagram is the seven node assembly we propose here, see fig.1 and Table 1.

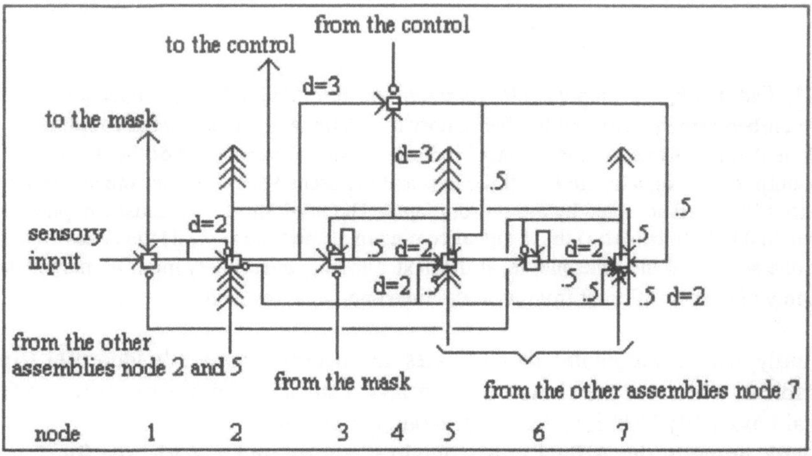

Fig. 1. *The seven node assembly.* The empty circles represent the absolute inhibitory connections. The single barbed arrows represent the facilitatory connections: their associated coupling coefficient value is one, unless the label indicates its value as one half of the threshold. Unless the label indicates a different value for the delay, it is one tau. Each of the nodes two, five and seven are Hebbian-in, Hebbian-out and non-Hebbian. The node two and five Hebbian out-connection, represented by the four barbed arrows, impinges onto the node two of all the other assemblies. The node seven Hebbian out-connection, represented by the two barbed arrow, impinges onto the nodes five and seven of all the other assemblies. The nodes five and seven Hebbian out-connections are frozen as soon as their associated coupling coefficients become equal or greater than one half of the threshold T. The seven node assembly inner working in the training sessions is presented in the Table 1. For a proper functioning, an adaptable Boolean net obtained by connecting seven node assemblies must contain two more subnets: a control, see the fig. 2, and a mask, see the fig. 3.

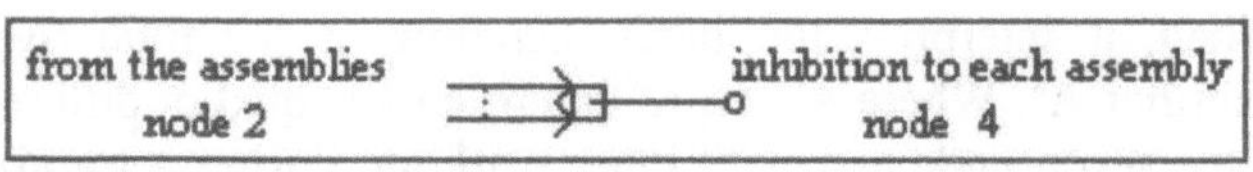

Fig. 2. *The control.* The node two activity in at least one assembly activates the control subnet. It inhibits all the assemblies node four.

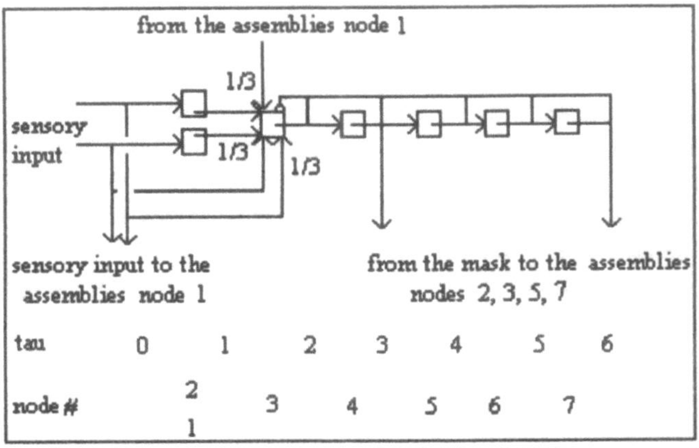

Fig. 3. *The mask.* The empty circles represent the absolute inhibitory connections. The single barbed arrows represent the facilitatory connections: their associated coupling coefficient value is one, unless the label indicate its value as one third of the threshold. The mask outputs two signals: the first three taus and the second six taus after the assembly A identifier has been accepted by the sensory input. The mask third node must compute the logical AND of the triplet: i) the A signal coming in at the time zero, ii) the A assembly node one activity at the time one, ie, at the next τ and iii) the sensory input A', namely an assembly identifier different from A, at the time one, ie, at the next τ.

Actually, the net automatically correlates the incoming assembly identifier triplets, see table 1, whenever they have been repeated a sufficient number of times to freeze the said assembly Hebbian, both in and out, connections.

To work properly the ABNN must contain a control and a mask, see fig. 2 and 3. The mask role is to simplify the net training. Actually, it is a subnet both inhibiting each assembly node two and three and contributing facilitatory connections, with associated a one half coupling coefficient, to each assembly node five and seven. Both the facilitatory and the inhibitory mask signals must be at each assembly input three and six taus after the first sensory input representing the identifier triplets. Accordingly, we have chosen the net wiring so to have a mask activated by the triplet: [i) an assembly identifier as sensory input at the time zero; ii) activation of an assembly node one at the time one, ie, after one τ; iii) an assembly identifier as sensory input at the time one, ie, after one τ]. The control task is to help an assembly to work properly in case only some, not all, of its Hebbian-out connections are frozen, once the training has been completed.

Once it has received, at the time zero, the first assembly identifier A, the net interprets the input, incoming at the time two, as the identifier A' of a different assembly. Then, it interprets the input incoming at the time five, as another assembly identifier A". Where A' and A" can be either equal or different.

We present the seven node assembly inner working, during a training session, in the Table 1. The trained assembly inner working is a simplified version of the latter.

To avoid the loop implementation in the net, the training procedures must satisfy the following condition [a]:

"In a seven node assembly the Hebbian-in connections can be frozen only if its node 2 and 5 Hebbian-out connections are frozen".
As it is easy to verify, a net organized in seven node assemblies can be trained so to respect the [a] condition. As a primitive is an assembly with some frozen Hebbian-out connections, the [a] condition is the same as to require the training to start associating, first, some assemblies with the primitives and, then, the former with new assemblies, not the other way round.

3 Conclusions

In the quest for a way to implement a Hebbian rule in a Boolean neural net, ie, to obtain an adaptable Boolean neural net, our proposal is to connect suitable Boolean subnets, the assemblies, by means of both Hebbian and non-Hebbian connections. We have discussed a seven node assembly wiring diagram, together with the resulting net properties. The three node assembly, discussed in [5], seems to be the smallest subnet we can connect to obtain an adaptable Boolean net. However, its disadvantages is the constraint imposed on the input sequences and, also if it is quite universal, it seems fit for some special purpose applications, only. By contrast the seven node assembly accepted input sequence set is universal, ie, there are no limitations on the sequences it stores autonomously as it interacts with the environment. Accordingly, it seems to be the smallest unconstrained wiring found so far to implement a Hebbian rule in a Boolean neural network.
Acknowledgements: this research was supported in part through the MPI/INFM 40% fund and the MPI 60% fund.

References

[1] Lauria, F. E. - On some sufficient conditions for the Caianiello-Hebb transform convergence - Rend. Acc. Sc. Fis. e Mat., Napoli, Serie IV- Vol. LXII (1995) 7-42.

[2] Lauria, F. E., M. Sette, S. Visco - Adaptable Boolean neural networks- Fridericiana Scientia, Liguori (Napoli)(1997) 212. ISBN 88-207-2676-9. http://www.na.infn.it/Gener/cyber/nnet.htm

[3] Rumelhart, D. E., J. L. McClelland and the PDP Research Group - Parallel distributed processing. Vol I Foundations -MIT Press, Cambridge (Mass.) USA (1986) 547 + xx.

[4] Caianiello, E. R. - Outline of a theory of thought processes and thinking machines- J. Theor. Biol. 2 (1961) 204-235.

[5] Lauria F. E., M. Sette - "Teaching macros through program execution examples" E. R. Caianiello ed. "Neural nets WIRN Vietri-92" World Sc. (Singapore), (1992) 204-209.

TABLE 1

NODE # \ TIME	**0**	**1**	**2**	**3**	**4**	**5**	**6**	**7**	**8**	**9**
1 non-Hebbian of		**A**	**A, A'**	**A, A'**	**A, A'**	**A, A''**	**A, A''**	**A, A''**	**A**	
2 Hebbian-in & out of			**A**	**A'**			**A''**			
3 non-Hebbian of				**A**						
4 non-Hebbian of										
5 Hebbian-in & out of						**A**				
6 non-Hebbian of							**A**			
7 Hebbian-in & out of					**A'**			**A''**	**A**	
INPUT \ TIME	**0**	**1**	**2**	**3**	**4**	**5**	**6**	**7**	**8**	**9**
from **the sensory input**, with delay equal to 1, to the node 1 of	**A**	**A'**			**A''**					
from the **mask,** with delay equal to 1, to the nodes 2,3,7 of				**A,A',A''**			**A,A',A''**			
from the **mask,** with delay equal to 2, to the nodes 5,7 of				**A,A',A''**			**A,A',A''**			
from the **control,** with delay equal to 1, to the node 4 of				**A,A',A''**	**A,A',A''**			**A,A',A''**		
from the Hebbian node 2 or 5 of A to the Hebbian node 2 of			**A'**			**A''**				
from the Hebbian node 7 of A' to the Hebbian node 5 of					**A**					
from the Hebbian node 7 of A'' to the Hebbian node 7 of								**A**		

Table 1. *The seven node assembly activity in the course of the training sessions*. The table gives the relative times for the seven assembly node activity. **A** is the *parent* assembly name, **A'** is the *first offspring* assembly name, **A''** is the *second offspring* assembly name. The node two Hebbian-out connections are frozen as soon as their associated coupling coefficients reach the threshold value T. The nodes five and seven Hebbian-out connections are frozen as soon as their associated coupling coefficients reach the value T/2. The Hebbian connection delay is one tau. The control activity inhibits the node four during the training sessions. It becomes active once the node two, five and seven Hebbian-out connections are frozen.

A Comparative Evaluation of Sequential Constructive Methods

Marco Muselli
Istituto per i Circuiti Elettronici - CNR
via De Marini, 6 - 16149 Genova, Italy

Abstract

A detailed comparison of the most promising algorithms in the class of sequential constructive methods is performed by employing artificial benchmarks and real-world problems.

Four of them incrementally build the hidden layer of the resulting neural network by subsequently adding threshold units. The last method considered makes use of neurons with a window-shaped activation function.

The performances obtained with each technique and their comparison with the widely employed back-propagation algorithm allow a thorough examination of the characteristics of these sequential constructive methods. In particular, the neural networks trained show good generalization ability also when the learning time is very low.

Keywords: Supervised learning, constructive methods, sequential learning, performance comparison, generalization ability.

1 Introduction

The solution of most real-world problems through the back-propagation algorithm [1] requires a high computational cost; in many cases, this prevents us to analyze training sets containing a sufficient number of samples. Consequently, the resulting neural network is often unable to catch the underlying input-output function associated with the given problem.

Another limitation inherent the back-propagation algorithm is the need to choose a priori the architecture of the multilayer perceptron to be trained. The determination of the number of layers and the number of nodes in each layer requires the execution of additional trials that increases the total learning time spent by the algorithm.

When we are facing with classification problems, a possible way to overcome these difficulties is provided by the class of *sequential constructive methods*: they subsequently add neurons to the hidden layer until all the input-output pairs contained in the given training set are satisfied. The insertion of a new neuron does not require the updating of the whole weight matrix, but only the training of the connections associated with the unit to be added.

Although the convergence properties of sequential constructive methods have been theoretically proved elsewhere [2], there has not yet been presented a detailed comparison which allows to establish the effective performances of this class of algorithms. This paper tries to fill this gap by examining the five most promising sequential constructive methods and by executing a series of common comparative trials. Furthermore, the application of the back-propagation algorithm on the same set of benchmarks allows to correctly evaluating the results obtained.

2 Training algorithms considered

Sequential constructive methods essentially differ in the algorithm employed for the addition of a new hidden neuron. In fact, the weights for the output layer are obtained through the application of simple algebraic equations and therefore do not need a proper learning phase. The following definition plays a fundamental role in the description of the algorithms examined.

Definition 1 *Let Q^+ and Q^- be two subsets of the n-dimensional input space D, containing a finite number of patterns q^+ and q^- respectively. A neuron will be called a* partial classifier *if it provides output $+1$ for all the patterns included in a non-empty subset R of Q^+ and output -1 for all the elements of Q^-.*

Neural networks generated by sequential constructive methods contain partial classifiers in the hidden layer. Thus, at any addition, the main goal is to build a partial classifier that provides the correct output for the greatest number of patterns of Q^+ and Q^-, so as to minimize the number of connections in the final neural network. Since the achievement of this result can lead to the solution of an NP-complete problem, sequential constructive methods only try to obtain near-optimal partial classifiers that approximate the desired configuration.

Now, we will give a brief description of the training algorithms considered in the comparative evaluation. Four of them employ threshold neurons for the construction of the hidden layer; their activation function is given by

$$\varphi(\mathbf{x}) = \operatorname{sgn}\left(\sum_{i=0}^{n} w_i x_i\right) = \begin{cases} +1 & \text{if } \sum_{i=0}^{n} w_i x_i \geq 0 \\ -1 & \text{otherwise} \end{cases} . \tag{1}$$

As usual we have added a component $x_0 = +1$ to the input pattern $\mathbf{x}$ so as to include the bias in the weights of the neuron.

The fifth method builds partial classifiers having the following window-shaped activation function

$$\psi(\mathbf{x}) = \begin{cases} +1 & \text{if } \left|\sum_{i=0}^{n} w_i x_i\right| \leq \delta \\ -1 & \text{otherwise} \end{cases} , \tag{2}$$

where the (small) real quantity δ is called *amplitude* and is meaningful only by an implementative point of view.

2.1 Irregular Partitioning Algorithm (IPA)

In this method [3] the subset R with highest size is incrementally built by subsequently adding to it the element $\mathbf{x} \in Q^+$ that maximizes the number of input-output pairs satisfied by the current partial classifier. It is then necessary to employ at every iteration a proper algorithm that verifies the existence of a threshold neuron separating R and Q^-. A possible choice is to adopt the Thermal Perceptron Learning Rule (TPLR) [4], because of its high convergence speed.

2.2 Carve Algorithm (CA)

A different approach to obtain the construction of a partial classifier for two sets of input patterns Q^+ and Q^- is to consider the convex hull generated by the elements of Q^-. In fact, it could be observed that only the points of Q^+ outside this convex hull can be separated from the elements of Q^- through a hyperplane. This is the basic idea of CA, but its direct application can lead to a total execution time which is exponential in the number n of inputs.

A near-optimal result can be obtained much faster by considering only a restricted number of faces of the convex hull; this can be done by applying a proper algorithm which has polynomial complexity [5].

2.3 Target Switch Algorithm (TSA)

To obtain a subset $R \subset Q^+$ with highest size, TSA [6] searches for a near-optimal threshold neuron by applying a proper training algorithm, e.g. TPLR [4]. If it does not correctly classify some patterns of Q^-, the more critical element of Q^+ (chosen according to a precise definition) is moved from Q^+ to Q^-. It can be shown that such a modification leads to a partial classifier for the given training set in a finite number of iterations.

2.4 Oil Spot Algorithm (OSA)

If the input patterns have binary components, they are placed at the vertices of an n-dimensional hypercube centered at the origin of the space $\mathcal{R}^n$. Thus, two patterns are *contiguous* if they are connected by a single edge of the hypercube. It can easily be seen that two edges are *parallel* if the corresponding pairs of vertices differ by the same component; moreover, two parallel edges are called *congruent* if their orientation coincides. Finally, according to the classification induced by the sets Q^+ and Q^- the edges that join two vertices belonging to opposite classes will be named *critical*.

By using these definitions, it is possible to show that the sets R and Q^- can be separated by a threshold neuron if the following two conditions are satisfied: (1) the patterns of R are the nodes of a connected subgraph of the hypercube $\{-1, +1\}^n$, and (2) two parallel critical edges are always congruent. This property is used by the oil spot algorithm [7] to construct partial classifiers in an incremental way.

Algorithm	Problem # 1			Problem # 2			Problem # 3		
	w	g	t	w	g	t	w	g	t
BPA	21	95.7%	8	42	84.3%	36	21	90.2%	4
IPA	31	87.8%	44	60	83.3%	163	29	87.8%	33
CA	30	90.7%	5	74	82.9%	13	35	88.7%	5
TSA	64	83.4%	59	116	82.3%	198	64	81.8%	37
OSA	350	82.2%	1	963	74.3%	4	341	80.8%	1
OSA & HC	13	99.7%	1	82	90.5%	1	39	91.1%	1
SWL	38	95.3%	15	16	100.0%	1	65	85.5%	27
SWL & HC	11	100.0%	1	18	98.7%	1	29	91.5%	1

Table 1: Number of nonnull weights w, generalization ability g and average CPU time t for the neural networks solving Monk's problems.

2.5 Sequential Window Learning (SWL)

The interest in window neurons (2) is motivated by the existence of a fast and efficient learning algorithm, based on the solution of systems of algebraic equations, which allows to find good partial classifiers for any pair of sets Q^+ and Q^- [8]. This method is used by SWL which incrementally builds the set R by adding to it the element $\mathbf{x} \in Q^+$ that maximizes the number of input patterns correctly classified by the current window neuron.

3 Experimental results

The sequential constructive techniques briefly described in the previous section have been extensively tested on five benchmark problems to evaluate their performance. The first of them (Monk's problems) has been artificially built while the other four groups of trials derive from a collection of data sets distributed by the machine learning group of the University of California at Irvine [9]. The influence of the procedure of Hamming Clustering (HC) [8] on the complexity and the generalization ability of the configurations generated will also be analyzed.

Thirty runs have been performed for every trial and every training algorithm so as to obtain a sufficient statistics of the measured quantities. In the tables of results we will denote with w the number of nonnull weights of the resulting configurations, with g their generalization ability (percentage of correctness on the test set), and with t the training time (sec.) needed for each construction. A reference value for each of these quantities has been obtained by the application of the Back-Propagation Algorithm (BPA) to every set of trials. All the execution times refer to a DECStation 3000/600 with 64 MB RAM under operating system Digital UNIX 4.0.

3.1 Monk's problems

Three artificial classification problems have been proposed in [10] as benchmarks for training algorithms. The first two Monk's problems are noise-free, whereas in the third case the outputs in the training set can undergo small

Algorithm	GL			IR			CH			VO		
	w	g	t	w	g	t	w	g	t	w	g	t
BPA	40	73.6%	143	10	96.1%	93	148	98.6%	799	34	94.9%	24
IPA	59	73.7%	71	35	91.0%	23	491	97.8%	9074	42	94.5%	45
CA	57	73.6%	43	30	89.7%	1	1297	92.5%	8452	69	93.7%	135
TSA	78	75.2%	33	32	93.2%	36	558	96.8%	4202	49	94.6%	3
OSA	4326	80.0%	41	1859	92.2%	6	7329	88.3%	15576	1036	90.7%	5
OSA & HC	81	79.0%	6	35	90.7%	2	156	98.3%	422	47	93.2%	3
SWL	393	52.2%	457	589	11.3%	104	371	93.1%	6146	139	86.4%	755
SWL & HC	37	74.6%	13	28	87.3%	3	102	98.0%	651	32	91.6%	7

Table 2: Number of nonnull weights w, generalization ability g and average CPU time t for the neural networks solving four real-world problems.

changes from their correct values. This last test can therefore provide a measure of the robustness of sequential constructive methods.

Since the inputs are discrete, a proper transformation must be applied to allow the employment of SWL and OSA; it leads to binary training sets containing 15 inputs. The results obtained in the simulations are shown in tab. 1.

The employment of HC allows the achievement of good performances both for SWL and OSA: the resulting neural networks show a high generalization ability and a discrete robustness (Monk's problem #3). Furthermore, the low computational cost of HC allows the reduction of the total CPU time in all the three tests. Interesting results are also offered by CA which is a good compromise between training speed and efficiency.

3.2 Real-world problems

The following four benchmarks maintained at the UCI machine learning repository [9] have been considered:

Glass identification (GL): Starting from 9 real parameters, we want to determine if a given piece of glass is float-processed or not.

Iris data set (IR): Three different types of iris plant, virginica, versicolor and setosa must be recognized by knowing the length and width of the sepal and petal.

Chess end-game (CH): In this problem, the goal is to determine whether or not a given chess board configuration is a winning position for the white player. The input pattern contains 35 binary components and one ternary component describing a board position of a chess end-game.

United States congressional voting record database (VO): From the results of 16 ternary key votes ('yea', 'nay' or '?') determine if a US House of Representatives Congressman is either Democrat or Republican.

In all these benchmarks 2/3 of the patterns (randomly chosen in each of the 30 trials) form the training set whereas the remaining 1/3 has been used to test the generalization ability of the resulting neural networks. The application of OSA and SWL needs a proper transformation of the input patterns: the corresponding binary training sets contain 256 (GL), 151 (IR), 37 (CH), and 48 (VO) inputs respectively. For real inputs (GL and IR) the thermometer code has always been used.

The results obtained in the simulations are reported in tab. 2.

In all these tests TSA generates two-layer perceptrons with good generalization ability in reasonable training times. As one can note, the performances of neural networks constructed by OSA and SWL does not suffer from data binarization. HC generally lowers the computational burden and the complexity of the resulting configurations. In many cases (particularly with SWL) it also allows to reach a better generalization ability.

References

[1] Rumelhart, D.E., Hinton, G.E., and Williams, R.J. Learning internal representations by error propagation. In *Parallel Distribute Processing* (Rumelhart D.E., and McClelland, J.L., eds.), Cambridge, MA: MIT Press (1986), 318–362.

[2] Muselli, M. A unified approach to sequential constructive methods. Contribution to *WIRN'98 - The 10-th Italian Workshop on Neural Nets.*

[3] Marchand M., and Golea, M. On learning simple neural concepts: From halfspace intersections to neural decision lists. *Network* **4** (1993), 67–85.

[4] Frean, M. A "thermal" perceptron learning rule. *Neural Computation* **4** (1992) 946-957.

[5] Young, S., and Downs, T. Improvements and extensions to the constructive algorithm CARVE. In *Artificial Neural Networks — ICANN 96* (Von der Malsburg, C., W. von Seelen, J.C.V., and Sendhoff, B., eds.), Berlin: Springer (1996), 513–518.

[6] Campbell C., and Perez Vicente, C. The target switch algorithm: A constructive learning procedure for feed-forward neural networks. *Neural Computation* **7** (1995), 1245–1264.

[7] Frattale Mascioli, F.M., and Martinelli, G. A constructive algorithm for binary neural networks: The oil-spot algorithm. *IEEE Transactions on Neural Networks* **6** (1995) 794-797.

[8] Muselli, M. On sequential construction of binary neural networks. *IEEE Transactions on Neural Networks* **6** (1995), 678–690.

[9] Merz, C.J., and Murphy, P.M. Uci repository of machine learning databases [http://www.ics.uci.edu/ mlearn/mlrepository.html]. Irvine, CA: University of California, Department of Information and Computer Science (1996).

[10] Thrun, S. et al. A performance comparison of different learning algorithms. Technical Report CMU-CS-91-197, Pittsburgh, PA: Department of Computer Science, Carnegie Mellow University (1991).

A Unified Approach to Sequential Constructive Methods

Marco Muselli
Istituto per i Circuiti Elettronici - CNR
via De Marini, 6 - 16149 Genova, Italy

Abstract

A general treatment of a particular class of learning techniques for neural networks, called *sequential constructive methods*, is proposed. They subsequently add units to the hidden layer until all the input-output relations contained in a given training set are satisfied.

Every addition involves the update of a small portion of the whole weight matrix and depends on a subset of samples whose size decreases with time. In most cases this leads to a large reduction of the computational cost.

General convergence theorems are presented that ensure the achievement of a good multilayer perceptron within a finite execution time. The output weights need not to be trained but are obtained by the application of simple algebraic equations.

Keywords: Supervised learning, constructive methods, sequential learning, convergence theorems, multilayer perceptron.

1 Introduction

The application of the back-propagation algorithm [1] suffers from some intrinsic limitations that make its employment difficult, particularly when the given training set contains a high number of input-output pairs. For example, the choice of the neural network architecture (the number of layers and the number of units in each layer) is often very critical. In fact, a configuration containing a small number of weights may lead to an excessive error in the training phase, whereas a too complex multilayer perceptron generally shows a poor generalization ability when new input patterns are considered.

In practice, to overcome this problem an initial architecture is found by applying heuristic rules and the optimal configuration is obtained after successive corrections that considerably increase the total learning time.

When classification problems are considered, a more efficient solution is provided by the class of *sequential constructive methods*. They build the resulting neural network by starting from an initial minimal configuration and subsequently adding units to the hidden layer. The insertion of a new neuron does not require the updating of the whole weight matrix, but only the training of the connections associated with the unit to be added.

This reduces the computational cost and allows to obtain optimal weight vectors for every neuron in a finite execution time. Moreover, the size of the training set considered for each insertion decreases during the algorithm thus allowing a further improvement of the convergence properties of sequential constructive methods. The application of simple algebraic equations gives the values for the output weights, which do not need a proper learning phase.

Several algorithms proposed in the literature belong to the class of sequential constructive methods [2, 3, 4, 5]. For each of them a proper convergence theorem ensures the achievement of a configuration that satisfies all the samples contained in a given consistent training set. As we will see in the next sections, these theoretical results can be collected in a unified approach that allows a common analysis of the convergence properties of sequential constructive methods. Such an approach is based on the concept of sequential decision list, which plays a fundamental role in the mathematical description of their functioning.

For the sake of brevity, the proofs of the convergence theorems will be only outlined; for a complete version see [6].

2 Sequential decision list

Consider a training set S containing a finite number s of input-output relations $(\mathbf{x}_j, \mathbf{y}_j)$, $j = 1, \ldots, s$, generated by an unknown function f and denote with $D \subset \mathcal{R}^n$ its domain, being n the number of inputs. Since sequential constructive methods have been expressly developed for the solution of classification problems, the range of the function f will be $\{-1, +1\}^m$, where m is the number of components of the output patterns $\mathbf{y}_j$.

In the case $m = 1$ the general technique employed for the construction of a neural network that approximates f can be easily described by introducing a generalization of the concept of decision list defined in [7].

Definition 1 *An ordered sequence of pairs* (L_j, d_j), $j = 1, \ldots, h+1$, *where* $L_j \subset D$ *and* $d_j \in \{-1, +1\}$ *will be called a* sequential decision list for a two-class problem, *or simply a* 1-SDL *if* $L_{h+1} = D$.

The first element L_j *of every pair will be called the* choice set *while the latter* d_j *will be called the* pertaining class.

Every 1-SDL is associated with a function $g : D \longrightarrow \{-1, +1\}$ given by $g(\mathbf{x}) = d_j$, being j the first (least) index for which $\mathbf{x} \in L_j$. Since $L_{h+1} = D$ the function g is defined on the whole domain D.

Now, consider a threshold neuron whose output y is defined as follows:

$$y = \operatorname{sgn}\left(\sum_{j=0}^{k} u_j z_j\right) = \begin{cases} +1 & \text{if} \sum_{j=0}^{k} u_j z_j \geq 0 \\ -1 & \text{otherwise} \end{cases} , \tag{1}$$

where $u_1, \ldots, u_k$ are the weights corresponding to the inputs $z_1, \ldots, z_k$ respectively. The bias u_0 is included in the summation by adding a new component $z_0 = +1$. The relevance of sequential decision lists for the incremental construction of neural networks is pointed out by the following basic result.

Theorem 1 *The function* g *associated with a given 1-SDL* (L_j, d_j), $j = 1, \ldots,$ $h+1$, *can always be realized by a perceptron containing a single hidden layer and an output threshold neuron.*

Proof. *(outline)* It is sufficient to put in the hidden layer h neurons whose output $z_j, j = 1, \ldots, h$, is given by

$$z_j = \varphi_j(\mathbf{x}) = \begin{cases} +1 & \text{if } \mathbf{x} \in L_j \\ -1 & \text{otherwise} \end{cases} , \tag{2}$$

Then, the following set of weights for the final threshold neuron leads to the realization of the given function g [2]

$$u_0 = \sum_{j=1}^{h} u_j + d_{h+1} \quad , \quad u_j = d_j 2^{h-j} \text{ for } j = 1, \ldots, h. \quad ; \quad \blacksquare \tag{3}$$

It is also possible to adopt a different architecture, called *cascade net* [3], in which every hidden layer contains a single unit that is fed by all the outputs of previous neurons together with the input pattern considered. As in the two-layer perceptron case, the final neuron is again a threshold unit whose activation function is given by Eq. (1).

The following theorem can be stated

Theorem 2 *The function g associated with a given 1-SDL (L_j, d_j), $j = 1, \ldots, h+1$, can always be realized by a cascade net containing h hidden neurons.*

Proof. *(outline)* The h hidden layers contain as many units whose output $z_j, j = 1, \ldots, h$, is given by

$$z_j = \psi_j(\mathbf{x}, z_1, \ldots, z_{j-1}) = \begin{cases} +1 & \text{if } \mathbf{x} \in L_j \text{ and } z_1 = \cdots = z_{j-1} = -1 \\ -1 & \text{otherwise.} \end{cases} \tag{4}$$

With this choice an input pattern $\mathbf{x} \in D$ can activate at most one of the hidden neurons by setting its output to the value +1. Then, a weight vector $\mathbf{u}$ for the output threshold unit that yields the value $g(\mathbf{x})$ generated by the given 1-SDL is [3]

$$u_0 = \sum_{j=1}^{h+1} d_j \quad , \quad u_j = d_j \text{ for } j = 1, \ldots, h. \quad \blacksquare \tag{5}$$

These results can be extended to the general case $m > 1$, by changing definition 1 in the following way.

Definition 2 *An ordered sequence of pairs $(L_j, \mathbf{d}_j)$, $j = 1, \ldots, h+1$, where $L_j \subset D$ and $\mathbf{d}_j$ is a nonnull vector with m components in $\{-1, 0, +1\}$, will be called* sequential decision list for a multiclass problem *or* m-SDL *if $L_{h+1} = D$ and $\mathbf{d}_{h+1}$ does not contain null components.*

The association of a function $g : D \longrightarrow \{-1, +1\}^m$ with any m-SDL is straightforward: for every input pattern $\mathbf{x} \in D$ the kth component $g_k(\mathbf{x})$ of the corresponding output $g(\mathbf{x})$ is given by $g_k(\mathbf{x}) = d_{jk}$, being j the first (least) index for which $\mathbf{x} \in L_j$ and $d_{jk} \neq 0$.

From definitions 1 and 2 it follows that an m-SDL $(L_j, \mathbf{d}_j)$ is equivalent to m 1-SDLs $(L_j, d_{jk}), k = 1, \ldots, m$, each of which contains only the pairs with $d_{jk} \neq 0$. Then, Theorems 1 and 2 can be extended to the multiclass case leading to the following results [6].

Theorem 3 *The function g associated with a given m-SDL $(L_j, \mathbf{d}_j)$, $j = 1, \ldots, h+1$, can always be realized by a perceptron containing a single hidden layer and m threshold neurons in the output layer.*

Theorem 4 *The function g associated with a given m-SDL $(L_j, \mathbf{d}_j)$, $j = 1, \ldots, h+1$, can always be realized by a cascade net containing h hidden units and m threshold neurons in the output layer.*

3 General procedure

Sequential constructive methods build the neural network associated with a classification problem by generating a sequential decision list starting from the given training set S. The kernel of these techniques is the learning algorithm for the hidden units which has the aim of constructing neurons with activation functions given by Eq. (2) or (4) for some choice sets L_j. The following definition will be used:

Definition 3 *Let Q^+ and Q^- be two subsets of the input space; a neuron will be called* partial classifier *if it provides output $+1$ for at least one pattern in Q^+ and output -1 for all the elements of Q^-.*

The general procedure for obtaining the multilayer perceptron associated with a finite training set S is common to every sequential constructive method. Its outline is reported in Fig. 1.

GENERAL SEQUENTIAL CONSTRUCTIVE METHOD

1. Set $h = 1$ *(number of hidden units)*
2. Choose a vector $\mathbf{d}_h$ in $\{-1, 0, +1\}^m$ containing the pertaining classes for some of the m outputs ($d_{hk} = 0$ if the hth hidden neuron does not affect the kth output).
3. Set
$$Q^+ = \bigcap_{\{k:d_{hk} \neq 0\}} D_{k,d_{hk}} \quad , \quad Q^- = \bigcup_{\{k:d_{hk} \neq 0\}} D_{k,-d_{hk}} \quad .$$
4. Train the hth hidden neuron so as to obtain a partial classifier for Q^+ and Q^-. Let R be the subset of Q^+ containing the input patterns for which the hth hidden units provides output $+1$.
5. Remove all the input patterns contained in R from the sets $D_{k,d_{hk}}$ for every k such that $d_{hk} \neq 0$. If $D_{k,+1} \neq \emptyset$ and $D_{k,-1} \neq \emptyset$ for some output k then set $h = h + 1$ and go to step 2.

Figure 1: General procedure followed by sequential constructive methods.

In this algorithm the following m pairs of sets $(D_{k,+1}, D_{k,-1})$, with $k = 1, \ldots, m$, are employed

$$\begin{aligned} D_{k,+1} &= \{\mathbf{x}_i \mid y_{ik} = +1, (\mathbf{x}_i, \mathbf{y}_i) \in S\}, \\ D_{k,-1} &= \{\mathbf{x}_i \mid y_{ik} = -1, (\mathbf{x}_i, \mathbf{y}_i) \in S\} \quad . \end{aligned}$$

They contains the input patterns of the given training set S whose kth output is $+1$ and -1 respectively and form at any iteration the current training set for each output. The sets Q^+ and Q^- for the learning of the hidden neurons will be extracted from the pairs $(D_{k,+1}, D_{k,-1})$ by following the method described at Step 3. It can easily be seen that, in this way, an m-SDL $(L_j, \mathbf{d}_j), j = 1, \ldots, h+1$, is constructed in which every choice set L_j contains some input patterns of the training set S.

The convergence properties of sequential constructive methods are ensured by the following result [6].

Theorem 5 *Every sequential constructive method provides, in a finite execution time, a neural network which satisfies all the samples contained in a given finite consistent training set S if the learning algorithm for the hidden neurons allows us to obtain within a finite computing time a partial classifier, whichever pair of sets Q^+ and Q^- is considered.*

The choice of the vectors $\mathbf{d}_h$ (Step 2) can be avoided by modifying the procedure in Fig. 1 and employing a proper algorithm for the training of the neurons in the output layer. The resulting method is reported in Fig. 2.

GENERAL SEQUENTIAL CONSTRUCTIVE METHOD
(Version with output training)

1. For each output $k = 1, \ldots, m$ choose a pertaining class d_k. Set
$$V_{k,+1} = D_{k,+1} \quad , \quad V_{k,-1} = D_{k,-1} \quad , \quad h = 0 \ .$$
2. Train the m threshold neurons in the output layer. Let $U_{k,+1}$ ($U_{k,-1}$) be the set of input patterns in $V_{k,+1}$ ($V_{k,-1}$) that are not correctly classified by the current neural network. Denote with k^+ (k^-) the index of the output associated with the set U_{k,d_k} ($U_{k,-d_k}$) having maximum size.
3. If $U_{k^+,d_{k^+}} \neq \emptyset$ then set $Q^+ = U_{k^+,d_{k^+}}$, $Q^- = D_{k^+,-d_{k^+}}$, otherwise if $U_{k^-,-d_{k^-}} \neq \emptyset$ then set $Q^+ = U_{k^-,-d_{k^-}}$, $Q^- = D_{k^-,d_{k^-}}$, otherwise the construction is complete.
4. Set $h = h + 1$. Train the hth hidden neuron so as to obtain a partial classifier for Q^+ and Q^-. Let R be the subset of Q^+ containing the input patterns for which the hth hidden units provides output $+1$.
5. If $U_{k^+,d_{k^+}} \neq \emptyset$, then set $V_{k^+,d_{k^+}} = V_{k^+,d_{k^+}} \setminus R$, otherwise set $V_{k^-,-d_{k^-}} = V_{k^-,-d_{k^-}} \setminus R$. Go to step 2.

Figure 2: Version with output training of sequential constructive methods.

This method can always construct a neural network that satisfies all the samples contained in a given (finite and consistent) training set S if the learning algorithm for the neurons in the output layer is able to obtain, within a finite execution time, an optimal configuration, i. e. a weight matrix that makes the minimum number of errors on the sets $D_{k,+1}$ and $D_{k,-1}$. A method of this kind is the pocket algorithm with ratchet [8, 9].

This property can be used to show the following result [6].

Theorem 6 *Under the hypotheses of Theorem 5 the version with output training of a sequential constructive method is able to provide, within a finite execution time, a neural network that satisfies all the samples contained in a given training set S.*

The algorithm employed to obtain a partial classifier (Step 4 of Figs. 1-2) changes from one sequential constructive method to another and determines its performances in the solution of classification problems. A comparative evaluation of the different choices introduced in the literature goes beyond the scope of the present work and forms the subject of a complementary paper [10].

References

[1] RUMELHART, D.E., HINTON, G.E., AND WILLIAMS, R.J. Learning representations by back-propagating errors. *Nature*, **323** (1988), 533–536.

[2] MARCHAND, M., GOLEA, M.,AND RUJÁN, P. A convergence theorem for sequential learning in two-layer perceptrons. *Europhysics Letters* **11** (1990), 487–492.

[3] MARCHAND M., AND GOLEA, M. On learning simple neural concepts: From halfspace intersections to neural decision lists. *Network* **4** (1993), 67-85.

[4] MUSELLI, M. On sequential construction of binary neural networks. *IEEE Transactions on Neural Networks* **6** (1995), 678–690.

[5] CAMPBELL C., AND VICENTE, C.P. The target switch algorithm: A constructive learning procedure for feed-forward neural networks. *Neural Computation* **7** (1995), 1245–1264.

[6] MUSELLI, M. Sequential constructive techniques. In *Neural Network Systems Techniques and Applications, vol. 2 - Optimization Techniques.* Ed. C. Leondes. (1998) New York: Academic Press, 81–144.

[7] RIVEST, R.L. Learning decision lists. *Machine Learning* **2** (1987), 229–246.

[8] GALLANT, S. I. Perceptron-based learning algorithms. *IEEE Transactions on Neural Networks* **1** (1990), 179–191.

[9] MUSELLI, M. On convergence properties of pocket algorithm. *IEEE Transactions on Neural Networks* **8** (1997), 623–629.

[10] MUSELLI, M. A comparative evaluation of sequential constructive methods. Contribution to *WIRN'98 - The 10-th Italian Workshop on Neural Nets.*

SOMLib: A Distributed Digital Library System based on Self-Organizing Maps

Andreas Rauber
Institut für Softwaretechnik, Technische Universität Wien
Resselgasse 3/188, A-1040 Wien, Austria
e-mail: rauber@ifs.tuwien.ac.at

Abstract

We describe an architecture for a distributed digital library system based on an unsupervised neural network model, namely the Self-Organizing Map. The system allows the clustering of text documents forming the basis for intelligent information retrieval. User profiles can be combined with full text queries or sample texts to locate documents within the library system. Contrary to conventional approaches consisting of one central library, the distributed SOMLib allows the combination and integration of several independent SOM-based library systems in a not necessarily hierarchical order. Furthermore, the system is not limited to a specific type or setup of the basic SOM architecture. Rather, variations of the basic architecture which might serve special needs of some users can be combined.

Keywords: Digital Libraries, Information Retrieval, Self-Organizing Map (SOM)

1 Introduction

The last years have seen an uninterrupted rise of the amount of information available in electronic form, with both the number as well as the size of such information collections continuing to grow. Similar to huge conventional libraries, which are interconnected by some organizational network, the combination of several independently managed information repositories must be facilitated, allowing to access and query a collection via one single interface. This requirement arises not only from the need for convenient interaction, but also from the very nature of the information repositories currently in use: Apart from the desire to combine specialized libraries, we are faced with the situation of publishers offering their annual publications in electronic form, be it online or on CD, which users want to combine into one collection. Finally, the mere size of such information repositories calls for their separation into smaller parts which are then again integrated into one single system, in order to keep them in manageable orders of magnitude. Most systems in use today rely on central databases to retrieve text documents based on keyword queries. However, with the ever increasing amount of information available, these systems face some severe limitations due to computational demands, since all queries have to be dealt with by a central search engine or its mirrors.

On the other hand research in neural networks has produced promising results in the the field of full text archives. A number of projects [1, 6] especially show the capabilities of one specific type of unsupervised neural network for dealing with the the problems of text document clustering, namely the Self-Organizing Map.[2] Apart from its applicability for query processing, this network architecture provides the advantage of intuitive visualization of the structure of the information repository [4], allowing the implementation of intuitive user interfaces for browsing collections. One of the most prominent examples in this application arena is the WEBSOM[1], providing access to over 1 million articles from 83 Usenet Newsgroups. However, the mere size of the map and the huge computational demands pose limitations from the technical point of view. Recent studies [3] successfully try to reduce the size of the maps and thus the computational demands by using a modified architecture, e.g. the Hierarchical SOM [5], consisting of a hierarchy of (small) maps being trained consecutively layer for layer top down. However, all training data must be processed centrally.

In this paper we propose a model of a distributed SOM-based library system (SOMLib), which combines both local text documents as well as remote library systems, which use a conforming feature vector representation. The system is not limited to a specific architecture of SOM nor does it need central parsing of all documents to create the training vector structure. The overall goal is the integration of several independently managed libraries using the learning capabilities of Neural Networks. Analogously, each user can create both a personal library built up of other library systems as well as create a user-profile covering her fields of interest to enhance keyword queries.

The rest of the paper is built up as follows: Section 2 describes the principles of our Neural Network approach for such a library system. Section 3 details the architecture of the system with a model setup and application scenario detailed in Section 4. Finally we present a short summary in Section 5.

2 A Neural Network Approach

The self-organizing map is an unsupervised neural network providing a mapping from a high-dimensional input space to a usually two-dimensional output space while preserving topological relations as faithfully as possible. Input signals $x \in \Re^n$ are presented to the map, consisting of a grid of units with n-dimensional weight vectors, in random order. An activation function based on some metric (e.g. the Euclidean Distance) is used to determine the winning unit (the 'winner'). In the next step the weight vector of the winner as well as the weight vectors of the neighboring units are modified following some learning rate in order to represent the presented input signal more closely.

Basically, the entries to be included in the library system are represented in the form of feature vectors, which are created by parsing the texts and processing the resulting word histograms to provide a compact and effective rep-

[1] http://nodulus.hut.fi

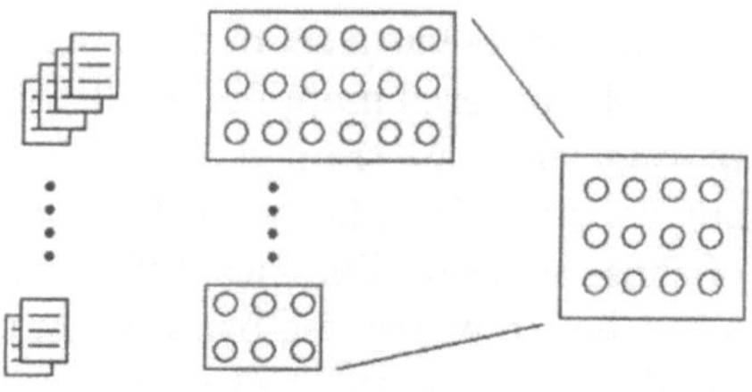

Figure 1: SOMLib-Architecture: Integration of several lower order SOMs

resentation of the texts, using a term frequency / inverse document frequency ($tf \times idf$) representation. These feature vectors are used as input vectors to train a standard self-organizing map, whereby the very architectures used for representing specific libraries do not necessarily have to be identical. Various modifications of the basic architecture, especially hierarchical SOMs, can be used to suit the differing needs of users concerning size, training time and map representation. This approach has proven successful in a number of applications on free form documents with much attention being paid to the design of intuitive user interfaces allowing easy browsing of large collections of documents and simple retrieval of similar pieces of texts. However, so far these systems require all processing, be it feature vector extraction or map training, to take place centrally. When building larger libraries including the documents of several independent repositories (e.g. university libraries, electronic archives of journals) limits are reached, be it the availability of all texts for training as well as the necessary computational power. Thus we need a method to combine different libraries which are based on differing feature vectors made up of the vocabulary of the documents used for training those individual libraries. By combining the feature vectors, the resulting new vectors will grow to form a kind of more universal alphabet as the number of included libraries increases. However, to keep training times low, we use the weight vectors of the SOMLib library maps as input vectors to the higher-order library instead of using all original vectors, resulting in a hierarchy of maps. Still it is not a strict hierarchy in the sense used in the hierarchical SOM model, where each node in a higher level map represents exactly one map at a lower level of the hierarchy. Rather, we train higher level maps bottom-up, obtaining a topographic ordering of all the weight vectors used in the various lower order maps. Thus one node in the higher order map can reference several nodes in different lower order maps. Furthermore, the maps need not necessarily be arranged in a hierarchical order, allowing mutual referencing.

3 SOMLib - The Architecture

Basically, there are two different types of SOMs in the SOMLib architecture. (Figure 1) First, there is a set of independent, small SOMs, referred to as *first order maps*, which are trained with the feature vectors obtained by parsing the documents. Thus every node represents a set of documents, with the whole map representing a topographically ordered mapping of all documents in the library. In a second step, *higher order maps* are trained using the weight vectors of those first order maps as input vectors. Note, however, that the vocabulary and thus the vector structures of those separate libraries differ from each other. Thus, a unique feature vector setup has to be created based upon the different vectors of the libraries to be included by merging the vector structures to train the higher-order map. The resulting map is conceptually identical to the various library maps it is based upon, with the nodes now representing a set of other nodes from the various lower order maps. Analogously, small SOMs trained with relevant documents can be used as user profiles to enhance keyword queries.

For any higher order map two different types of referencing must be considered: with the *static referencing scheme* each node passes on the documents it represents to any higher order map that uses its weight vector for training. In this case every node in any map directly represents a set of documents. With the *node referencing scheme*, higher order maps contain references to the nodes of the maps they are built upon, in which case a node represents a certain topic-area of any lower order map. Static referencing allows direct access to the represented documents while avoiding problems that arise if any of the lower order maps is retrained, in which case the assignment of files to nodes changes. With the node-referencing scheme this requires either the 'old' map to remain available for referencing higher order maps or a re-routing of the corresponding referenced nodes by providing appropriate 'substitute nodes' has to be implemented. This can easily be achieved by mapping all nodes of the 'old' map onto the new, retrained map. An advantage provided by the node-referencing scheme is the fact, that it allows the retrieval of documents that were added to the lower order library document collection after the higher order map was trained. Such documents are not automatically added to the list of referenced documents in the static referencing scheme, but require an update process instead. A combined referencing scheme might provide all the benefits while avoiding the shortcomings: static referencing allows a fast browsing of a high order library map and - with appropriate information given by the static referencing scheme - fast access to the desired up-to-time information on the corresponding lowest order library map in an interactive drill down manner.

4 SOMLib at Work

From the user's point of view, SOMLib facilitates both intuitive library browsing interfaces as well as query processing tools. The topographic mapping provided by a SOM locates similar documents close to each other on neighboring nodes. Thus, SOMLib proves to be a good analogy to what we are used to find in conventional libraries, where books are sorted by topical areas. Furthermore, the implicit hierarchy - which in fact is rather a web of sub-libraries interconnected by reference - allows quick and intuitive access to relevant information by choosing relevant sub-libraries when manually selecting library section after library section. To use a conventional analogy, a user interested in neural networks will start from a high-level map, where he finds (by query or simply by looking at the labels of the nodes) a node covering that topic and several related ones in the close surroundings. Opening this node (section) he either finds a list of documents covering that topic, which are available at different libraries (i.e. static referencing) or a list of sections of other libraries containing documents on that topic. These links can be followed to finally reach the very sections of first order libraries, where access to the desired documents is provided (node referencing).

A query to the library system can consists of a sample text which is parsed using the vector structure of the library system. The resulting query vector is then presented to the library to calculate the winning node. In a first order SOM, the documents mapped onto the winning node are returned as query results. The same applies for higher order maps implementing only static referencing. If both referencing schemes are implemented, in higher order maps the statically referenced documents are returned as first hit query results together with the next order referenced nodes from the node-referencing scheme, i.e. the nodes of the lower order maps that are mapped onto the winning node. In an interactive search the user can now browse through the library 'hierarchy' either manually or by passing the query vector on to desired lower order maps with appropriate conversion of the query vector to match the lower order libraries' vector structure. For automatic retrieval, all references are followed to retrieve documents that are mapped on winning nodes in the lowest order SOM.

From the library administrator's point of view there are two different situations to be considered. On the one hand there are the first level SOMs which are trained with the feature vectors created by parsing (a subset of) existing documents. The resulting maps are relatively small since they only need to represent the very documents present in the library. New documents can be added to the map by parsing them using the previously extracted vector structure and mapping the resulting feature vectors. As long as the general scope of the library does not change extensively, new documents can be added without destroying the topology preserving mapping. As new topics emerge, the small first level libraries need to be retrained. The 'old' SOMLib map can either be retained to serve other referencing higher order maps, or the nodes of the old map can be mapped onto the corresponding nodes in the new SOMLib map by determining the winning node on presentation of the (modified to match the new vector structure) weight vectors of the old map's nodes. If a first order map tends to grow too big, one can choose to split the underlying documents into groups to create separate first order SOMLibs, which are then combined in a higher order map.

Secondly, there are higher order maps to be administered. These are based on several lower order maps, the structure vectors of which are merged to create a new vector. The modified weight vectors of the lower order maps are then used to create the higher level SOMLib map. In many cases a natural hierarchy will evolve in institutional arenas, say several university departments will have their own SOMLibs as first order maps, which are then integrated in a single second order map at university level, which in turn may be combined at a national level and so on. Others may choose to combine first or higher order SOMLibs of institutions covering a certain topic of interest, with the possibility for mutual referencing, to create their personal library system.

5 Conclusion

We have presented a model of a new library organization tool called SOMLib, which is based on a prominent unsupervised neural network architecture, the self-organizing map (SOM). The applicability of SOM and modified architectures to the problem of library organization has been evaluated and shown by many projects so far. SOMLib goes one step further by integrating several independent SOM-based libraries to form higher order library maps. This allows the maintenance of the very libraries locally where training requirements can be met since the maps are relatively small and only a comparably small amount of documents has to be considered. Higher order maps only use the trained lower order maps as basis for their training data to keep both map size and training time low, creating a kind of 'super library'. These higher order libraries do not necessarily need to form a strict hierarchy but can include any other SOMLib system, forming a network of library maps. Users can browse these libraries utilizing the intuitive user interfaces facilitated by the topology preserving mapping of SOMs. Furthermore, documents can be retrieved based on queries, which can either be presented as sample texts, or as keyword queries which in turn can be enhanced by merging the query with a user profile SOM. Due to its modular structure, management of the various libraries is greatly simplified, since they can be independently enlarged and retrained. Finally the system does not pose limitations on architecture variation, so local libraries can be built to best suit the needs of the individual users.

References

[1] T. Honkela, S. Kaski, K. Lagus, and T. Kohonen. Newsgroup exploration with WEBSOM method and browsing interface. Report A32, Helsinki Univ. of Technology, Lab. of Computer and Information Science, Otaniemi, Finland, 1996.

[2] T. Kohonen. *Self-Organizing Maps.* Springer Verlag, Berlin, Germany, 1995.

[3] D. Merkl. Exploration of text collections with hierarchical feature maps. In *Proc. of the 20th Int'l ACM SIGIR Conference on Research and Development in Information Retrieval*, pages 186 – 195, Philadelphia. PA, 1997.

[4] D. Merkl and A. Rauber. Alternative ways for cluster visualization in self-organizing maps. In *Proc. of the Workshop on Self-Organizing Maps (WSOM97)*, Helsinki, Finland, 1997.

[5] R. Miikkulainen. Script recognition with hierarchical feature maps. *Connection Science*, 2:83 – 101, 1990.

[6] A. Rauber and D. Merkl. Finding structure in text archives. In *Proc. European Symp. on Artificial Neural Networks (ESANN98)*, Bruges, Belgium, 1998.

Incremental Min-Max Network. Part 1: Continuous Spaces

A. Rizzi, M. Biancavilla, & F.M. Frattale Mascioli
INFO-COM Dpt., University of Rome "La Sapienza"
Via Eudossiana 18, 00184 Rome, Italy
mascioli@infocom.ing.uniroma1.it

Abstract

A very important and active research topic are Fuzzy Min-Max neural networks. In the present paper an incremental Min-Max based learning algorithm is discussed in comparison with traditional and optimized ones, with the aim of obtaining a better decision region in the input space and an optimal structural complexity. Experiments show promising results.

1 Introduction

The Min-Max neural network proposed by Simpson [1,2] is the basis of a family of fuzzy neural nets particularly efficient in solving classification and clustering problems. This efficiency is clearly illustrated in [3], where a comparison is carried out among 10 classical AI algorithms, 3 statistical routines and 10 neural based algorithms. The version of Min-Max proposed in [3] aims to handle mutually non-exclusive output classes. Several other versions can be developed paying attention to specific inconveniences and performances. Some of the present authors proposed in [4] a version characterized by the application of Learning Theory paradigms to improve the generalization capability of the model. Further inconveniences still remain to be removed in order to improve the classification performances of the resulting Min-Max version. One of them, i. e. the fixed covering resolution of the input space, is particularly dangerous and merits to be removed in order to improve the classification performance.

The version of Min-Max proposed in the present paper aims to overcome such problem. It will be described in Sect. 3. In order to understand what is proposed, Sect. 2 will present a short review of original Min-Max algorithm and of the version described in [4]. The paper will be concluded by simulation tests reported for illustrating the effectiveness of the proposed Min-Max version, called Incremental Min Max net (IMM net).

2 The Min-Max classifier

In this section we shortly review the original Min-Max algorithm for classification [1] and its optimized version with respect to the generalization capability, proposed in [4]. Min-Max technique consists in partitioning the input space in hyperboxes with the sides parallel to the coordinate axes and with appropriate membership function controlled by a parameter γ, wich defines its slope outside the hyperbox. Patterns are first normalized and thus they are contained in the unitary hypercube of the input space. The location of each hyperbox is completely defined by two extreme vertices: the 'Min' and the 'Max' vertices. The algorithm is characterized by a three-step process:

- Expansion of hyperboxes, to include a pattern of the training set;
- Overlap test, to determine overlaps among hyperboxes associated with different classes;
- Contraction, to eliminate overlaps.

Expansion process is controlled by a parameter $\theta \in (0,1]$ that establishes the maximum size allowed to each hyperbox.

The modified version [4] optimizes the generalization capability of the net by applying a simple formulation of the learning theory. According to this theory, the best architecture is obtained by minimizing an objective function coincident with the sum of two terms, respectively measuring the accuracy and the complexity of the net, i.e.:

$$F=(1-\lambda)E+\lambda C \qquad (1)$$

where: E is the percentage of misclassification on the training set; C is the percentage of hyperboxes with respect to the cardinality of the training set and $\lambda \in [0,1]$ is a weight to be determined taking into account that small values of λ yield nets more complex but less subject to errors (the opposite situation is characterized by large values of λ).

The choice of λ value depends on the application at hand; however, λ is much less critical than θ (λ=0.5 can be chosen as the default compromise value).

3 The proposed classifier

A careful investigation of the mechanism underlying the Min-Max algorithm points out that it operates with a fixed resolution in the input space. This lack of flexibility cannot be balanced by the choice of a suitable value θ. The only effect of θ is to modify the complexity of the net without remedying the said dishomogeneity. Moreover, the hyperboxes (HBs in the following) obtained by training procedure strongly depend on the presentation order of training examples.

In order to remove the previous dishomogeneity, it is necessary to match the sizes of the HBs to the clusters of points representing the training prototypes in the input space. In this manner, it is possible to describe details of the decision region boundaries by small HBs and, at the same time, to use larger HBs to cover input regions far away from them.

The learning process is iterative and based on the constructive approach of learning theory. After each iteration we determine the value of the objective function (1). The resulting classifier will be denoted as Incremental Min-Max Network (IMM), by referring to the incremental resolution gained during the process.

The iteration is characterized by four steps:

1. Expansion;
2. Overlap test;
3. Contraction;
4. Merging.

Let's examine them in detail:

1) During the first step, each point of the set to be clusterized is presented to the system:

- It does not produce any changes if it belongs to a corresponding class HB already created;
- It causes the expansion of a HB of its own class if it does not belong to any cluster of the corresponding class already created and the expanded HB does not surpass the limit θ;
- It creates a new HB if it does not belong to any cluster of the corresponding class already created and any HB of the corresponding class cannot be expanded without surpassing the limit θ.

When the overall set has passed through the net, the resulting HBs could overlap each other.

2) In the second step, we determine the overlaps among HBs obtained in Step 1. Two HBs of different classes are ovelapped if each their projection onto the coordinate axes produces two superimposed segments. The HBs which result to be overlapped are assigned to a list denoted as *mixed.* They contain patterns of two or more different classes. The remaining HBs, which contain patterns of a unique class, are assigned to a list denoted as *pure.*

3) The third step aims to cover the input space with pure HBs. For this purpose, we consider each couple of HBs of the mixed list. If the two considered HBs overlap, we reduce them by cutting the corresponding overlap region. The latter is assigned to a new list: the *overlap* list. At the end of the previous procedure the mixed list will be empty. In fact, the reduced HBs contain only patterns of a same class and consequently they are assigned to the pure list. The overlap list, so obtained, will be handled in the next iteration, as explained in the following.

4) During the merging step, the clusters in the pure list are divided taking into account of the corresponding output classes. Each HB of the list is merged with the others of the same class, if this merging does not involve any overlap with clusters belonging to other classes. This process permits to reduce at a minimum the complexity of the network allowing wide HBs in larger area of the input space and keeping small the HBs in narrower areas. In this way, the system is able to reproduce much more faithfully the distribution of the points in the input space, describing details using smaller clusters and backgrounds using bigger ones.

After the last step, a new iteration starts considering recursively each overlap region as a new input space. It is interesting to note that the learning process carries out a progressive zoom, reducing the portion of the input space to be examined after each iteration. This characteristic is the main reason for the name IMM of the algorithm.

The learning session comes to the end when no overlaps are obtained after the contraction phase of the current iteration. According to learning theory, after each iteration, we determine the value of the objective function F defined in (1). F is a function of the preset value θ and of the iteration I:

$$F=F(\theta,I) \tag{2}$$

The optimal configuration C_{OPT} of the classifier is that which minimizes F with respect to θ and I, i. e.:

$$C_{OPT}=\min_{\theta}(\ \min_{I}(F(\theta,I))) \tag{3}$$

The proposed algorithm achieves a good flexibility with a good capability of adjusting the HBs on the basis of the actual location of input prototypes. This capability is paid by the necessity of following an iterative process in place of a unique pass of the training set required by the original Min-Max learning algorithm. However, this increased computational cost does not seem a serious obstacle, since it is always not very large.

3.1 Saturated version of IMM classifier

In the introduction, we remarked the existence of inconveniences affecting Min-Max algorithm. The version proposed above aims to eliminating one of the most dangerous. There is however another inconvenience: the excessive dependence on the presentation order of the prototypes. The effect of this presentation is controlled by the parameter θ. In particular, if θ takes on its maximum value, i.e. it is saturated, at the end of the first iteration, there will be as many HBs as output classes (one HB per class). Consequently, each point of the training set expands the corresponding class cluster. Given a set of points in the input space belonging to the same class, it univocally remains determined the minimum HB containing them. This determination is obtained independently of the presentation order (it is a direct consequence of the commutative property of min and max operators).

We note that the new version of the IMM net is less flexible, because function F depends only on iteration number (θ is definitely set to 1). In some cases, the resulting HBs take on forms too narrow and long with a loss of classification performance. However, it is much simpler than the previous version. For this reason it will be preferred in the case of hybrid input data classification, as it will be discussed in [5].

A critical situation which can arise with this saturated version of IMM net happens when two HBs of different classes coincide. The previous IMM algorithm is unable to manage this case. A simple method for overcoming it, adopted in [5], consists in executing a cut of the HB in the middle of its longest side. Then, the two halves are added to the overlap list and the coincident HBs are eliminated.

4 Simulation tests

The effectiveness of the IMM classifier was tested on well known benchmarks. Its performance depends on the membership function used. We investigated this influence in depth. The best results were obtained with the following one:

$$\mu(\mathbf{x}) = \prod_{i=1}^{N} \mathbf{Max}\left(0, \mathbf{Min}\left(1, 1+\gamma \cdot \mathbf{Min}\left(\mathbf{x}_i - \mathbf{v}_i, \mathbf{w}_i - \mathbf{x}_i\right)\right)\right) \tag{4}$$

where: X is the input pattern; x_i is the i-th input component (i = 1,..., N); v_i is the i-th component of the Min vertex of the HB under consideration; w_i is the corresponding component of the Max vertex. The previous μ function is a modification of the Simpson's original one, introduced for eliminating its bias. The value of γ is chosen with the aim of covering a large part of the hypercube. A reasonable value is obtained by imposing:

$$\prod_{i=1}^{N} (1-\gamma)^N = 0.1 \tag{5}$$

The resulting value of γ can be used as a default.

Iris benchmark

The IMM net performance in correspondence to the Iris problem is illustrated in Tab. 1 by the confusion matrix of the 3 output classes. This table must be compared with that related to Simpson's original Min-Max (Tab. 2). The better performance of IMM version is evident, taking also into account the reduced structural complexity (8 versus 14 Hbs).

Classes	Class 1	Class2	Class 3
Class 1	50	0	0
Class 2	0	49	1
Class 3	0	1	49

Tab. 1: confusion matrix of IMM for the iris data set (8 HBs).

Classes	Class 1	Class 2	Class3
Class 1	49	1	0
Class 2	0	48	2
Class 3	0	11	39

Tab. 2: confusion matrix of Simpson's Min-Max (θ= 0,1) for iris data set (14 HBs).

Cancer, Diabetes and Vowels benchmarks

The performance of the IMM net is described in Tab. 3. The classification results, compared with those reported in [6], prove the effectiveness of the IMM net.

Problems	Error in the test set (%)	Number of rules
Cancer	1.5	28
Diabetes	24	65
Vowels	47	59

Tab. 3: performances of IMM in some benchmark problems.

Spirals

The data sets for training and test were chosen as in [8]. In the present case we have used for training the examples of both the training and validation sets, since we apply the minimization of (1). In Tab. 4 there are reported the best results obtained with the IMM net and with other approaches. The latter are quoted from [8]. The accuracy performance of IMM is similiar to that obtained by the Min-Max version proposed in [8], which is better than those obtained with other methods. Some intersting comparisons can be achieved taking into account for the results published in [3].

However, it is already possible to say that IMM is a very promising system, mainly considering the scene of modified Min-Max algorithms, for its simple way of functioning, its remarkable flexibility and its contained computational cost. Work in progress regards the possibility of using structures different from HBs to cluster the data space. In particular, a PCAs based technique could be used to form generalized HBs, avoiding the limitation of having sides parallel to the coordinate axes.

Model of the neural net	Complexity (number of neurons)	Classification error on test set (%)
IMM	135	3.8
Original Min-Max [1]	136	7.9
Modified Min-Max [8]	133	3.7
Fuzzy Artmap [9]	169	7.2
FBF [10]	108	15.3
AOFLS [11]	144	8.1

Tab. 4: Spiral data set results

References

[1] P.K. Simpson. Fuzzy min-max neural networks – Part 1: Classification. *IEEE Trans. on Neural Networks*, Vol. 3, No. 5, pp. 776-786, 1992.

[2] P.K. Simpson. Fuzzy min-max neural networks – Part 2: Clustering. *IEEE Trans. on Fuzzy System*, Vol. 1, No. 1, pp. 32-45, 1993.

[3] A. Joshi, N. Ramakrishman, E.N. Houstis, J.R. Rice. On neurobiological, neuro-fuzzy, machine learning, and statistical pattern recognition techniques. *IEEE Trans. on Neural Networks*, Vol. 8, No. 1, pp. 18-31, Jannuary 1997.

[4] F.M. Frattale Mascioli, G. Martinelli. A constructive approach to neuro-fuzzy networks. *Signal Processing*, Vol. 64, No. 3, pp. 347-358, 1998.

[5] F.M. Frattale Mascioli, M. Biancavilla, and A. Rizzi. Incremental Min-Max Network - Part 2: Continuous an Discrete Spaces. WIRN Vietri 1998.

[6] L. Prechelt. Proben1 - A Set of Neural Network Benchmark Problems and Benchmarking Rules. Technical Report 21/94, September 30, 1994.

[7] M. Prakash, M. N. Murty. Growing subspace pattern recognition methods and their neural-network models. *IEEE Trans. on Neural Networks*, Vol. 8, pp. 161-168, Sept. 1997.

[8] M. Menganti, F. Saviello, R. Tagliaferri. Fuzzy neural networks for classification and detection of anomalies. *IEEE Trans. on Neural Networks*, special issue on "Neural Networks and Hybrid Intelligence Models: Foundations, Theory and Applications". In Press.

[9] G. A. Carpenter, S. Grossberg, N. Markuzon, J. Reynolds, D. B. Rosen. Fuzzy artmap: a neural network architecture for incremental supervised learning of analog multidimensional maps. *IEEE Trans. on Neural Networks*, Vol. 3, pp. 698-713, 1992.

[10] H. M. Kim, J. M. Mendel. Fuzzy basis functions: comparison with other basis functions. *IEEE Trans. on Fuzzy Systems*, Vol. 3, No. 2, pp. 158-168, 1995.

[11] L. X. Wang. Adaptive fuzzy systems and control. Englewood Cliffs, NJ, Prentice-Hall, 1993.

Tangent Models: Comparing the TD-Neuron Constructive Algorithm versus SVD Based Algorithms

Diego Sona Alessandro Sperduti Antonina Starita
Dipartimento di Informatica, Università di Pisa
Corso Italia, 40, 56125, Pisa, Italy

Abstract

In this paper we present comparative results of our constructive algorithm versus the HSS. Specifically, we tested the HSS algorithm using both the original version based on the two-sided tangent distance, and a new version based on the one-sided tangent distance. Empirical results over the NIST-3 database show that the TD-Neuron is superior to SVD based algorithms since it reaches a better trade-off between error and rejection.

1 Introduction

In several pattern recognition systems the principal and most desired feature is the robustness against transformations of patterns. Different approaches to the problem aiming at the reduction of the classification time and space requirements, using tangent distance theory [SLD93], were studied by some authors.

Specifically, Hastie et al. [HSS95] developed rich models for representing large subsets of the prototypes through a Singular Value Decomposition (SVD) based algorithm, while Schwenk & Milgram [SM95b, SM95a] proposed a modular classification system (*Diabolo*) based on several auto-associative multi-layer perceptrons, which use tangent distance as the error reconstruction measure. Finally, Sona et al. [SSS97] devised a constructive algorithm based on the definition of a neuron (TD-Neuron) where the input net is computed by using the one-sided tangent distance instead of the standard dot product.

In this paper we compare the performance of Sona et al. [SSS97] constructive algorithm versus the HSS algorithm. In particular, we tested the HSS algorithm using both the original version, and a new version based on the one-sided tangent distance. The one-sided version of the HSS algorithm was derived in order to have a fair comparison against the TD-Neuron, which exploits the one-sided tangent distance. Empirical results over the NIST-3 database of handwritten digits show that the TD-Neuron is superior to both SVD based algorithms since it reaches a better trade-off between error and rejection.

2 Tangent Distance Overview

2.1 HSS models

Given a set of patterns $\{\boldsymbol{X}_1, \ldots, \boldsymbol{X}_N\}$, all of the same class, Hastie et al. [HSS95] proposed the *tangent subspace model*, $\boldsymbol{W}(\boldsymbol{\theta}) = \boldsymbol{W} + \sum_{i=1}^{n} \boldsymbol{T}_i \theta_i$, where $\boldsymbol{W}$ is the *centroid* and the $\boldsymbol{T}_i$ constitute the associated invariant subspace. The model is computed minimizing over $\boldsymbol{W}$ and $\boldsymbol{T}_i$ the error function

$$\sum_{i=1}^{N} \min_{\boldsymbol{\theta}_i\ \boldsymbol{\alpha}_i} ||\boldsymbol{W}(\boldsymbol{\theta}_i) - \boldsymbol{X}_i(\boldsymbol{\alpha}_i)||^2. \tag{1}$$

The above problem can be solved for a fixed subspace dimension by an iterative algorithm based on Singular Value Decomposition, proposed by Hastie et al. [HSS95]. The problem can also be formulated using the one-sided tangent distance, in this way the equation (1) becomes

$$\sum_{i=1}^{N} \min_{\boldsymbol{\theta}_i} ||\boldsymbol{W}(\boldsymbol{\theta}_i) - \boldsymbol{X}_i||^2, \tag{2}$$

which is a way of representing the principal component analysis of the $\boldsymbol{X}_i$, also called *Karhunen-Loéve Expansion*.

2.2 TD-Neuron

The TD-Neuron is so called since it can be considered as a neural computational unit which receives an input vector $\boldsymbol{X}_k$, and computes (as input net) the square of the one-sided tangent distance of the input vector with respect to a set of internal parameters (weights). This set of parameters are organized in such a way to form a tangent model. Formally, we have

$$net_k = \min_{\boldsymbol{\theta}} ||\boldsymbol{W}(\boldsymbol{\theta}) - \boldsymbol{X}_k||^2 + \beta, \tag{3}$$

where β is the offset. Under the condition that the tangent vectors constitute an orthonormal basis, equation (3) can exactly and easily be computed by using the projections of the input vector over the model subspace (see Figure 1)

$$net_k = ||\underbrace{\boldsymbol{X}_k - \boldsymbol{W}}_{\boldsymbol{d}_k}||^2 - \sum_{i=1}^{n}[(\boldsymbol{X}_k - \boldsymbol{W})^t\, \mathbf{T}_i]^2 + \beta = \boldsymbol{d}_k^t \boldsymbol{d}_k - \sum_{i=1}^{n}[\underbrace{\boldsymbol{d}_k^t\, \boldsymbol{T}_i}_{\gamma_{ik}}]^2 + \beta. \tag{4}$$

The output of the TD-Neuron is then computed by transforming the *net* through a nonlinear monotone function f. In our experiments, we have used the symmetric sigmoidal function $o_k = \frac{2}{1+e^{net_k}} - 1$.

2.2.1 Training the TD-Neuron

Given a training set $\{(\boldsymbol{X}_1, t_1), \ldots, (\boldsymbol{X}_N, t_N)\}$, where $t_i \in \{-1, 1\}$ is the i-th desired output, and N is the total number of patterns in the training set, an

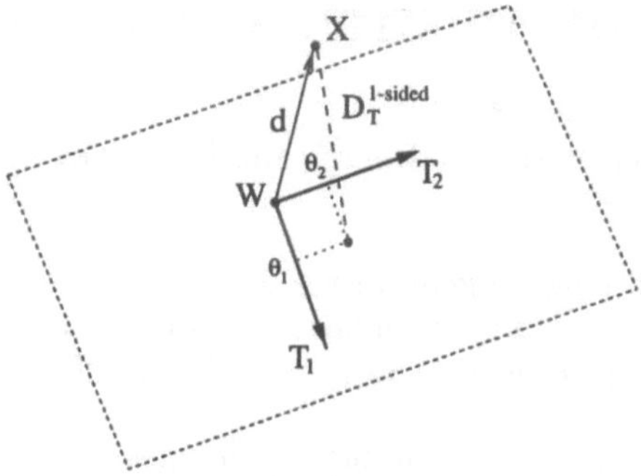

Figure 1: Geometric interpretation of equation (4). Note that $net = (D_T^{1-sided})^2$.

error function can be defined as $\mathcal{E} = \frac{1}{2}\sum_{k=1}^{N}(t_k - o_k)^2$, where o_k is the output of the TD-Neuron for the k-th input pattern.

Using equation (4) with the sigmoidal function, it is trivial to compute the changes for the centroid, the tangent vectors, and the offset, by using a gradient descent approach. Before training the TD-Neuron by gradient descent, however, the tangent subspace dimension must be decided. To solve this problem we have developed a constructive algorithm which adds tangent vectors one by one, according to the computational needs. This idea is also justified by the observation that a typical run of gradient descent, over equation (4), leads to the sequential convergence of the tangent vectors according to their relative importance. This behavior suggests starting the training using only the centroid and then adding tangent vectors as needed. Under this learning scheme (see Figure 2), the changes to the parameters can be computed as

$$\Delta \boldsymbol{W} = -\eta \left(\frac{\partial \mathcal{E}}{\partial \boldsymbol{W}}\right) = -2\,\eta \sum_{k=1}^{N} [(t_k - o_k)\, f' \boldsymbol{d}_k] \tag{5}$$

$$\Delta \boldsymbol{T}_i = -\eta \left(\frac{\partial \mathcal{E}}{\partial \boldsymbol{T}_i}\right) = -2\,\eta \sum_{k=1}^{N} [(t_k - o_k)\, f'\, \gamma_{ik}\, \boldsymbol{d}_k] \tag{6}$$

$$\Delta \beta = -\eta_\beta \left(\frac{\partial \mathcal{E}}{\partial \beta}\right) = \eta_\beta \sum_{k=1}^{N} [(t_k - o_k)\, f'\,] \tag{7}$$

where η and η_β are learning parameters.

On the basis of empirical evidence we have concluded that the learning phase of the centroid can considerably be reduced by initializing the centroid with the mean value of the patterns belonging to the positive class. We have also devised a "good" initialization algorithm for tangent vectors (see Figure 3) which tries to minimize the drop in the input net for all the patterns due to the increase in the tangent subspace dimension. This is obtained by introducing a new tangent vector which mainly spans the residual subspace between the patterns in the positive class and the current model. In this way, patterns which are in the negative class will only be mildly affected by the new introduced tangent vector.

CONSTRUCTIVE ALGORITHM

Initialize the centroid $\boldsymbol{W}$
Update β and $\boldsymbol{W}$ with equations 7 and 5 till they converge
Freeze $\boldsymbol{W}$
REPEAT
    Initialize a new tangent vector $\boldsymbol{T}_i$
    Update $\boldsymbol{T}_i$ and β with equations 6 and 7, and orthonormalize
        $\boldsymbol{T}_i$ with respect to $\{\boldsymbol{T}_1, \ldots, \boldsymbol{T}_{i-1}\}$ till it converges
    Freeze $\boldsymbol{T}_i$
UNTIL new $\boldsymbol{T}_i$ gives little accuracy changes

Figure 2: The constructive algorithm for the TD-Neuron.

3 Results

We have tested our constructive algorithm versus the HSS algorithm (with the one-sided and the two-sided tangent distance) using 10704 binary digits taken from the NIST-3 dataset. The binary 128x128 digits were transformed into 64-grey level 16x16 format by a simple local counting procedure. The only one preprocessing transformation performed was the elimination of empty borders.

The training set consisted of 5000 randomly chosen digits, while the remaining digits were used in the test set. For each algorithm, a single tangent model for each class of digit was computed (for the two-sided version of the HSS algorithm we used 6 transformations for each pattern). The classification of the test digits was performed using the label of the closest model for HSS or the highest output of the TD-Neurons. We have performed also a classification using the Nearest Neighbor rule (1-NN) with the Euclidean distance as the classification metric. In Table 1 we have reported the results obtained for different numbers of tangent vectors. From the results it can be noted that while the two-sided HSS algorithm does overfit the data after the 9th tangent, this is not true for the remaining two algorithms. Nevertheless all the models reach a similar performance with the same amount of parameters (the two-sided version of HSS with 9 tangents uses 6 additional tangents for input patterns), which is slightly below the performance attained by the 1-NN classifier using Euclidean distance. However the tangent models have the advantage of being less demanding both in space and response time.

TANGENT VECTOR INITIALIZATION

for each class c of patterns compute the mean value of differences
    between patterns and model: $\boldsymbol{d}_c = \frac{1}{n}\sum_{i=1}^{n}(\boldsymbol{X}_i - \boldsymbol{W}(\boldsymbol{\theta}))$;
orthonormalize the mean value of differences $\boldsymbol{d}_c$ of all negative
    classes ($c \in \{x | x \text{ is a negative class}\}$);
orthonormalize the mean value of the positive class with respect to the
    other classes.

Figure 3: Initialization procedure for the tangent vectors.

# Tang.	**HSS 1-sided** % Cor	**HSS 2-sided** % Cor	**TD-Neuron** % Cor
0	83.19	85.33	92.41
1	88.50	89.59	94.01
2	89.45	91.36	94.65
3	92.13	93.92	95.04
4	93.16	95.25	95.18
5	94.06	95.81	95.46
6	94.36	96.07	95.72
7	94.86	96.20	95.97
8	95.50	96.48	95.93
9	95.85	**96.60**	96.13
10	95.88	96.41	96.23
11	95.71	96.20	96.30
12	95.78	96.39	96.32
13	95.86	96.42	96.44
14	96.13	96.28	96.46
15	**96.42**	96.18	**96.51**
1-NN with Euclidean distance			**96.84**

Table 1: The results obtained with models generated by both versions of HSS, the TD-Neuron, and Euclidean 1-NN algorithms.

Although from Table 1 it seems that the generated tangent models are equivalent, when introducing a rejection criterion (threshold over the difference between the first and the second best outputs) the model generated by the TD-Neuron outperforms the other two models, due to its discriminant capability (see Figure 4). Furthermore, introducing the rejection criterion leads also to the surprising result that the one-sided version of HSS performs better than the two-sided version.

4 Discussion and Conclusion

In this paper we compared the TD-Neuron constructive algorithm versus two different versions of the HSS algorithm. The obtained results over the NIST-3 database of handwritten digits show that the TD-Neuron is superior to the HSS algorithms based on Singular Value Decomposition since it reaches a better trade-off between error and rejection.

It must be pointed out that, during the model generation, for a fixed number of tangent vectors, the HSS algorithm is faster than our, because it needs only a fraction of the training examples (only one class). However, our algorithm is remarkably more efficient when a family of tangent models, with an increasing number of tangent vectors, must be generated.

An additional advantage of TD-Neuron model is that, because the training algorithm is based on a gradient descent technique, several TD-Neurons can be arranged in a feed-forward network, which can be trained by a trivial extension of back-propagation. This may leads to a remarkable increase in the transformation invariant features of the system.

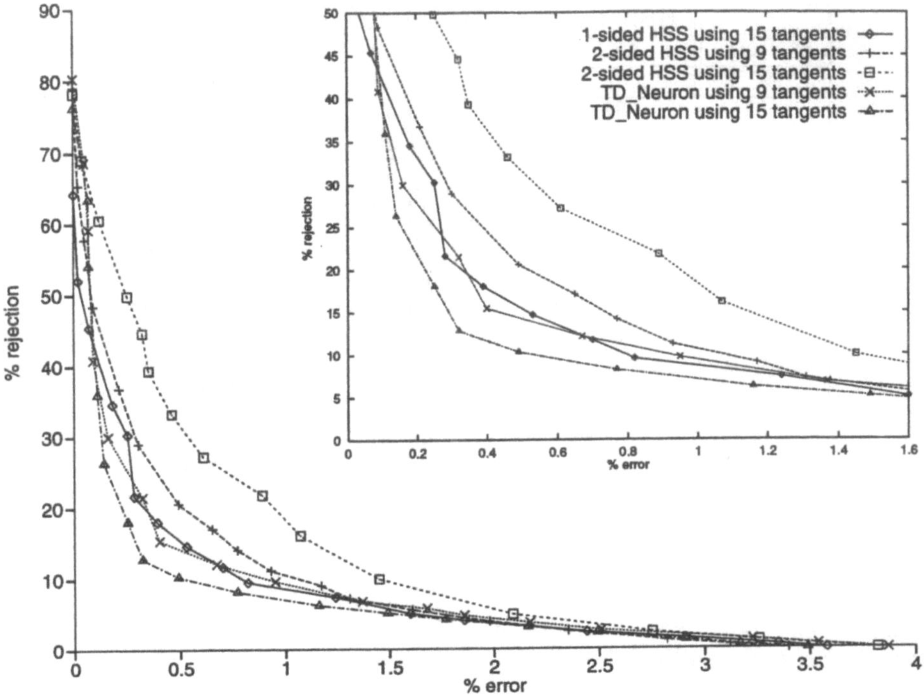

Figure 4: Error-rejection curves for the tangent models.

References

[HSS95] T. Hastie, P.Y. Simard, and E. Säckinger. Learning prototype models for tangent distance. In G. Teasauro, D.S. Touretzky, and T.K. Leen, editors, *NIPS*, volume 7, pages 999–1006, Cambridge MA, 1995. MIT Press.

[SLD93] P.Y. Simard, Y. LeCun, and J. Denker. Efficient pattern recognition using a new transformation distance. In S.J. Hanson, J.D. Cowan, and C.L. Giles, editors, *NIPS*, volume 5, pages 50–58, San Mateo CA, 1993. Morgan Kaufmann.

[SM95a] H. Schwenk and M. Milgram. Learning discriminant tangent models for handwritten character recognition. In *International Conference on Artificial Neural Networks*, pages 985–988. Springer-Verlag, 1995.

[SM95b] H. Schwenk and M. Milgram. Transformation invariant autoassociation with application to handwritten character recognition. In G. Teasauro, D.S. Touretzky, and T.K. Leen, editors, *NIPS*, volume 7, pages 991–998, Cambridge MA, 1995. MIT Press.

[SSS97] D. Sona, A. Sperduti, and A. Starita. A constructive learning algorithm for discriminant tangent models. In M.C. Mozer, M.I. Jordan, and T. Petsche, editors, *NIPS*, volume 9, pages 786–792, Cambridge MA, 1997. MIT Press.

A Neural Multi-Agent Architecture for an Autonomous Mobile Robot Control

S.Vitabile, F.Torterolo, F.Sorbello
Department of Electrical Engineering - University of Palermo
90128 Palermo, Italy

Abstract

The aim of this paper is to introduce a multi-agent architecture for an autonomous robot control oriented to obtain high design modularity and high parallelism of decision processes. A multi-agent approach allows us to design several specialized agents and each of them is optimized in order to solve a single part of the whole task. Each agent is able to cooperate with other agent exchanging information, so the robot is able to make the appropriate policy of management to reach a prefixed goal. We tested the proposed architecture with a simple application: an autonomous mobile robot performs object retrieval tasks into unknown structured environments. The robot is not given a priori map of the environments and it can explore the unknown environments, equipped with a simple b/w camera sensor.

1 Introduction

The main architectural approaches proposed by the literature may be subdivided in two big categories: the *functional* approach and the *behavioral* approach. At the basis of the functional approach[1] there is the idea that the architecture of an autonomous agent may be decomposed in a hierarchy of functional layers. The bottom layers are related to the processing of data flowing from sensors and effectors, whereas the top layers process information coming out from the bottom layers in order to find a strategy to satisfy the goal of the robot. However this approach has shown some limitations in real-world situations, since the world changes in unpredictable manner, and a complex plan may be continuously rebuilt by adapting itself to new situations.
Instead, the behavioral approach is based on the idea that the control architecture of an agents my be well decomposed as a set of behavioral modules arranged in parallel. Each module receives a sensory input and produces a control signal responsible for a specific behavior of the robot. The main claims of this approach is that complex tasks are not programmed in a central control module as in the top layers of the functional approach, but they emerge from the interactions of these simple modules[8].

More recently works in the robot control architecture design tries to mix the functional and the behavioral approach. For example, Brady and Hu [3], proposed a robot architecture where the control module is able to react in a fast way with some low level behavioral modules, but the module has also reasoning capabilities. In Gaglio et al. [2] a layered architecture with more behavioral modules hierarchically guided by a cognitive module is proposed.

Following the last approaches, we propose an hybrid, modular architecture in which each task is an autonomous agent that in a complex system can communicate, can perform actions, and above all, it can acts autonomously [7]. With a modular approach we obtain the typical advantages of the functional approach in complex reasoning tasks, and the typical behavioral advantages as the rapidity of response when the environment changes in a sudden and fast way. A modular approach allows us to design independent, specialized agent in order to solve a particular sub-process of the whole task. So we can exploit the peculiarity of different expert system (as neural networks, bayesian networks, genetic algorithms) in order to optimize each sub-process. If we develop a common agents language, the designed modules can become a part of an architecture that is able to manage and synchronise each process, obtaining both high design modularity and a high parallelism of decision processes.

We have tested the proposed approach implementing the multi-level mental process of an autonomous robot that performs object retrieval tasks into unknown structured environments. The robot has only a background knowledge of the adopted environment geometrical rules and it can explore the unknown environments, equipped with a simple b/w camera sensor. In section 2 we describe both the proposed multi-agent architecture and its modules, in the section 3 we describe the implemented experimental task, the control architecture configured for the proposed task and the obtained experimental results. Finally section 4 reports conclusions and future developments of our work.

2 An Hybrid Multi-Agents Architecture

We adopt a hybrid, modular multi-agent architecture with both reactive and deliberative capabilities to design the robot control system. In fact, we need a reactive behavior for fast responses to sudden and unpredictable environment evolutions, whereas cognitive characteristics are also given to the robot in order to perform intelligent and complex reasoning processes. So the transduction problem (i.e. how translate external real world into an accurate, adequate symbolic description) and the representation/reasoning problem (i.e. how to symbolically represent information about complex real-world entities and processes and how to get agents to reason with this information) can be investigated and dealt with independent and specialized agents, taking into account both environment description and representation accuracy and real time constrains.

Our architecture is constituted by a set of agents that may be grouped into four categories (see Fig.1). The first agent categories, called sensor agents (SA), is composed by agents that translate the external world perceptions into accurate and adequate symbolic environment description just in time for that description to be useful. A second agent class, called navigation agents(NA), is a groups of agents performing robot effectors and robot sensors control skills. A third group, called decision agents (DA), is composed by cognitive agents that

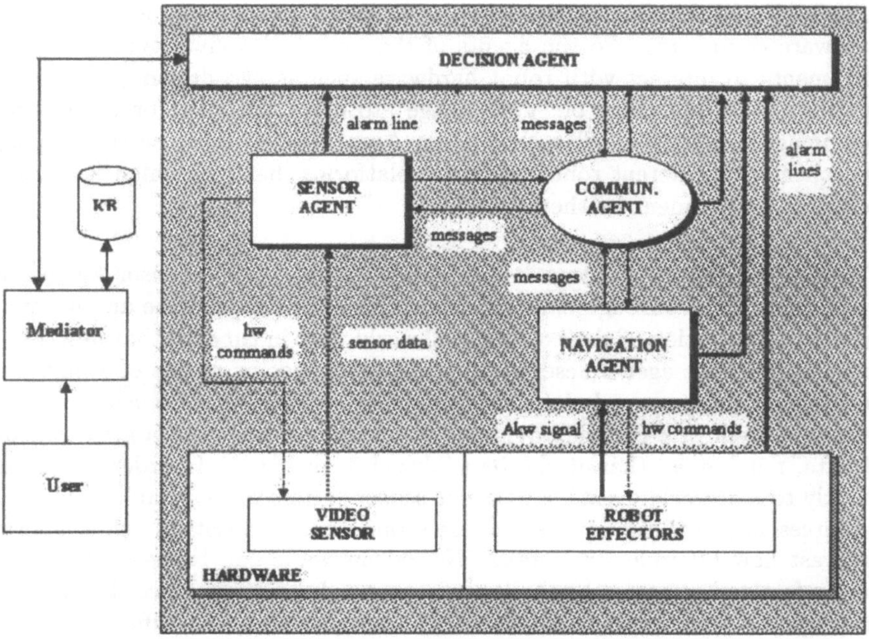

Figure 1: A survey of the multi-agent architecture.

cooperate, on the basis of SA and NA supplied information, in planning and reasoning actions in order to achieve a desired goal. Messages between agents are routed by the Communicator Agent that can be seen as the messages dispatcher of the architecture. In our approach we have a Mediator Module whose purpose is to interact with both the user and the multi-agent architecture. According to the specified mission and through the access to the own knowledge base the module specifies the architecture competence which will give to the autonomous mobile robot the ability to execute the particular mission.
In what follow we give further information about each group of agent and the related partial goal.

The Decision Agents A group of deliberative agents, called Decision Agents, are the highest decision level of the proposed multi-agent architecture. The agents store and merge the supplied agents information about environment structure, robot position and agents' status in a dynamic KB. Consequently, they plan the best strategy for environment exploration taking the properly decisions to reach a prefixed goal. The agents are able to interact and manage each reactive process (agent) in order to reach one or more prefixed goals, following their vocation and their behavioral.

The Navigation Agent Robot sensors and effectors control has been implemented with reactive agents, called Navigation Agents, that are able to construe each high level command in order to breed one or more low level robot hardware commands. So the agents of the proposed architecture use simple statements to interact with robot hardware such as "go drawing a circle" or "perform wall following" or "grab image" to drive the robot or to acquire a new image respectively. The Navigation Agent allows us to use the proposed approach with different robot hardware platforms, being the high level agent language independent by them.

The Sensor Agents Some "transduction" agents, called Sensor Agents, acquire information through external sensors and give an accurate and adequate symbolic description of the external real world, just in time for that description to be useful. The agents describe the external scenes by means of a high-level language, joining a symbol to each analyzed external environment zone. A sensors system in a complex dynamic world has to deal with a great amount of data, much more than it can be analyzed in real time. In order to respond quickly to real world events, a dynamic sensor system must be able to focus its resources on just those aspects that are significative and critical, while ignoring the vast flow of irrelevant details. Neural networks can be used in order to have a fast and punctual symbolic environment description, since they are very strong and attractive for their generalization capabilities with different external environments condition.

Communication Agent and Agent's Language Each agent of the proposed architecture can communicate using a designed communication language in order to perform a simple speech act. In our task communication is the intentionally exchange of information about environment configuration and the properly robot action to reach a prefixed goal. We adopt a high level language derived from the KQML language [6] with fixed grammar rules in which only the dynamic change of the sentences builds robot knowledge. Messages have different priority level in order to manage agents interactions and to arbitrate agents conflicts. Agent messages are routed by the Communicator Agent that can be seen as the messages dispatcher of the architecture.

3 A Real Task for a Vision Guided Mobile Robot

We have tested the proposed control architecture implementing the multi-level mental process of an autonomous robot that performs object retrieval tasks into unknown structured environments. The robot is not given a priori map of the environments and it can explore the unknown environments, equipped with a simple b/w camera sensor.

Environment and Task Description Our structured environments are built as hardware stores with corridors and stands or shelves (see Fig.2). Store's corridors are identified by black square zones with a white edge, whereas the

Figure 2: The robot (a RWI B-12 prototype) inside a structured environment.

store's shelves are identified by white square zones. A store is divided in sectors and each sector is identified by one or more markers. Instruments are on the shelves and their placement in the store is linked to a related store sector, i.e. to the related markers. Robot's tasks consist to retrieve one or more objects in the described structured environments. The robot is not given a priory knowledge about environment configuration and it can autonomously explore the environment in order to retrieve the objects. In the first phase the robot explores the structured environment without a prefixed path to retrieve the required object. The exploration phase goes on even if the object is retrieved. When a new object retrieval request happened, the robot performs the best path-planning to reach the object position if the object is in the learned KB. Otherwise the robot restarts the environment exploration. In this way, the robot is able to perform very fast object retrieval tasks after the first learning phase.

The Architecture Configuration In the first phase the Mediator Module specifies the architecture competence through the interaction between the User and Deliberative Agents. In this phase are defined the objects sequence for the robot retrieval task, the typology and the number of the environment objects and markers, the available robot sensors and so on. In the experimental tasks the robot is equipped with a simple b/w CCD camera, so the Sensor Agents can be seen as a collection of modules performing vision processes. Vision processes are based on algorithms of digital image processing to classify a square environment zone and to segment markers and instruments in the environment. The extracted symbols are sent to two neural networks for their classification. In particular we use Multilayer Perceptrons for classification problems: the capability to approximate boundary surfaces of arbitrary complexity makes MLP the universal classifiers. The supplied information are stored in a dynamic KB. The KB is organized as a tree and it is inferred in order to obtain the shortest path to retrieve an object in the environment. We use the backtracking search algorithm to obtain it.

Experimental Results We have implemented and tested the proposed architecture on a RWI B-12 autonomous robot in the CSAI Lab of the University of Palermo. The robot is equipped by a visiòn system composed of a CCD video camera mounted on pan-tilt unit. Robot mounts an Intel 80486 processor on board running Linux and it is linked with a network of Unix workstation by a Local Radio Ethernet, too. Different environment configuration was built in order to test the proposed multi-agent architecture [4, 5]. The designed agents work in the same time on robot in order to perform a very fast object retrieval task with complex environments, too.

The multi-agent software architecture was developed using a full C software code for each agent of the architecture and a software exploiting the potentiality of Prolog language to design to dynamic KB.

4 Conclusions

The adopted strategy is a powerful means to face real robotics tasks: it allows us to split the robot cognitive processes in most simple cognitive sub-process, designing independent and specialised agent for each sub-task. Each agent is able to cooperate with other agent exchanging information, so the robot is able to make the appropriate policy of management to reach a prefixed goal.

Future developments will aim to apply the proposed approach to several robotics tasks in unstructured and unknown environments.

References

[1] R.E. FIkes, P.E.Hart, and N.J.Nilsson,Learning and Executing Generalized Robot Plans,Artificial Intelligence,1972,(**3 251-288**)

[2] A. Chella, S. Gaglio, G.Sajeva, F. Torterolo,A Hibrid Architecture for Autonomous Agents,1997,AIIA.

[3] H.HU and M.Brady,A parallel processing architecture for sensor-based control of intelligent mobile robots,Robotic and Automation,1996,(**17 235-237**)

[4] S. Vitabile, F. Torterolo, F. Sorbello, A Multi-Agent Architecture based on Neural Networks and Knowledge Bases for Automatic Object Retrieval Tasks by an Autonomous Mobile Robot, International ICSC Symposium on Engineering of Intelligent Systems, 1998, (**568-574**)

[5] A New Multi-Agent Architecture for the Distribuited Artificial Intelligence in a Real Robotics Experiment", Proc. of 16th Iasted International Conference Applied Informatics, 1998.

[6] T. Finin , Y. Labrou,A Semantic Approach for KQML, a General Purpose Communication Language for Software Agents,Third International Conference on Information and Knowledge Management (CIKM'94),1994

[7] M. Wooldridge, N. R. Jennings,Intelligent agents: Theory and Practice,The Knowledge Engineering Review,1995,(**10:2 115-152**)

[8] R. A. Brooks, Intelligence Without Representation, Artificial Intelligence,1991,(**47, 139-159**)

Author Index

MIX
Papier aus verantwortungsvollen Quellen
Paper from responsible sources
FSC® C105338

If you have any concerns about our products,
you can contact us on
ProductSafety@springernature.com

In case Publisher is established outside the EU,
the EU authorized representative is:
Springer Nature Customer Service Center GmbH
Europaplatz 3, 69115 Heidelberg, Germany

Printed by Libri Plureos GmbH
in Hamburg, Germany